MICROALGAL BIOTECHNOLOGY

Bioprospecting Microalgae for Functional Metabolites towards Commercial and Sustainable Applications

Innovations in Biotechnology

MICROALGAL BIOTECHNOLOGY

Bioprospecting Microalgae
Metabolites towards Com
Sustainable Applica

Edited by

Jeyabalan Sangeetha, PhD
Svetlana Codreanu, PhD
Devarajan Thangadurai, PhD

AAP | APPLE ACADEMIC PRESS

First edition published 2023

Apple Academic Press Inc.
1265 Goldenrod Circle, NE,
Palm Bay, FL 32905 USA

760 Laurentian Drive, Unit 19,
Burlington, ON L7N 0A4, CANADA

CRC Press
6000 Broken Sound Parkway NW,
Suite 300, Boca Raton, FL 33487-2742 USA

4 Park Square, Milton Park,
Abingdon, Oxon, OX14 4RN UK

Library and Archives Canada Cataloguing in Publication

Title: Microalgal biotechnology : bioprospecting microalgae for functional metabolites towards commercial and sustainable applications / edited by Jeyabalan Sangeetha, PhD, Svetlana Codreanu, PhD, Devarajan Thangadurai, PhD.
Names: Sangeetha, Jeyabalan, editor. | Codreanu, Svetlana, editor. | Thangadurai, Devarajan, 1976-
Series: Innovations in biotechnology (Series)
Description: First edition. | Series statement: Innovations in biotechnology | Includes bibliographical references and index.
Identifiers: Canadiana (print) 20220400172 | Canadiana (ebook) 20220400199 | ISBN 9781774912379 (hardcover) | ISBN 9781774912386 (softcover) | ISBN 9781003332251 (ebook)
Subjects: LCSH: Microalgae—Biotechnology. | LCSH: Microalgae—Industrial applications.
Classification: LCC TP248.27.A46 M48 2023 | DDC 660.6—dc23

Library of Congress Cataloging-in-Publication Data

Names: Sangeetha, Jeyabalan, editor. | Codreanu, Svetlana, editor. | Thangadurai, Devarajan, 1976- editor.

Title: Microalgal biotechnology : bioprospecting microalgae for functional metabolites towards commercial and sustainable applications / edited by Jeyabalan Sangeetha, PhD, Svetlana Codreanu, PhD, Devarajan Thangadurai, PhD.

Description: First edition. | Palm Bay, FL : Apple Academic Press ; Boca Raton, FL, USA : CRC Press, 2023. | Series: Innovations in biotechnology | Includes bibliographical references and index. | Summary: "Microalgae are a valuable resource of carbon materials that may be used in biofuels, pharmaceuticals, cosmetics, and health supplements. Still, there are tremendous challenges in the microalgae production process, such as mass cultivation, strain improvement, biomass disruption, and reprocessing of nutrients and water that have been the encumbering microalgal industry in several ways. Microalgal biotechnology has the capability to introduce remarkable breakthroughs and innovations. This volume brings together current advancements in the field of microalgal biotechnology to provide understanding of the fundamentals and progress of the industry-oriented technologies. The volume covers biofuel production and its technical challenges; current trends of microalgae in agriculture, aquaculture and food biotechnology; market space; biosafety and other regulatory issues; and sustainability of microalgal products. The volume first provides insight into desmids and other conjugating algae along with prospective trends of diatoms. It then goes on to explore the commercial applications of microalgae that take advantage of their bioactive compounds, natural colorants, and functional metabolites. The role of microalgae in agriculture, aquaculture, and the food sector are discussed, and tools from genome manipulation to bioprocess engineering are studied. Several chapters focus on future prospects of biofuel and biodiesel production for environmental and economic viability. The key features of the book: Presents the role of microalgae in various industries including food, agriculture, aquaculture, biofuel, and metabolites Shows the historical and prospective use of microalgae elements for economic and ecological benefits Explains the integrated technologies for massive production of microalgae-derived products Includes the various industrial case studies to improve the sustainable production of microalgae products Discusses current developments and confronts in microalgae bioprocessing Microalgal Biotechnology: Bioprospecting Microalgae for Functional Metabolites towards Commercial and Sustainable Applications will be a useful reference for academicians, researchers, industry professionals, postgraduate students, and policymakers who are interested in future prospects of microalgae commercialization"-- Provided by publisher.

Identifiers: LCCN 2022037400 (print) | LCCN 2022037401 (ebook) | ISBN 9781774912379 (hbk) | ISBN 9781774912386 (pbk) | ISBN 9781003332251 (ebk)

Subjects: LCSH: Microalgae--Biotechnology.

Classification: LCC TP248.27.A46 M525 2023 (print) | LCC TP248.27.A46 (ebook) | DDC 660.6--dc23/eng/20220829

LC record available at https://lccn.loc.gov/2022037400

LC ebook record available at https://lccn.loc.gov/2022037401

ISBN: 978-1-77491-237-9 (hbk)
ISBN: 978-1-77491-238-6 (pbk)
ISBN: 978-1-00333-225-1 (ebk)

BOOK SERIES
Innovations in Biotechnology

Series Editor
Devarajan Thangadurai, PhD
Professor, Karnatak University, Dharwad, India

Books in the Series

- **Fundamentals of Molecular Mycology**
  Authors: Devarajan Thangadurai, PhD, Jeyabalan Sangeetha, PhD, and Muniswamy David, PhD
- **Biotechnology of Microorganisms: Diversity, Improvement, and Application of Microbes for Food Processing, Healthcare, Environmental Safety, and Agriculture**
  Editors: Jeyabalan Sangeetha, PhD, Devarajan Thangadurai, PhD, Somboon Tanasupawat, PhD, and Pradnya Pralhad Kanekar, PhD
- **Phycobiotechnology: Biodiversity and Biotechnology of Algae and Algal Products for Food, Feed, and Fuel**
  Editors: Jeyabalan Sangeetha, PhD, Devarajan Thangadurai, PhD, Sanyasi Elumalai, PhD, and Shivasharana Chandrabanda Thimmappa, PhD
- **Biogenic Nanomaterials: Structural Properties and Functional Applications**
  Editors: Devarajan Thangadurai, PhD, Saher Islam, PhD, Jeyabalan Sangeetha, PhD, and Natália Martins, PhD
- **Seaweed Biotechnology: Biodiversity and Biotechnology of Seaweeds and Their Applications**
  Editors: Jeyabalan Sangeetha, PhD, and Devarajan Thangadurai, PhD
- **Microalgal Biotechnology: Bioprospecting Microalgae for Functional Metabolites towards Commercial and Sustainable Applications**
  Editors: Jeyabalan Sangeetha, PhD, Svetlana Codreanu, PhD, and Devarajan Thangadurai, PhD
- **Algal Metabolites: Biotechnological, Commercial, and Industrial Applications**
  Editors: Jeyabalan Sangeetha, PhD, Devarajan Thangadurai, PhD, Saher Islam, PhD, and Ravichandra Hospet, PhD

About the Editors

Jeyabalan Sangeetha, PhD
Assistant Professor, Central University of Kerala, Kasaragod, Kerala, India

Jeyabalan Sangeetha, PhD, is an Assistant Professor at Central University of Kerala, Kasaragod, South India. She has edited/co-edited several books in her research areas, which include environmental toxicology, environmental microbiology, environmental biotechnology, and environmental nanotechnology. She earned her BSc in Microbiology and PhD in Environmental Science from Bharathidasan University, Tiruchirappalli, Tamil Nadu, India. She holds an MSc in Environmental Science from Bharathiar University, Coimbatore, Tamil Nadu, India. She is the recipient of a Tamil Nadu Government Scholarship and a Rajiv Gandhi National Fellowship of the University Grants Commission (UGC), Government of India, for her doctoral studies. She served as Dr. D.S. Kothari Postdoctoral Fellow and UGC Postdoctoral Fellow at Karnatak University, Dharwad, South India, during 2012–2016 with funding from the University Grants Commission, Government of India, New Delhi.

Svetlana Codreanu, PhD
Associate Professor, Institute of Microbiology and Biotechnology, Chisinau, Republic of Moldova

Svetlana Codreanu, PhD, is an Associate Professor at the Institute of Microbiology and Biotechnology in Chisinau, Republic of Moldova, is co-founder of Moldovan Society for Microbiology and FEMS Delegate. She studied Biology and obtained her PhD in Microbiology from Moldova State University. Her current scientific research focuses on Applied Phycology and Biotechnology. She is regularly involved as an evaluator of scientific projects and research infrastructures, reviewer of scientific journals, and abstract selections in conferences and PhD defenses. She is the author and co-author of over 100 scientific publications and 17 patents. She received numerous awards, including the WIPO Award and Gold Medal.

Devarajan Thangadurai, PhD
Professor, Karnatak University, Dharwad, India

Devarajan Thangadurai, PhD, is a Professor of Botany at Karnatak University in South India. He has authored/edited over 30 books with national and international publishers and has visited 24 countries in Asia, Europe, Africa, and the Middle East for academic visits, scientific meetings, and international collaborations. He received his PhD in Botany from Sri Krishnadevaraya University in South India as a CSIR Senior Research Fellow with funding from the Ministry of Science and Technology, Government of India. He served as a Postdoctoral Fellow at the University of Madeira, Portugal; University of Delhi, India; and ICAR National Research Center for Banana, India. He is the recipient of a Best Young Scientist Award with a Gold Medal from Acharya Nagarjuna University, India, and the VGST-SMYSR Young Scientist Award of the Government of Karnataka, Republic of India.

Contents

Contributors

Alok Adholeya
TERI Deakin Nanobiotechnology Center, Sustainable Agriculture Division,
The Energy and Resources Institute, New Delhi, India

Malik Badshah
Department of Microbiology, Faculty of Biological Sciences, Quaid-i-Azam University,
Islamabad – 45320, Pakistan

Aavany Balasubramanian
Anna University Regional Campus, Coimbatore – 641046, Tamil Nadu, India

Paramasivan Balasubramanian
Agriculture and Environmental Biotechnology Laboratory, Department of Biotechnology and
Medical Engineering, National Institute of Technology, Rourkela, Odisha – 769008, India

Usman Ali Chaudhry
Infection Control and Disease Prevention Center, Ministry of Health, Tabuk,
Kingdom of Saudi Arabia

Rayanee Chaudhuri
Agriculture and Environmental Biotechnology Laboratory, Department of Biotechnology and
Medical Engineering, National Institute of Technology, Rourkela, Odisha – 769008, India

Laurent Dufossé
Université de la Réunion, CHEMBIOPRO Chimie et Biotechnologie des Produits Naturels,
ESIROI Département Agroalimentaire, Sainte-Clotilde F – 97490, Ile de La Réunion,
Indian Ocean, France

Suchitra Gaur
TERI Deakin Nanobiotechnology Center, Sustainable Agriculture Division,
The Energy and Resources Institute, New Delhi, India

Jyothish Jayakumar
Anna University Regional Campus, Coimbatore – 641046, Tamil Nadu, India

Srijay Kamat
Department of Biotechnology, Goa University, Goa – 403206, India

Amritpreet Kaur
TERI Deakin Nanobiotechnology Center, Sustainable Agriculture Division,
The Energy and Resources Institute, New Delhi, India

Rupali Kaur
Center of Biotechnology, Nehru Science Center, University of Allahabad, Prayagraj,
Uttar Pradesh – 211002, India

Samiullah Khan
Department of Microbiology, Faculty of Biological Sciences, Quaid-i-Azam University,
Islamabad – 45320, Pakistan

Mahesh Khot
Laboratorio de Recursos Renovables, Centro de Biotecnología, Universidad de Concepción, Concepción – 4030000, Chile

Manjita Mishra
Advanced Laboratory for Phycological Assessment (ALPHA Plus), Center of Biotechnology, University of Allahabad, Prayagraj, Uttar Pradesh – 211002, India

Aneela Nawaz
Department of Microbiology, Faculty of Biological Sciences, Quaid-i-Azam University, Islamabad – 45320, Pakistan

Pavle Pavlović
Department of Ecology, Institute for Biological Research "Siniša Stanković," University of Belgrade, Bulevar Despota Stefana 142, Belgrade – 11060, Serbia

Md. Akhlaqur Rahman
Department of Biotechnology, S.S. Khanna Girls' Degree College, Prayagraj, Uttar Pradesh – 211003, India

Gouri Raut
Bioenergy Division, Agharkar Research Institute, Pune – 411004, Maharashtra, India

Pankaj Kumar Singh
Diatom Research Laboratory, Amity Institute of Biotechnology, Amity University, Noida, Uttar Pradesh, India

Marija Stamenković
Department of Ecology, Institute for Biological Research "Siniša Stanković," University of Belgrade, Bulevar Despota Stefana 142, Belgrade – 11060, Serbia

Shanthy Sundaram
Advanced Laboratory for Phycological Assessment (ALPHA Plus), Center of Biotechnology, Nehru Science Center, University of Allahabad, Prayagraj, Uttar Pradesh – 211002, India

Pavithra Suresh
Anna University Regional Campus, Coimbatore – 641046, Tamil Nadu, India

Archana Tiwari
Diatom Research Laboratory, Amity Institute of Biotechnology, Amity University, Noida, Uttar Pradesh, India

Chidambaram Kulandaisamy Venil
Department of Biotechnology, Anna University, Regional Campus, Coimbatore – 641046, Tamil Nadu, India

Matheswaran Yaamini
Department of Biotechnology, Anna University, Regional Campus, Coimbatore – 641046, Tamil Nadu, India

Abbreviations

ABA	abscisic acid
AChE	acetylcholinesterase
AFA	*Aphanizomenon flos-aquae*
AFM	atomic force microscopy
AGPs	arabinogalactan proteins
ALA	α-linolenic acids
APC	allophycocyanin
ARA	arachidonic acid
ATP	adenosine triphosphate
BHT	butylated hydroxytoluene
BiP	binding protein
CBZ	carbamazepine
CDVs	cardiovascular diseases
CN-N	cyanovirin-N
CNTs	carbon nanotubes
CO_2	carbon dioxide
CYN	cylindrospermopsin
DE	diatomaceous earth
DH	digital holography
DHA	docosahexaenoic acid
DMS	dimethylsulfide
DMSP	dimethylsulfoniopropionate
EMS	early mortality syndrome
EOT	enhanced optical transmission
EPA	eicosapentaenoic acid
EPS	exopolysaccharides
EPS	extracellular polymeric substances
EST	expressed sequence tag
FA	fatty acid
FAEEs	fatty acid ethyl esters
FAO	Food and Agriculture Organization
FDFs	fast death factors
FEM	finite element method

GA	gibberellic acid
GHGs	greenhouse gases
GLA	γ-linolenic acid
GTX	gonyautoxins
HBV	hepatitis B virus
Hsps	heat shock proteins
IAA	indole acetic acid
ISKNV	infectious spleen and kidney corruption infection
KCS	3-ketoacyl-CoA synthase
LCPAs	long-chain polyamines
LPSs	lipopolysaccharides
MAAs	mycosporine-like amino acids
MC-LR	microcystin-LR
MCs	microcystins
MDH	molasses hydrolysate direct medium
MHL	molasses hydrolysate nitrogen-limited medium
MSNs	mesoporous silica NPs
NADPH	nicotinamide adenine dinucleotide phosphate-oxidase
NF-kB	nuclear factor-kB
NNV	nervous necrosis virus
NO_2	nitrogen dioxide
NOS	nitrogen reactive species
ORFs	open reading frames
OVAT	one variable at a time
PAR	photosynthetically active radiation
PBPs	phycobiliproteins
PC	phycocyanin
PC	phytochelatins
PCE	photo conversion efficiency
PE	phycoerythrin
PG	propyl gallate
PGRs	plant growth regulators
PHs	protein hydrolysates
POD	peroxide dismutase
PP1	phosphatases 1
PP2A	phosphatases 2A
PSPs	paralytic shellfish poisons
PUFA	polyunsaturated fatty acids

QDs	quantum dots
ROS	reactive oxygen species
RSIV	red sea bream iridovirus
RSM	response surface methodology
SAD	stearoyl-ACP desaturase
SCMT	supercritical methanol transesterification
SDA	stearidonic acid
SDV	silica deposition vesicle
SEM	scanning electron microscope
SITs	silicic acid transporters
SOC	soil's organic carbon
SOD	superoxide dismutase
SOM	soil organic matter
STX	saxitoxins
TAG	triacylglycerol
TEM	transmission electron microscopy
TL	twin-layer
TNF	tumor necrosis factor
UGC	University Grants Commission
WSSD	white spot syndrome disease
YHD	yellow head disease

Preface

Microalgae are a distinct group of unicellular photosynthetic algae that produce diverse valuable products, including bioactive molecules, biodiesel feedstock, human and animal food. Microalgae have the potential to offer a sustainable source of industrially relevant nutraceuticals, biofuels, food supplements, and biochemicals. Microalgal biotechnology has become a rapid-paced field of research and coming interminably closer to commercial viability as this field consolidates the cutting-edge research and concurrently looks into policy considerations and market potential. With the great concern in renewable energy resources, microalgae biotechnology has been subjected to a great leap in prominence in recent times. The impact of this area of research on society and the environment has been outlined as a desirable outcome of scientific advancement.

Highlighting the promising role of microalgae being commercial commodities, this book *Microalgal Biotechnology* covers recent progress on a number of fronts, including biofuel production and its technical challenges, current trends of microalgae into agriculture, aquaculture, and food biotechnology, market space, biosafety, and other regulatory issues and sustainability of microalgal products. This book deals with integrated tools to bring the great potential of microscopic algae into application, accelerate the progress of working production practices, and put final products into the market. The book explicates the comprehensive and authoritative overview of microalgae-based studies, practices, and products. Distributed into 10 discrete chapters, this book covers the most promising applications of microalgae. Chapter 1 provides insight into desmids and other conjugating algae. Prospective trends of diatoms are reviewed in Chapter 2. Chapters 3 to 5 reveal the commercial applications of microalgae, including bioactive compounds, natural colorants, and functional metabolites. The role of microalgae in agriculture, aquaculture, and the food sector have been discussed in Chapters 6 to 8; thus, ultimate enabling tools from genome manipulation to bioprocess engineering are studied. The last two chapters focus on the future prospects of biofuel and biodiesel production for environmental and economic viability.

Contributions from academic circles and industrial applications, and case reports make the book a broad survey of recent progress in the field of microalgal biotechnology. So, this book would be of great interest to researchers in biology, bioengineering, and biotechnology for sustainable and mass production of great value microalgal products. Further, the book will permit the protagonists of industry and academia, in addition to decision-makers, to get a clear image of existing opportunities and future possibilities in microalgal biotechnology.

— *Editors*

CHAPTER 1

Applied Studies on Desmids and Other Conjugating Algae (*Zygnematophyceae, Streptophyta*)

MARIJA STAMENKOVIĆ and PAVLE PAVLOVIĆ

Department of Ecology, Institute for Biological Research "Siniša Stanković," University of Belgrade, Bulevar Despota Stefana 142, Belgrade – 11060, Serbia

ABSTRACT

The conjugating algae (*Zygnematophyceae*, *Streptophyta*) have been primarily used as a model and test organisms for the study of many biological processes. Recent investigations demonstrated that they might have fair potential to be used for the commercial production of various metabolites as well as for the bioremediation of wastewaters. Due to the adaptation to high light intensities, *Zygnematophyceae* typically have high amounts of photosynthetic pigments, among which zeaxanthin and lutein have a precious role in medicine and agriculture. Furthermore, under certain conditions, some conjugating algae may produce astaxanthin, gallotannins, heat shock proteins (hsps), and antioxidant enzymes, which have nutraceutical and pharmaceutical significance. High amounts of several fatty acids (FAs), such as palmitic, oleic, linoleic, and α-linolenic acid, have been recorded in desmids (order *Desmidiales*), and these FAs may have an important role in cosmetics and industry as well as for biodiesel production. Some conjugating algae (e.g., *Spirogyra* sp.) have high biomass productivity and may accumulate large amounts of carbohydrates,

Microalgal Biotechnology: Bioprospecting Microalgae for Functional Metabolites towards Commercial and Sustainable Applications. Jeyabalan Sangeetha, PhD, Svetlana Codreanu, PhD, & Devarajan Thangadurai, PhD (Eds.)

representing a suitable substrate for bioethanol and hydrogen production. Furthermore, fresh and dried biomass of several representatives of the families *Desmidiaceae* and *Zygnemataceae* appeared efficient substrate for the biosorption of metals (As, Cd, Pb, Cu, Mn, Zn, Cr, Sr, Ba, Pb, Ni, and Hg), toxic substances, remedy residues, and nutrients, thus rendering this group interesting for the purification of various types of agricultural and industrial wastewaters.

1.1 INTRODUCTION

The conjugating green algae comprise an almost exclusively freshwater group of *Streptophyta* organisms, many of extraordinary beauty. They are referred to a class generally known as the *Conjugatophyceae* ('Conjugatae') in Central Europe and the *Zygnematophyceae* ('*Zygnemophyceae*') elsewhere in the world (Guiry, 2013). A class *Zygnematophyceae* is characterized by conjugation-mediated sexual reproduction, absence of flagellated life cycle stadia, and lack of centrioles, suggesting that putative flagella were not secondarily lost in the course of evolution (Gerrath, 1993; Guiry, 2013). There is considerable disagreement as to the classification of conjugating green algae, both within the group and with respect to their relationship to other green algae. Traditionally, the class *Zygnematophyceae* includes two orders: *Zygnematales* (families *Zygnemataceae* and *Mesotaeniaceae* – saccoderm, i.e., 'false' desmids) and *Desmidiales* – placoderm, i.e., 'true' desmids (*Closteriaceae*, *Desmidiaceae*, *Gonatozygaceae*, *Peniaceae*) (Mix, 1972; Růžička, 1977; Coesel and Meesters, 2007; Guiry, 2013). However, some authors consider that *Zygnematophyceae* contains the order *Zygnematales*, whereas the *Desmidiales* emerged within the *Zygnematales* (Lemieux et al., 2016). Accordingly, the *Zygnematophyceae* are supposed as sister clade of the *Mesotaenium*, together forming the sister clade of the land plants (Wickett et al., 2014; de Vries et al., 2016; Gitzendanner et al., 2018).

The 'true' desmids are named after the Greek word *desmos* (bond or chain), since cells of the majority of taxa are transversally carved by a constriction (*sinus*) into two symmetrical semicells connected by an *isthmus* (Brook, 1981; Stamenković and Hanelt, 2017). This applies to the group of placoderm desmids which possess a complex two-layered cell wall, perforated by a system of pores (Brook, 1981; Guiry, 2013).

The name 'saccoderm desmids' was coined to refer to the possession of a cell wall consisting of a single piece and lacking vertical pores in the cell wall. The *Zygnemataceae* represent a family of filamentous or unicellular, uniseriate (unbranched) green algae. The family is notable for its diversely shaped chloroplasts, such as stellate in *Zygnema*, helical in *Spirogyra*, and flat in *Mougeotia*. *Zygnematophyceae* are considered to have a cosmopolitan freshwater distribution; however, numerous ecological and taxonomic investigations revealed that many desmid taxa (at the level of genus, species, and variety) are capable of occupying specific geographic zones, characterized by particular climatic attributes (Stamenković and Hanelt, 2017). Recent investigations demonstrated a clear relationship between the climatic factors (temperature, photosynthetically active radiation (PAR) and ultraviolet radiation (UVR)) and the distributional potential of conjugating algae, taking into account their photosynthetic, physiological, and ultrastructural adaptations which had been revealed during and after certain temperature and irradiation treatments (Stamenković and Hanelt, 2011, 2013a, b, 2014; Stamenković et al., 2014; Holzinger and Pichrtová, 2016).

Zygnematophyceae can be found in nearly every aquatic environment, and they are important as primary producers, while filamentous forms may be essential habitats for invertebrates, fish, and other algae (Hoshaw and McCourt, 1988; Hall and McCourt, 2015). Filamentous *Zygnematales* (especially *Mougeotia* sp. and *Spirogyra* sp.) can often be found in abundance and typically exist as floating mats in stagnant water in ditches and ponds. *Spirogyra* sp. tends to develop the floating masses which can foul intake pipes and filter beds, as frequently seen in water treatment plants (Hainz et al., 2009; Stancheva et al., 2013). In contrast, blooms of *Mougeotia* and *Zygnema* often occur in littoral zones of softwater lakes undergoing early stages of acidification (pH 5.0–6.0; Turner et al., 1991), and several species of these taxa have been used as acidification bioindicators (Schneider and Lindstrøm, 2011; Hall and McCourt, 2015). The genus *Zygogonium* seems to prefer acid habitats and may be the dominant taxon found in low pH waters (Lynn and Brock, 1969). The placoderm desmids are generally most common and diverse in oligotrophic lakes and ponds as well as in acidic, highly colored, dystrophic lakes (Gerrath, 1993).

Diverse communities of *Zygnematophyceae* have been used in indices of water condition as well as for assessments of the conservation value of aquatic habitats (Coesel, 2001, 2003; Ngearnpat and Peerapornpisal, 2007; Stancheva and Shaeth, 2016). Yet, many desmid taxa increased their tolerance threshold to various pollutants, and so 'mesotrophic' desmids can be found in habitats heavily enriched with organic biodegradable compounds (Stamenković and Cvijan, 2008a, b, c, 2009; Stamenković et al., 2008; Fužinato et al., 2011). Furthermore, *Zygnematophyceae* may dominate certain marginal, rare, and extreme habitats. The species *Ancylonema nordenskioeldii* and *Mesotaenium berggrenii* are cryophilic and can be found on glaciers, snow, and their meltwater. Occasionally, phytoplankton studies report that desmids are dominant or subdominant phytoplankton organisms. Some of the records include a bloom of *Staurastrum pingue* in an Austrian lake (Lenzenweger, 1980), the dominance of *Staurodesmus* species in some lakes in Greenland (Gerrath, 1993), a bloom of *Staurastrum tetracerum* or *Cosmarium variolatum* var. *rotundatum* in Indian cistern (Venkateswarlu, 1983; Jyothi et al., 1989) and in a eutrophic lake in Mid Wales (Brook, 1982). In accordance with the cosmopolitan distribution of this group, there have been numerous floristic and ecological investigations which directly or indirectly involved conjugating algae (Brook, 1981; Palamar-Mordvintseva, 1982; Gerrath, 1993; Lenzenweger, 1996, 1997, 1999, 2003; Coesel and Meesters, 2007; Hall and McCourt, 2015). Desmids are increasingly mentioned in the limnological literature (even if only identified to the level of genus), and are themselves often included in experiments related to basic ecological and physiological research (Gerrath, 1993; Domozych et al., 2016; Lütz-Meindl, 2016; Stamenković and Hanelt, 2017).

It is important to emphasize that microalgae are receiving considerable attention due to their ability to synthesize valuable compounds (e.g., pigments, and enzymes), accumulate high-energy compounds (e.g., lipids, carbohydrates) and sequester carbon. They are therefore considered as a 'third generation' feedstock for biofuel production and have a great potential as a renewable source (Hu et al., 2008). In the recent years investigations have been dedicated to the few conjugating algae which have become experimental organisms in various laboratories, where they have been found to be useful for the study of certain biological processes (e.g., Kasai and Ichimura, 1986, 1990; Pichrtová, 2014a, b; Stancheva et al., 2014, 2016; Stamenković and Hanelt, 2017). Certain studies also showed that zygnematophycean algae might have a fair potential to be used for the purposes of wastewater bioremediation (Vogel and Bergmann, 2018) as well as for the production of commercially interesting products such as lipids, carbohydrates, proteins, and pigments (Remias et al., 2012a; Pacheco et al., 2017; Pinto et al., 2018; Stamenković et al., 2019, 2020).

Several review papers regarding the ecophysiological responses of *Zygnematophyceae* to climate conditions and abiotic stress appeared rather recently (Holzinger and Karsten, 2013; Holzinger and Pichrtová, 2016; Lütz-Meindl, 2016; Stamenković and Hanelt, 2017). However, there have been no detailed surveys on the possibilities of exploitation this algal group for the production of valuable metabolites and for the purification of freshwater systems. Taking into account the cosmopolitan distribution of conjugating algae, and that some of the taxa thrive in polluted waters and may produce high values of biomass (Vogel and Bergmann, 2018; Stamenković et al., 2019), investigations on their commercial value have multiplied (Ge et al., 2018). Hence, this chapter intended to summarize and discuss the most important results regarding the metabolic products and applied aspects of *Zygnematophyceae* and to estimate possible commercial attributes of this group.

1.2 PHOTOSYNTHETIC AND NON-PHOTOSYNTHETIC PIGMENTS

A photosynthetic pigment is a pigment that is present in chloroplasts or photosynthetic bacteria and have functional and structural role in the photosynthetic mechanism (Lawlor, 2000). Non-photosynthetic pigments in *Zygnematophyceae* include phenolic compounds (gallotannin derivatives) in *Zygnemataceae* and *Mesotaeniaceae* as well as astaxanthin in *Spirogyra* sp. (Remias et al., 2012a, b; Pacheco et al., 2015; Stamenković and Hanelt, 2017). Composition of photosynthetic pigments in *Zygnematophyceae* is comparable to that of embryophytes, including chlorophylls *a* and *b* and carotenoids with violaxanthin xanthophyll cycle (Herrmann, 1968; Züllig, 1982; Fawley, 1991; Van Heukelem et al., 1992; Lütz et al., 1997), concomitantly with their possible mutual origin (McCourt et al., 2000). Under the relatively low light applied, four desmids (*Closterium acerosum*, *Cosmarium botrytis*, *Micrasterias americana* and *Staurastrum orbiculare*) were uniform in possessing β-carotene, lutein, violaxanthin (Vx), antheraxanthin (Ax), loroxanthin, and neoxanthin and in lacking canthaxanthin and echinenone, while small amounts of zeaxanthin

(Zx) were found (Züllig, 1982). The detailed pigment analysis of four *Cosmarium* strains (*C. crenatum* var. *boldtianum*, *C. punctulatum* var. *subpunctulatum* 570 and 571, and *C. beatum*) cultivated at optimum growth conditions and under high-light stress revealed that their pigment composition corresponded to that of sun plants or sun-acclimated plants (Demmig-Adams and Adams, 2006), taking into account relatively high amounts of Vx, Zx, and β-carotene, and a high xanthophyll cycle pool size (over 25% of all carotenoids) (Stamenković et al., 2014a). Considering photosynthetic parameters and ultrastructural characteristics of the *Cosmarium* strains treated at conditions which imitated those in nature (Stamenković and Hanelt, 2011, 2013a, b, 2014; Stamenković et al., 2014b) as well as physiological characteristics of saccoderm desmids and other zygnematophycean algae studied (Holzinger et al., 2009; Remias et al., 2012a, b; Pichrtová et al., 2013), conjugating algae can be regarded as an algal group adapted to high light intensities (Stamenković and Hanelt, 2017). According to the high de-epoxidation rates and as well as relatively high Zx amounts in control and UV-treated samples, Arctic, Antarctic, and temperate *Zygnema* sp. strains showed high-light adaptation of their pigment composition (Holzinger et al., 2018).

As *Zygnematophyceae* are inhabitants of shallow freshwater environments, the 'sun-type' adaptations of the photosynthetic pigment composition are regarded as characteristics that may enable them to cope with rapid and large changes in light intensity. The larger the xanthophyll cycle pool is, the greater is the capacity to form Zx, and presumably the capacity for the photoprotective processes associated with Zx (Demmig-Adams and Adams, 2006). Remarkably, the xanthophyll cycle pool size of the *Cosmarium* strains studied (except *C. crenatum*), when grown under relatively low light regime of a climate chamber, ranges approximately 72–98 mmol (mol Chl a)$^{-1}$ which is slightly below the range of several high-light adapted plants treated under strong light (89–150 mmol (mol Chl a+b)$^{-1}$) (Demmig-Adams and Adams, 1992a, b; Demmig-Adams, 1998). When stressed under photoinhibitory light intensities (700–1,200 μmol photons m^{-2} s^{-1}), the *Cosmarium* strains produced a markedly high amount of Zx (66–88 mmol (mol Chl a)$^{-1}$) which corresponded to that of sun-adapted plants (e.g., perennial shrubs, vines, and some crop species) grown under 2,000 μmol photons m^{-2} s^{-1} (70.4–95.9 mmol (mol Chl a+b)$^{-1}$) (Demmig-Adams and Adams, 1992a). Overall xanthophyll pool size increased during high-light treatments confirming the *de novo* synthesis of xanthophylls;

the tropical species (*C. beatum*) was attributed by the highest xanthophyll pool size – 113.9 mmol (mol Chl a)$^{-1}$ (Stamenković et al., 2014a).

The high production of Zx in the *Cosmarium* strains investigated was found in some biotechnologically important algae such as *Scenedesmus* sp. and *Dunaliella tertiolecta*, but also in a full-sunlight grown culture of *Spongiochloris spongiosa* from the Canadian Arctic (Casper-Lindley and Björkman, 1998; Masojídek et al., 2004). Additionally, the amount of lutein in the *Cosmarium* strains was very high both at low light conditions and after high light treatments (200–260 mmol (mol Chl a)$^{-1}$), which all indicate a fair potential for the possible production of these valuable carotenoids (Stamenković et al., 2014a). The amount of lutein in Arctic, Antarctic, and temperate *Zygnema* strains was around 5 nmol mg^{-1} CDW (i.e., 3 μg mg^{-1} CDW) (Holzinger et al., 2018) which was slightly below the range of some commercially grown microalgae for the lutein production (del Campo et al., 2001). Zeaxanthin and lutein are forceful antioxidants and play a critical role in the prevention of age-related macular degeneration, the leading cause of blindness. Both pigments are used as a feed additive and colorants, and they may also be protective against cataract formation (Yeum et al., 1999), while zeaxanthin has cancer-preventive properties and may help slowing atherosclerosis progression (Sajilata et al., 2007).

Unlike primary carotenoids which constitute structural and functional components of the photosynthetic apparatus, astaxanthin is a secondary carotenoid accumulating in microalgal cytosolic lipid bodies under environmental stress or adverse culture conditions (Han et al., 2013). Astaxanthin has been used as a nutraceutical and a pharmaceutical, to fight against free-radical associated diseases like oral, colon, and liver cancers, cardiovascular diseases (CDVs), and degenerative eye diseases. Astaxanthin is also a common coloring agent in aquaculture to impart red pigmentation in salmon and rainbow trout (Lorenz and Cysewski, 2000; Guerin et al., 2003). With the exception of the well-known commercial sources of astaxanthin, such as *Haematococcus pluvialis* (up to 4% astaxanthin by cell dry weight, CDW) and *Chlorella zofinigiensis* (0.7% CDW) as well as *Chlamydomonas nivalis*, the occurrence of astaxanthin appears species- and strain-specific (Orosa et al., 2000; Han et al., 2013). Interestingly, a significantly high amount of astaxanthin was produced from a strain of *Spirogyra* sp. collected from Ireland (SAG 170.80), which appeared orange in stress conditions due to the presence of astaxanthin (Pacheco et al., 2015; Pinto et al., 2018). Pacheco et al. (2015) obtained 0.12 g 100 g^{-1} CDW of total pigments with a composition of 56% free astaxanthin, 16% β-carotene and 5% lutein and canthaxanthin. The pigment extraction may significantly decrease the cost of biorefinery handling, for the hydrogen-producing from total carbohydrates (Pacheco et al., 2015; Pinto et al., 2018).

Non-photosynthetic pigments visible as UVR and PAR-screening compounds were evidenced in saccoderm taxa typically living on bare ice surfaces of polar and alpine areas – *Mesotaenium berggrenii*, *Ancylonema nordenskioeldii* and *Cylindrocystis brebissonii* f. *cryophila*, which periodically cause the appearance of dark purple, brownish or gray ice and snow surfaces (Nedbalová and Sklenár, 2008; Remias et al., 2009, 2012a, b). The electron-dense vacuoles and compartments covering the chloroplasts are of high ecophysiological significance due to their possible storage of large amounts of polyphenolics and hydrolyzable tannins which have screening effects, hence protecting organelles in the center of cells (Remias et al., 2009, 2012a, b). Sporangia of *Zygogonium ericetorum* contain brownish residual cytoplasmatic content which apparently has an UVR-protective and/or antioxidative role due to high amounts of soluble phenolic compounds, and this feature is regarded a taxonomic trait of the genus *Zygogonium* (Stancheva et al., 2014, 2016).

While phenolics are less common in non-streptophycean green algae, these have been reported repeatedly in *Zygnematophyceae* (Cannell et al., 1987, 1988a, b; Han et al., 2007; Remias et al., 2012a, b; Stancheva et al., 2012; Pichrtová et al., 2013). A gallotannin derivative (galloylglucopyranose, i.e., purpurogallin carboxylic acid-6-*O*-β-D-glucopyranoside) is a main component responsible for giving vacuoles of the alpine *M. berggrenii* a brownish-purple color (Remias et al., 2012b). It was also found in *Zygogonium ericetorum* (Newsome and van Breemen, 2012; Aigner et al., 2013) while its derivatives were identified in *Spirogyra varians* (Han et al., 2009). The broad absorption capacity of these phenolic compounds (the entire UVA and UVB range and a large part of PAR) indicates their important role as cellular protectants against excessive irradiation. Polyphenolics in *Zygnematophyceae* may also play an antioxidative and/or antimicrobial role or may represent a 'physiological sink' for a surplus of reductive energy (Remias et al., 2012a; Aigner et al., 2013).

The benzotropolone derivatives such as gallotannins can likely be used as UV absorbers and antioxidants in sunscreen or cosmetic compositions, as well as antimicrobial agents (Wagner et al., 2009, 2011). Cannel et al. (1987, 1988a) investigated inhibitors of the glycosidases (α-glucosidase, α-amylase and β-galactosidase) and revealed that several closely related conjugating algae, namely *Spirogyra varians*, *Mougeotia* sp., *Zygnema cylindricum* and *Mesotaenium caldariorum* all produced inhibitors of α-glucosidase. Further, an α-glucosidase inhibitor/antibiotic was purified from *Spirogyra varians* and was determined to be the pentagalloylglucose 3-*O*-digalloyl-1,2,6-trigalloylglucose (Cannel et al., 1988b), which was previously found in *Spirogyra arcta* (Nakabayashi et al., 1954a, b, 1955). It has long been known that hydrolysable tannins are potent enzyme inhibitors, acting by their tendency to precipitate proteins (Cannel et al., 1988b). The inhibition of α-glucosidase by gallotannins may lower the rate of glucose absorption through delayed carbohydrate digestion and extended digestion time, hence, gallotannins could be possibly used as anti-diabetic drugs (Benalla et al., 2010; Firdaus and Prihanto, 2014). As α-glucosidase is involved in the pathway for N-glycans for viruses such as HIV and human hepatitis B virus (HBV), the application of inhibitors of α-glucosidase can prevent fusion of HIV and secretion of HBV (Mehta et al., 1988). These facts may indicate a wide possible application of zygnematophycean gallotannins in medicine.

1.3 LIPIDS AND FATTY ACIDS (FAS)

Microalgae that have the ability to produce substantial amounts (e.g., 20 to over 50% CDW) of triacylglycerols (TAGs) as a storage lipid under photo-oxidative stress or other adverse environmental conditions are referred to as oleaginous algae (Hu et al., 2008). The earlier reports on a few desmid taxa (*Mesotaenium* sp., *Staurastrum* sp., *Cosmarium* sp., *C. bioculatum* and *C. botrytis*, *Xanthidium subhastiferum*) demonstrated rather low total lipid/fatty acid (FA) contents, which may be explained by influences of inadequate growth conditions or sampling time (Cranwell et al., 1990; Hempel et al., 2012; Aminul Islam et al., 2013; Song et al., 2013; Wacker et al., 2016).

A detailed study on the fatty acid methyl esters (FAME) content and FAME production in *Cosmarium* and *Staurastrum* strains demonstrated that the strains had the initial FAME content around 50 mg g^{-1} CDW (~ 5% CDW) at the start of cultivation in optimal growth conditions. However, majority of the desmid strains investigated showed an increase in total FAME content during the growth reaching over 100 mg g^{-1} CDW (Stamenković et al., 2019). The polar oligotrophic taxon, *C. crenatum* var. *boldtianum*, had the highest total FAME content (308.1 mg g^{-1} CDW) as well as the highest FAME productivity at stationary phase (11.1 mg L^{-1} day^{-1}). Furthermore, the distinctly high FAME contents (> 200 mg g^{-1} CDW, > 20% CDW) were noted for *C. meneghinii* 59 (259.9 mg g^{-1} CDW), *S. boreale* 631 (231.5 mg g^{-1} CDW), *C. regnellii* 2792 (226.5 mg g^{-1} CDW), *C. leave* 508 (209.7 mg g^{-1} CDW), and *S. punctulatum* 501 (207.3 mg g^{-1} CDW). Strikingly, these total FAME quantities of the mentioned strains were twofold higher than amounts recorded for *Chlorella vulgaris*, *C. oleofaciens*, *Ellipsoidion parvum*, and *Tetradesmus obliquus*, which had a maximum total FAME content around 100 mg g^{-1} CDW when grown in media with higher nitrate concentrations and enriched with CO_2 (Abomohra et al., 2013, 2017, 2018; El Sheekh et al., 2017). These six desmid strains had the total FAME quantity in the range of the marine species *Phaeodactylum tricornutum* and *Nannochloropsis oculata* (187.3 and 267.1 mg g^{-1} CDW, respectively), more than twofold higher than in *Scenedesmus dimorphus*, *Franceia* sp. (Aminul Islam et al., 2013), *Chlorella vulgaris* (Nascimento et al., 2013), and manifold (6–10 times higher) compared to that of the chlorophycean strains *Ankistrodesmus falcatus*, *A. fusiformis*, *Kirchneriella lunaris*, *Chlamydomonas* sp., *Tetradesmus obliquus*, and *Pseudokirchneriella subcapitata* (Griffiths and Harrison, 2009; Griffiths et al., 2012; Nascimento et al., 2013). However, due to the moderate biomass productivity the desmid strains investigated had rather modest FAME productivity which ranged 8–11.1 mg L^{-1} day^{-1} at late stationary phase (Stamenković et al., 2019). The FAME productivity of the desmid strains was slightly lower than productivities reported for *Tetraselmis elliptica*, *Chlorella vulgaris*, *Chlorococcum oleofaciens*, and *Tetradesmus obliquus*, ranging between 14 and 18 mg L^{-1} day^{-1} (Abomohra et al., 2013, 2018).

A study on FA profiles of 29 *Cosmarium* and *Staurastrum* strains revealed that the dominant FAs in desmids were palmitic (C16:0), linoleic (LA, C18:2*n-6*), α-linolenic acids (ALA, C18:3*n-3*), and hexadecatrienoic acid (C16:3*n-3*) (Stamenković et al., 2019). Among the saturated FAs, palmitic acid comprised around 30% at the start of the logarithmic phase in most desmids investigated. The highest proportions of palmitic acid were found in the tropical species, *C. beatum* 533 (42.5%), and in eutrophic strains from moderate climate, *C. formosulum* 536 and *C. impressulum* 58. Monounsaturated FAs comprised a rather low and constant proportion of the total FAs – around 5% of total FAs. However, the proportion of oleic acid was particularly high in polar strains, reaching the average up to 24% in *C. nasutum* 566, and 22.7% in *Staurastrum monticulosum* 521, respectively, as well as in the new isolates: *S. boreale* (631), *S. polymorphum* (628) and *S. punctulatum* (501) (21%, 15.5% and 13.8%) (Stamenković et al., 2019, 2020).

Desmid strains investigated had over 50% polyunsaturated fatty acids (PUFA) at the start of cultivation, and in most strains the proportion increased during the growth, being the highest in the polar taxon, *C. crenatum* 561 (70%). Linoleic acid was the most abundant PUFA, comprising 30–40% of total FAs; the highest proportions were found in the polar taxon *C. crenatum* 561 (48.4%) and also in *S. crenulatum* 597 (42.4%). Hexadecatrienoic acid and ALA were especially abundant FAs in tropical-subtropical strains at the start of cultivation; the maximum ALA percentage was 33.8% in *C. humile* 1879 (Stamenković et al., 2019, 2020). Previous investigations also showed that the majority of desmid strains had higher ALA proportions than most green microalgae, while *C. lunula*, *Arthrodesmus convergens*, *C. botrytis* and *Triploceras gracile* had > 40% ALA of total FAs (Lang et al., 2011; Wacker et al., 2016). Very high levels of trienoic FAs are especially met in tropical desmids and can be of importance because free radicals that are byproducts of the intensive photosynthetic light reactions stimulate oxidation of PUFAs (Browse, 2009). Because this oxidation might be expected to mediate against a high degree of unsaturation, it has been inferred that there is a strong selective advantage to have such high levels of trienoic FAs in the thylakoid (Allakhverdiev et al., 2009). Interestingly, stearidonic acid (SDA) was recorded in a high proportion in strains cultivated > 35 years such as *C. impressulum* 58 (22.4%). The appearances of high levels of SDA in these desmids may be explained as a genotypically fixed response to stress under long-term suboptimal cultivation conditions (Stamenković et al., 2020).

In general, the qualitative composition of FAs of the desmids investigated in detailed studies by Lang et al. (2011); and Stamenković et al.

(2019, 2020) corresponded largely to that of safflower, flaxseed, sunflower, corn, sesame, and cottonseed (Jones and King, 1996; Gunstone, 2011; Luciana et al., 2014), taking into account the high amounts of LA and palmitic acids and the low amount of oleic acid in most desmids. As both palmitic acid and LA are utilized to produce cosmetics, and they have an anticancer potential (Ando et al., 1998; Darmstadt et al., 2002; Pascual et al., 2017), while LA is extensively used in the fabrication of quick-drying oils (Porter et al., 1995), the FA extracts from desmids may have a wide economic significance. Interestingly, the strain characterized by the highest biomass productivity among all desmids, *S. boreale* 631, also had a high proportion of oleic acid – around 21% of total FAs, within the range of green microalgae used for the biodiesel investigations (Abomohra et al., 2013, 2018; Nascimento et al., 2013; Song et al., 2013; Talebi et al., 2013; Vidyashankar et al., 2015). Hence, this strain can be proposed for further studies which may lead to the commercial biodiesel production. Considering that the proportion of C18:3 acids decrease at the end of the stationary phase in most desmids (< 12% CDW; Stamenković et al., 2019, 2020), this may positively influence oxidative stability of biodiesel, according the international standards (ASTM D6751-08, EN 14214). In addition, several investigations demonstrated that energy production from *Spirogyra* biomass is feasible, and that this alga is suitable for the production of biogas (Ortigueira et al., 2015; Ramaraj et al., 2015; Vogel and Bergmann, 2018), biodiesel (Hossain et al., 2008), and bioethanol (Eshaq et al., 2010). Although Hossain et al. (2008) showed that biodiesel production appeared higher in *Oedogonium* than *Spirogyra* sp., a strain of *Spirogyra* sp. obtained 8.09 g of dry weight and 1.8 g lipids.

1.4 CARBOHYDRATES AND MUCILAGINOUS ENVELOPES IN ZYGNEMATOPHYCEAN ALGAE

Polysaccharides are of interest for the applications in fossil fuels, in the production of biochemicals for food and feed, pharmaceuticals, and cosmetics. Apart from their useful water-binding and gel-forming properties, they can additionally have anti-inflammatory, antitumor, antiviral, antibacterial, antiadhesive or antioxidant activities (Xiao and Zheng, 2016). Their high structural complexity makes chemical production unfeasible and thus biological sources are used, as well-known commercially available polysaccharides synthesized by algae demonstrate (agar, alginate, and carrageenan) (Jiao et al., 2011). Based on the moisturizing and hydrating properties, polysaccharides are interesting ingredients in cosmetic formulations (Wang et al., 2015). Some polysaccharides contain uronic acids and/or sulfate esters, which confer a negative charge to the molecule. This distinctive characteristic opens the possibility of binding cations, especially heavy metals, and using the polysaccharides for bioremediation purposes (Ekelhof and Melkonian, 2017a).

Zygnematophycean algae secrete the polysaccharides through their cell wall as extracellular polymeric substances (EPS) which form envelopes around cells (Kiemle et al., 2007). The presence of mucilaginous sheaths covering cells and/or filaments is regarded as an intrinsic characteristic of conjugating algae (Mix, 1972; Surek, 1983; Coesel, 1994; Paulsen and Vieira, 1994), with the exception of few eutrophic taxa (Stamenković and Hanelt, 2011). The EPS of *Zygnematophyceae* contain seven to nine different monosaccharides as well as sulfate esters, with high percentages of xylose, fucose, and glucuronic acid (Domozych et al., 2005; Kiemle et al., 2007). From a biotechnological viewpoint, the high fucose content may be of particular interest because fucose contributes to emulsifying properties, it is reported to have anti-aging effects (Péterszegi et al., 2003), and may suppress allergic contact dermatitis (Hasegawa et al., 1980). The biological functions of the EPS in *Zygnematophyceae* range from protection against radiation (Eder and Lütz-Meindl, 2010) and desiccation (Pichrtová et al., 2014a) to nutrient capture (Freire-Nordi et al., 1998, 2006), and EPS may reduce the sinking rate (Coesel, 1994). From the applied aspect, the high fucose content and the strong water binding capacity make the EPS interesting ingredients in future skin-care products (Ekelhof and Melkonian, 2017a). Furthermore, microalgal EPS retain their stable matrix structure and form a 3-D polymer network for cells to interact with each other, and mediate their adhesion to surfaces. Their superior rheological properties make EPS particularly useful in mechanical engineering (e.g., biolubricants and drag reducers) and food science/engineering (e.g., thickener and preservatives) uses (Xiao and Zheng, 2016).

Ekelhof and Melkonian (2017a) grew *N. digitus* in immobilized culture using lab-scale porous substrate bioreactors, so-called twin-layer (TL) systems, for the EPS production, which makes this alga interesting for biotechnological applications with a focus on cosmetics and food additives. It is shown that the cell as well as the EPS dry weight content is increased at least sixfold in immobilized compared to suspension culture. The EPS amount increased with time during the cultivation, and the medium exchange resulted in a significantly high final EPS concentration of 20.8 g m^{-2}. The major EPS component was xylose with 39.6% in suspension culture EPS and 37.2% in TL EPS, whereas the fucose content was higher in the TL samples (21.4%). The average of 14.6% of glucuronic acid, gives the polysaccharides their characteristic anionic properties (Ekelhof and Melkonian, 2017a).

In comparison to other algae, the finally produced EPS concentration in *N. digitus* (0.79 g L^{-1}) appeared high (Ekelhof and Melkonian, 2017a). The other Zygnematophyceaen species, *Penium margaritaceum* produced 0.75 g EPS L^{-1}, under non-stressed conditions (Domozych, 2007). Diatoms only produce EPS in the range of milligrams per liter (Wolfstein and Stal, 2002; Da Silva Maria et al., 2016), while *Chlamydomonas mexicana* reached about two-thirds of the concentration determined in *N. digitus* (Kroen, 1984). Comparable or higher amounts were observed for two *Porphyridium* species (Adda et al., 1986; Fuentes-Grünewald et al., 2015) as well as in the cyanobacterium *Cyanothece* sp. (Mota et al., 2013). Therefore, Ekelhof and Melkonian (2017a, b) demonstrated that the relatively slow-growing, but excessively EPS producing microalgal species *N. digitus* can be grown in porous substrate bioreactors and that this culturing technique is a promising alternative to suspension culture for the *Zygnematophyceae*. Formerly, Domozych (2007) reported that substitution of $NaNO_3$ with NH_4Cl and NH_4NO_3 resulted in noticeable decreases in cell number but increases in EPS production from *P. margaritaceum*. Ha et al. (1988) reported that the EPS of *C. turpinii*, which consisted of fucose, xylose, galactose, glucose, and glucuronic acid, had pseudoplastic properties due to the high viscosity, indicating the ability to function as a thickening agent.

The EPS composition determined in the study by Ekelhof and Melkonian (2017a, b) compares well with that of other desmids (Kattner et al., 1977; Paulsen and Vieira, 1994; Kiemle et al., 2007; Eder and Lütz-Meindl, 2010), even though differences in the abundance of the individual monosaccharides were observed. This underlines that EPS composition in conjugating algae is strain-specific, as suggested by Kiemle et al. (2007). *Closterium* sp., *Penium margaritaceum* and *Spondylosium panduriforme* also produced a large amount of EPS rich in xylose, fucose, galactose, and glucuronic acid and such EPS functioned by linking cells and substrate together, and contributing to initial adhesion, capsule formation and gliding (Domozych and Rogers-Domozych, 1993; Paulsen and Vieira, 1994; Domozych et al., 2005, 2007; Domozych and Domozych, 2014).

The high biomass productivity of the microalga *Spirogyra* sp. and its capacity to accumulate high amounts of carbohydrates, make this microalga attractive as substrate for bioethanol and bio H_2 production (Elsharnouby et al., 2013). The biofuel production by *Spirogyra* sp. is still in development and so far, few studies regarding bio-H_2 production using this alga have been published (Pacheco et al., 2015; Pinto et al., 2018). *Spirogyra* sp. represents as suitable source of high carbohydrate content for several reasons: (i) the outer cell wall is composed of pectose while the inner wall contains mainly cellulose; (ii) *Spirogyra* sp. cells produce a large mucilaginous layer on the outside; (iii) the chloroplasts contain pyrenoids around which starch accumulates (Pinto et al., 2018). Eshaq et al. (2011) and Wibowo et al. (2013) also reported *Spirogyra* sp. as one of the most efficient microalgae in carbohydrate accumulation, and consequently very adequate as feedstock for biofuel production.

The fermentation of the *Spirogyra* sp. biomass produced a peak in the H_2 yield (47 mL g^{-1} alga CDW or 156 mL g^{-1} total sugars) (Pacheco et al., 2015). The hydrogen percentage reached 10% (v/v) in the biogas produced, and this yield was comparable to that obtained by other authors with different microalgal biomass. It is closely dependent on the sugar content (30% w/w) of the *Spirogyra* sp. biomass used, since sugar is the main source for hydrogen production through dark fermentation (Pacheco et al., 2015). Pinto et al. (2018) applied macronutrient depletion as stress to increase the carbohydrate content in *Spirogyra*, and they gained a 36% (w/w) accumulation of carbohydrates by N depletion whereas P- or S-limitation produced no significant effect.

1.5 PROTEIN AND ENZYME INVESTIGATIONS IN ZYGNEMATOPHYCEAN ALGAE

Algae are generally regarded as a viable protein source, with composition meeting the Food and Agriculture Organization of the United Nations (FAO) requirements and they are regarded as good as other protein sources, such as soybean and egg (Bleakley and Hayes, 2017). Some microalgal sources have protein content higher than conventional animal or plant sources, e.g., the protein content of *Arthrospira* (*Spirulina*) *platensis* is 65%, higher than that of dried skimmed milk (36%), soy flour (37%), chicken (24%), fish (24%), beef (22%) and peanuts (26%) (Moorhead et al., 2011). Nevertheless, cysteine, tryptophan, and lysine are often limiting amino acids in most algae species, and bioavailability is considered lower than that of animal products (Barka and Blecker, 2016). It is thought that tannins and high polysaccharide content are the main factors which may negatively impact the digestibility of algal proteins (Bleakley and Hayes, 2017).

Up to date, there have been no detailed previous studies on the possible utilization of proteins from *Zygnematophyceae*. Since algal cell walls make them indigestible to humans and other animals, the isolation of protoplasts appeared a logical solution to the problems of increasing the availability of usable proteins (Berliner and Wenc, 1976a). Accordingly, a fast-growing desmid, *C. turpinii*, was incubated in a mineral medium + 0.4 M mannitol + 0.5% cellulysin and it was transformed into protoplasts in 4 h. Berliner and Wenc (1976a, b) concluded that the protoplasts obtained by this method appear to be good protein sources as they need not be harvested immediately and are not contaminated by cell wall products.

During the ecophysiological study on the heat response of *Micrasterias denticulata*, the occurrence and distribution of heat shock proteins (hsps) have been investigated (Weiss and Lütz-Meindl, 1999). Heat shock proteins hsp70 and BiP (binding protein) were detected by means of immunoblotting, in cells grown at different cultivation temperature. Densitometric measurements revealed an increase in hsp70 of about 45 to 155% after continuous heat treatment at different temperature levels, and of about 40 to 115% after cyclic heat exposure. Both the duration of heat exposure and the preceding cultivation temperature were decisive for the intensity of the hsp response (Weiss and Lütz-Meindl, 1999). The results of these authors indicate that at least hsp70 may be functional in protecting algae from the impact of temperature changes as its synthesis is triggered by temperature elevation, in accordance with the fact that accumulations of hsps in cells lead to increased tolerance and resistance (Hendrick and Hartl, 1993). The heat-induced inhibition of normal protein synthesis via block of pre-rRNA synthesis in plants usually coincides with the synthesis of a new set of proteins (hsps) which are either not present or present at a low level in untreated cells (Kregel, 2002). Stamenković et al. (2014b) suggested the presence of hsps in several *Cosmarium* strains treated at temperatures > 35°C by the observation of heat shock granules (stress granules) in the cytoplasm, similarly to what was observed for heat-stressed *M. denticulata* cells and several plant cells cultures (Neumann et al., 1984; Nover and Sharf, 1984; Meindl, 1990). In addition to the RNA preservation, hsgs represent multifunctional complexes, simultaneously combining enhanced chaperone activity by themselves as well as storage and release of hsps and other essential compounds during stress response and recovery (Smýkal et al., 2000; Kregel, 2002).

Hsps may have a wide and significant role in medicine; it is speculated that they are involved in BiP fragments from dead malignant cells presenting them to the immune system (Nishikawa et al., 2008). Hsps may be useful for increasing the effectiveness of cancer vaccines (Binder, 2008), whereas small molecule hsps, especially Hsp90 show promise as anticancer agents (Solit and Rosen, 2006). Hsps (especially hsp60 and hsp70) are used in clinical studies to treat rheumatoid arthritis and type I diabetes (Jansen et al., 2018). Therefore, some *Zygnematophyceae* may represent an interesting potential source of hsps; yet, comprehensive studies are needed to reveal the feasibility of the utilization of hsps from microalgae.

Among *Zygnemataceae*, *Spirogyra* was found to contain a large quantity of proteins (approx. 23.2% CDW) (Ontawong et al., 2013). Some species of *Spirogyra* have recently drawn attention due to their antioxidant activity *in vivo* and *in vitro* as well as renoprotective effects (Ontawong et al., 2013; Thumvijit et al., 2013a, b), which suggests that *Spirogyra* could be used as a source of natural antioxidants (Dash et al., 2014; Kumar et al., 2015). Whereas the antioxidant benefits of several terrestrial plant foods are established, much less is known about whether algal foods provide similar benefits (Wells et al., 2016). Macro- and microalgae peptides with antioxidant or preservative properties can prolong food shelf life either by delaying or inhibiting oxidation (Caporgno and Mathys, 2018).

The foremost enzymes that restrict oxidative damage in algae include the superoxide dismutases (SOD) that remove superoxide radical anions, and catalases and peroxidases, which convert hydrogen peroxide to water (Wells et al., 2016). De Jesus Raposo et al. (1989) attempted to correlate the occurrence of the copper-zinc form of SOD (Cu/Zn-SOD) with postulated evolution of land plants from green algae, and found Cu/Zn-SOD only in land plants and *Streptophyta* (including *Zygnematales* and *Desmidiales*). Apparently, all other groups of eukaryotic algae lack the Cu/Zn-SOD (Asada, 1977; Gerrath, 1993). The authors stated that the development of Cu/Zn-SOD occurred probably as oxygen levels on earth were increasing, and it was one of the crucial events which allowed the primitive *Streptophyta* to invade the land (Gerrath, 1993). In general, SOD has powerful antinflammatory activity, and it is a well-known remedy in the treatment of osteoarthritis, rheumatoid arthritis, pulmonary fibrosis, colitis, and urinary tract inflammatory disease (Bafana et al., 2011). It may also prevent radiation-induced dermatitis and reduce free radical damage to skin (Campana et al., 2004). However, thorough biochemical and clinical analyzes are needed to demonstrate if SOD extracts from microalgal sources are suitable to be used in medicine and cosmetics.

In addition to the phylogenetic significance of SOD, the enzyme studies in *Zygnematophyceae* dealt with the distribution patterns of enzymes which contribute to phylogenetic theories concerning the origin of land plants, such as urease/urea amidolyase and glycollate oxidase/dehydrogenase as well as arginase and arginine deiminase (Stewart and Mattox, 1978; Syrett and Al-Houty, 1984; Laliberté and Hellebust, 1991). Further, a recent analysis of adhesion mechanisms in *Zygnematophyceae* revealed that arabinogalactan proteins (AGPs) in the extracellular matrix are most likely key adhesion molecules (Palacio-Lopez et al., 2019). AGPs are suggested to perform in cell-to-cell adhesion in algae forming thalli, and cell to surface adhesion in the filamentous forms. These findings enabled a broader evolutionary understanding of the function of AGPs in green algae (Palacio-Lopez et al., 2019).

1.6 USE OF ZYGNEMATOPHYCEAN ALGAE IN BIOREMEDIATION OF WASTEWATERS

1.6.1 RELATIONSHIPS WITH METALS AND BIOSORPTION OF HEAVY METALS

Some earlier investigations showed that several desmids (*Cosmarium pachydermum*, *Desmidium swartzii*, *Euastrum oblongum*, *E. verrucosum*, *Pleurotaenium truncatum*) had high tolerance to zinc, manganese, vanadium, and chromium, but not to copper (Url, 1955). Desmids appeared sensitive to aluminum at concentrations in the range 100–200 μg L^{-1}, with the exception of *Xanthidium octocornis*, *Staurodesmus indentatus*,

Staurastrum arachne var. *curvatum* and *S. pentacerum* (Pillsbury and Kingston, 1990). Sodium arsenate slightly stimulated growth of *C. leave* and *C. obtusatum* at low concentration (5 μg mL^{-1}) but higher concentration (20 μg mL^{-1}) caused the inhibition of cell yield (Jayaraman and Sarma, 1985). Cadmium chloride (up to 0.25 μg mL^{-1}) inhibited growth in *C. lanceolatum*, *C. moniliferum*, *C. leave* and *C. obtusatum* (Jayaraman and Sarma, 1985).

Lorch (1978, 1986) studied the effect of lead (Pb) and manganese (Mn) on five desmids. Pb had little effect on dry weight yield of *Netrium digitus*, however, it decreased significantly growth of other desmids. Concentrations of 7 mg L^{-1} Mn affected morphogenesis in *Micrasterias rotata*, *Gonatozygon aculeatum*, and *Penium spirostriolatum*. Smaller Mn concentrations (1–2 mg L^{-1}) increased growth of *P. spirostriolatum* and *C. ehrenergii*, indicating the nutritional requirement for Mn in these species (Lorch, 1978). Mercury (0.08 ppm) caused partly an inhibition of CO_2 fixation and an inhibition of thymidine incorporation in *C. moniliferum* (Sastry and Chaudhary, 1989). Sodium selenate at concentrations > 20 μg mL^{-1} was lethal to *C. lanceolatum* and *C. obtusatum* in sulfate-free medium; yet, the addition of sulfate increased the rate of survival the desmids investigated (Jayaraman and Sarma, 1985). Interestingly, *Cylindrocystis brebissoni* and *C. subarctoum* were found in all regions of zinc-polluted British streams, including those areas with the distinctly high levels of Zn in water (20–30.2 mg L^{-1}) (Sathaiah et al., 1984).

With the exception of *Oocardium stratum*, peculiar colonial desmid inhabiting brooks rich in lime (Rott et al., 2012), desmids were regarded as calciphobes (Url, 1955). Moss (1972) examined the growth responses of desmids associated with both eutrophic and oligotrophic waters and looked at the influence of Ca^{2+} levels and the ratios of monovalent to divalent cations as possible controlling factors. He reported that 0.1 Ca^{2+} mg L^{-1} appeared to be an adequate concentration for all the desmids tested, except *Desmidium swartzii* and *Roya* sp. for which even 1.0 mg Ca^{2+} L was insufficient. However, Brook (1965) and Kovask (1973) found a considerable number of desmids at high Ca^{2+} concentrations (40–60 mg L^{-1}), and questioned whether Ca^{2+} is a limiting factor in desmid distribution. Experiments to modify Bold's Basal Medium for growth of *C. subtriordinatum* showed that optimum concentrations of Ca^{2+}, Mg^{2+} and Na^{+} were 25 mg L^{-1} and that of K^{+} was 100 mg L^{-1} (Rajalakshmi, 1986), revealing that optimal Ca^{2+} values were similar to those found in studies on other desmids (Brook, 1981).

Increasing environmental pollution due to progressive traffic and industrial as well as agricultural production leads to the release of heavy metals into air, soil, and water. As a typical inhabitant of unpolluted freshwater ecosystems, *Micrasterias denticulata* appeared a suitable model organism for the investigation of influences of heavy metals on physiology and anatomy of microalgae (Lütz-Meindl, 2016). Morphogenesis and cell development of *M. denticulata* are negatively affected by Zn, Al, Cd, Cr, and Pb when applied at the highest concentrations that still allowed cell growth (Volland et al., 2011, 2012, 2014; Andosch et al., 2012). It was shown that aluminum is only bound to the cell wall of *Micrasterias* when applied in long-term experiments, and it was not found intracellularly. Similar results have been obtained after incubation of *Micrasterias* cells with lead that leads to severe cell shape malformation but is neither found in the cell wall nor in any intracellular compartment (Volland et al., 2011, 2014). In this way, the cell wall may act as a kind of filter by accumulating the metals thus preventing more severe intracellular damage. The result that Pb does not enter *Micrasterias* cells does not correspond to findings in higher plants, where Pb was frequently found to be taken up into the cytoplasm and to affect intracellular components (Lütz-Meindl, 2016). Zinc and copper were identified in cell wall precipitations of *Micrasterias* after long-term exposure, and both metals were found in mucilage vesicles which are secreted steadily in *Micrasterias* (Volland et al., 2011). Elimination of metals from the cytoplasm by using mucilage vesicles as fast vehicles seems to represents an important detoxification mechanism in *Micrasterias*. Zn is additionally compartmentalized in vacuoles of *Micrasterias* which become electron-dense up on continuing Zn influence. Among all metals investigated Cu is compartmentalized the best in *Micrasterias*. Besides sequestration in the cell wall and in mucilage vesicles, Cu was also found as precipitates in starch grains where it may help to avoid toxic effects as long as starch is not catabolized (Volland et al., 2011). Whereas cell development and pattern formation is almost completely suppressed by 1 mM solutions of both Cr^{3+} and Cr^{6+}, only Cr^{6+} evoked a complete arrest of cell divisions even when applied in the low concentration (5 μM). Furthermore, electron-dense precipitations in bag-like structures were found along the inner side of the cell wall under Cr^{6+} impact. As these Cr-containing bags were located outside of the plasma membrane, these results indicate that Cr is extruded from the *Micrasterias* cell in the form of an iron-oxygen compound (Volland et al., 2012).

Among all metals tested on *Micrasterias* cadmium was the only one that, though extremely toxic to physiology and ultrastructure of the cells, was not compartmentalized intracellularly at all (Volland et al., 2011; Andosch et al., 2012). Cd is highly water-soluble and enters aquatic ecosystems and soils mainly as a consequence of anthropogenic activities such as disposal of electronic components. It was shown that Cd induces the formation of phytochelatins in *Micrasterias*, which were neither detected in control cells nor in Cu-exposed *Micrasterias* cells (Volland et al., 2013). Phytochelatins are known to be involved in detoxification of metals in higher plants, fungi, and green algae such as *Chlamydomonas* and others (Lütz-Meindl, 2016). Filtering by the extraplasmatic matrix (mucilage and/or cell wall), excretion by mucilage vesicles and intracellular compartmentalization seem to be the most important detoxification strategies of desmids that allow survival of the cells within certain concentration and duration limits (Lütz-Meindl, 2016).

The attribute that some heavy metals can be retained within desmid cells is used for the bioaccumulation of strontium using *Closterium moniliferum* (Krejci et al., 2011a, b). *Closterium moniliferum* belongs to a small number of organisms that form barite ($BaSO_4$) or celestite ($SrSO_4$) biominerals. Wilcock et al. (1989) demonstrated, but did not quantify, Sr incorporation into desmid crystals in a culture medium with a high ratio of Sr^{2+} to Ba^{2+}. This Sr incorporation is a consequence of Sr^{2+} substitution for Ba^{2+} in the barite crystal lattice to form a $(Ba,Sr)SO_4$ solid solution. Krejci et al. (2011a) have shown that it is possible to create growth conditions under which the desmid *C. moniliferum* precipitates crystals with up to 45 mol.% of the Ba lattice positions replaced by Sr. The ability to sequester Sr in the presence of an excess of Ca^{2+} is of considerable interest for the remediation of ^{90}Sr from the environment and nuclear waste. Based on elevated levels of sulfate detected in the terminal vacuoles, "sulfate trap" model was proposed, where the presence of dissolved barium leads to preferential precipitation of $(Ba,Sr)SO_4$ due to its low solubility relative to $SrSO_4$ and $CaSO_4$ (Krejci et al., 2011b). Therefore, desmids represent attractive candidates for bioremediation of low-level radioactive effluents as they are robust in culture, needing only sunlight and a few nutrients, and with a small amount of Ba^{2+} in the environment could act as Sr^{2+} sinks through the sulfate trap mechanism (Krejci et al., 2011a, b).

In general, green algae are the most commonly used algal group to adsorb metals, and they have considerable potential to treat metal-contaminated wastewaters. Carbohydrates and EPS are believed to play an important role in metal biosorption by algae and major influential factors include EPS composition, metal species, solution chemistry and operating conditions (Mehta and Gaur, 2005; Xiao and Zheng, 2016). The accumulation of heavy metals in algae involves two processes: an initial rapid (passive) uptake followed by a much slower (active) uptake (Gadd, 1988). During the passive uptake, metal ions adsorb onto the cell surface within a relatively short span of time (a few seconds or minutes), and the process is metabolism independent. Active uptake is metabolism-dependent, causing the transport of metal ions across the cell membrane into the cytoplasm (Mehta and Gaur, 2008). The use of dried, nonliving or chemically pretreated microalgae seems to be a preferred alternative to the use of living cells in industrial applications for the removal of heavy metal ions from wastewater. The use of dead cells offers the following advantages over live cells: the metal removal system is not subject to toxicity limitations, there is no requirement for growth media and nutrients, the biosorbed metal ions can be easily desorbed, and biomass can be reused (Kratochvil and Volesky, 1998). Conventional methods of metal removal from wastewater, such as by activated carbon, are expensive and not always effective for metals in low concentrations (Ince and Ince, 2017).

Among *Zygnematophyceae*, *Spirogyra* sp. is naturally abundant throughout the world and ease of harvesting. In recent years, many studies have applied both living and dead specimens of *Spirogyra* sp. to nutrient removal (De Busk et al., 2004), as well as tannery and textile wastewater treatment (Mohan et al., 2002; Özer et al., 2006; Khalaf, 2008; Onyancha et al., 2008). Many studies have reported that dried biomass of *Spirogyra* sp. has a very high capacity for binding with metals due to the presence of polysaccharides, proteins, or lipids on the surface of cell walls. These contain functional groups such as aminos, hydroxyls, carboxyls, and sulfates, which can act as binding sites for metals (Vogel and Bergmann, 2018).

Lee and Chang (2011) showed that the adsorption rate of Pb^{2+} and Cu^{2+} by dry *Spirogyra* biomass was extremely high during the first 30 min of application, comprising approximately 95% of the total adsorption, due to the adsorption on the surface of the algae powder. Slower adsorption that followed may have involved other mechanisms, such as complexation, micro-precipitation, and binding site saturation. The capacity of *Spirogyra* to adsorb Pb^{2+} and Cu^{2+} was 87.2 mg g^{-1} and 38.2 mg g^{-1}, respectively. However, when pH exceeded 5, the capacity to absorb Pb^{2+} and Cu^{2+} decreased; this may have been due to the precipitation of copper hydroxides and lead hydroxides (Gupta et al., 2006; Lee and Chang, 2011). In another study, *Spirogyra* biomass had maximum adsorption capacity of Pb^{2+} around 140.8 mg metal g^{-1} of biomass with initial Pb^{2+} concentration of 200 mg L^{-1}. The value of Pb uptake was significantly higher than reported for other biosorbents (Gupta and Rastogi, 2008). Romera et al. (2007) demonstrated that brown macroalgae appeared better metal absorbers than green algae; yet, *Spirogyra insignis* absorbed rather a high amount of Pb^{2+} (51.5 mg g^{-1}). Further, Singh et al. (2000) obtained a larger uptake of Pb using *Spirogyra* as biosorbent, compared to Cd^{2+}, Zn^{2+} and Cu^{2+} uptake. The sorption values for nickel, copper, and zinc were very similar and the general sequence of the biosorption maximum in *Spirogyra* biomass was Pb > Cd ≥ Cu > Zn > Ni (Romera et al., 2007; Matei et al., 2014). Melčáková and Růžovič (2010) used dry biomass of *Spirogyra* to bind up Zn^{2+} ions, and concluded that *Spirogyra* sp. had slightly higher adsorption capacity for zinc (17.8 mg g^{-1}) than leaves of *Reynoutria japonica* (17.0 mg g^{-1}).

Gupta et al. (2001, 2006) revealed that dried biomass of *Spirogyra* had the maximum biosorption capacity of 133.3 mg Cu^{2+} g^{-1} at an optimum pH of 5. Desorption studies were conducted with Cu-loaded biomass using different desorption agents including HCl, EDTA, H_2SO_4, NaCl, and H_2O. The maximum desorption of 95.3% was obtained with HCl in 15 min. Furthermore, maximum removal of Cr^{6+} from wastewater was around 14.7×10^3 mg metal kg^{-1} of dry weight at pH 2 (Gupta et al., 2001). Removal of Cr^{3+} was more than 70% using differently treated and untreated *Spirogyra* biomass with the initial concentration 30 mg L^{-1} (Bishnoi et al., 2007). All these results indicate that *Spirogyra* can be used as an efficient and economic biosorbent material for the removal and recovery of toxic heavy metals from polluted waters (Gupta et al., 2001, 2006; Bishnoi et al., 2007).

Mercury is generally considered to be one of the most toxic metals found in the environment, and various treatment technologies such as precipitation, ion exchange, and adsorption have been employed to remove metal pollutants from aqueous solutions. One of the promising techniques for the removal of Hg is the biosorption using living or nonliving microalgae (Rezaee et al., 2008). The authors used dried biomass of *Spirogyra* sp. which accumulated over 60–70% of Hg at lower pH and low temperatures (Rezaee et al., 2008). Biosorption of arsenic (III) and (V) from aqueous solutions by living and dried biomass of freshwater microalgae was investigated using five strains belonging to *Chlorella*, *Oscillatoria*, *Scenedesmus*, *Spirogyra*, and *Pandorina* (Sibi, 2014). The dried biomass of *Spirogyra* absorbed up to 25 mg As^{3+} g^{-1} and it was found slightly more effective for arsenic sorption than the living biomass (Sibi, 2014). Dried *Spirogyra* biomass absorbed also Se in the range 70–98.2%, for metal concentrations 5–10 mg L^{-1} (Mane et al., 2011).

In addition, biomass of *Zygnema* sp. showed a fair potential to remove Cr^{6+}, Hg, and Ni (Khoramabadi et al., 2008; Soni and Gupta, 2011; Sivaprakash et al., 2015). Maximum absorption of Cr with formaldehyde-treated alga was 80% at pH 2, and 63% using untreated biomass (Soni and Gupta, 2011). Remarkably, dry biomass of *Zygnema fanicum* achieved the maximum of Hg absorption rate (80%) at pH 8.5, and when the initial Hg concentration is increased, absorption rate also increased (Khoramabadi et al., 2008). Using dried biomass of *Zygnema* sp. maximum nickel removal of 76.4% at a dosage of 7.5 g L^{-1} was achieved (Sivaprakash et al., 2015). The authors concluded that *Zygnema* biomass was found to be a valuable material for the removal of Ni from industrial wastewater, and a better substitute for the conventional adsorbents.

Beside the absorption of various heavy metals, biomass of *Spirogyra* was used to estimate fluoride sorption capacity (Venkata Mohan et al., 2007). The highest fluoride biosorption values were observed at pH 2 (62%) where the overall surface charges on the algal cells should be positive, which facilitated the binding of negatively charged fluoride ion. The fluoride sorption efficiency at pH 7 was found to be 54%, which was suitable for practical reasons with respect to the upscaling of the technology (Venkata Mohan et al., 2007). Özer et al. (2006) achieved almost complete removal of Acid Red 274 dye from synthetic wastewater using dried biomass of *Spirogyra rhizopus*, whereas *Spirogyra* sp. exhibited maximum removal of reactive dye (Synazol) of 85% at pH 3 (Khalaf, 2008). Alaguprathana and Poonkothai (2017) revealed that *Spirogyra gracilis* could have decolorized the textile effluent up to 74% at pH 4.

1.6.2 BIOSORPTION OF NUTRIENTS

In general, desmids are recognized as algae thriving in freshwater environments characterized by low amounts of nitrogen (N) and phosphorus

(P) compounds (Brook, 1981). Most desmids utilize nitrate as a nitrogen source, but some desmids may utilize ammonia instead. Desmids requiring ammonia include *S. tetracerum* (Venkateswarlu, 1983) and *C. aciculare* (Coesel, 1991), both of which are known to occur in eutrophic water bodies. The inability of *C. aciculare* to grow in media with nitrate as the sole nitrogen source was traced to the complete lack of nitrate reductase activity in this desmid (Coesel, 1991). This finding was along with the presence of dense populations of *Cl. aciculare* in eutrophic lakes having high ammonium concentrations (30–50 μM). An experiment designed to study the effect of external phosphorus depletion on growth and intracellular phosphorus storage in *Peridinium cinctum* and *Cosmarium* sp. demonstrated that the *Cosmarium* showed a significant decrease of biomass yield in the lowest phosphate concentration (Elgavish and Elgavish, 1980; Elgavish et al., 1982). Returning phosphate deficient cells to higher phosphate levels (8.3 μM) restored biomass yield and intracellular phosphate values to normal levels for cells grown with higher phosphate levels. *Cosmarium* cells grown in excess phosphate (200 μM) took up more than 50% of the phosphorus in the culture, whereas only about 7% of the total phosphorus was in the *Peridinium* cells. When phosphate-deficient *Cosmarium* cells from stationary phase of growth were transferred to a lower phosphate level (0.6 μM), within 8 h almost all the phosphate in the culture medium was found within the cells. Similar rapid uptake has been reported for phosphate-deficient cells of other algae (Healey, 1979; Powel, 2011) as well as for desmids from eutrophic habitats, which is considered their strategy to insert phosphorus quickly during temporary pulses (Spijkerman and Coesel, 1996). Recent floristic investigations demonstrate that many desmids known as indicators of oligotrophic level have been commonly found in mesotrophic to eutrophic waters, while eutrophic taxa have been recorded in effluents from agricultural complexes (Fehér, 2003; Stamenković and Cvijan, 2008; Ferragut and Bicudo, 2009, 2012; da Silva et al., 2018). This indicated that desmids increased their tolerance threshold to increased concentrations of nutrients and various pollutants, and they could possibly be used as biosorbents of excess nutrients.

The presence of nutrients in the form of nitrate, nitrite, ammonia/ammonium, or phosphorus in wastewater may lead to eutrophication. Phytoremediation is the use of algae for the removal of pollutants from wastewaters since algal species are relatively easy to grow, adapt, and manipulate within a laboratory setting and appear to be ideal organisms for use in remediation studies (Sen et al., 2013). In addition, phytoremediation has advantages over other conventional physicochemical methods, such as ion exchange, reverse osmosis, dialysis, and electro-dialysis, membrane separation, activated carbon adsorption, and chemical reduction or oxidation, due to its better nutrient removal efficiency and the low cost of its implementation and maintenance (Thomas et al., 2016).

Among zygnematophycean algae *Spirogyra* sp. received a considerable attention to be used for wastewater purification (Ge et al., 2018). Kumar et al. (2016) used a freshly collected *Spirogyra* sp. for the phytoremediation of sugar mill effluent. Their study showed that the maximum removal of total dissolved solids (24.92%), biological oxygen demand (BOD) (47.82%), chemical oxygen demand (COD) (15.73%), total N (40%), PO_4^{3-} (44.44%), Ni (42.79%), Fe (24.78%) and Mn (34.92%) was recorded after 60 days of the experiment. Naturally isolated *Spirogyra* sp. was grown in three different types of municipal wastewater – primary, secondary, and centrate wastewaters (Ge et al., 2018). Nitrogen and phosphorus removal efficiencies ranged from 50.6–90.6% and 60.4–99.1%, respectively. Based on ultimate analysis, the biomass showed relatively consistent protein (16.7–19.5% of the dry mass fraction), carbohydrate (41.5–55.0%) and lipid (2.8–10.0%) contents (Ge et al., 2018). The study indicated the feasibility of using *Spirogyra* sp. to recover nutrients from multiple municipal wastewater sources with the simultaneous production of biomass that contains valuable biochemical components for energy applications. *Spirogyra* sp. utilized efficiently both NO_3^- and NH_4^+, possibly with a slight preference for ammonia nitrogen (Ge et al., 2018).

Barnard et al. (2017) used *Spirogyra grevilleana* in an experimental biofiltration system (Algal Filtration Device) to reduce levels of *Escherichia coli*, nitrates, and phosphates from a lake. *Spirogyra grevilleana* reduced *E. coli* by 100% and significantly reduced nitrate concentrations (30%) and phosphate concentrations (23%) while maintaining dissolved oxygen and pH at normal levels. Initial results indicate that the use of *S. grevilleana* in conjunction with an algal filtration device is potentially capable of creating potable water (Barnard et al., 2017). Furthermore, microalgal consortium consisting of *Actinastrum*, *Scenedesmus*, *Chlorella*, *Spirogyra*, *Nitzschia*, *Chlorococcum*, *Closterium*, and *Euglena* removed 96% ammonium and >99% orthophosphate from a diary wastewater, and achieved 99% removal of these compounds from municipal wastewater with the addition of CO_2 (Woertz et al., 2009). Interestingly, fresh biomass of *Spirogyra* showed high removal rates for several pharmaceuticals from wastewater: 31% carbamazepine (CBZ), 99% caffeine and acetaminophen, 89% propranolol, 92% ibuprofen, 54% diclofenac, 35% clofibric acid, and 95% bisphenol A (Garcia-Rodríguez et al., 2015). Beside *Spirogyra* sp., *Zygnema sterile* was grown in high-rate culture system and appeared to have a high biosorption rates for nutrients (Kong et al., 2007).

1.7 CONCLUSIONS AND FUTURE PROSPECTS

The class *Zygnematophyceae* is a cosmopolitan and widely spread group of algae in freshwater ecosystems. Members of this algal group are known as valuable model organisms for the study of many biological processes as well as for the investigation of ecophysiological attributes in microalgae. Some of these investigations pointed those desmids and other conjugating algae might have a fair potential to produce metabolites that can be used for commercial and economic purposes.

Being algae adapted to high light intensities, *Zygnematophyceae* have high amounts of photosynthetic pigments, among which zeaxanthin and lutein have precious role in medicine and agriculture. Furthermore, studies showed that astaxanthin and gallotannins in *Zygnemataceae* could possibly be utilized from some strains and they have medical relevance. A high quantity of FAs was found in several desmids grown in standard conditions, which rendered them as oleaginous algae. Several FAs, such palmitic, linoleic, and α-linolenic acid, are found in abundance in desmids, and they may have an important role in cosmetics, medicine, and industry. Furthermore, new desmid isolates have high proportions of oleic acid, which is particularly suitable for biodiesel production. EPS in the form of mucilaginous sheaths contain a large proportion of fucose, and may also have a role in biosorption of heavy metals. *Spirogyra* sp. has a high biomass productivity and it accumulates large amounts of carbohydrates, rendering this alga as a suitable substrate for bioethanol and hydrogen production. Some desmids (*Cosmarium* spp. and *Micrasterias denticulata*) produce an abundance of hsps and antioxidant enzymes, and so further studies should deal with the extraction and utilization of these compounds.

Finally, both *Desmidiaceae* and *Zygnemataceae* appeared efficient microalgae to be used for biosorption of heavy metals and nutrients. Dried biomass of *Spirogyra* sp. is especially effective in Pb, Hg, and

Cd uptake, and it may possibly be used for the bioremediation of larger quantities of industrial wastewaters. *Closterium moniliferum* appeared promising alga for the bioaccumulation of radioactive strontium, when small amounts of barium and calcium are also provided in medium. Therefore, the conjugating algae are not only fascinating algae concerning their taxonomy and physiology; they may also have a precious role in the production of useful products as well as in the purification of various types of wastewaters.

ACKNOWLEDGMENTS

This work was supported by the Ministry of Education, Science, and Technological Development of Serbia, Grant No. 173018. M. Stamenković was a scholarship holder of the German Academic Exchange Service (DAAD) foundation during her PhD studies (DAAD No. A/08/91041). A postdoctoral research grant of the Swedish Institute (SI No. 02390/2016) was provided to M. Stamenković.

KEYWORDS

- **bioremediation**
- **biosorption**
- **carbohydrates**
- **desmids**
- **fatty acid**
- **fucose**
- **gallotannin**
- **heat shock protein**
- **lutein**
- **xanthophyll**
- **zeaxanthin**
- ***Zygnematophyceae***
- **α-linolenic acid**

REFERENCES

Abomohra, A. E. F., Eladel, H., El-Esawi, M., et al., (2018). Effect of lipid-free microalgal biomass and waste glycerol on growth and lipid production of *Scenedesmus obliquus*: Innovative waste recycling for extraordinary lipid production. *Bioresour. Technol., 249*, 992–999.

Abomohra, A. E. F., El-Sheekh, M., & Hanelt, D., (2017). Screening of marine microalgae isolated from the hypersaline Bardawil lagoon for biodiesel feedstock. *Renew. Energ., 101*, 1266–1272.

Abomohra, A. E. F., Wagner, M., El-Sheekh, M., & Hanelt, D., (2013). Lipid and total fatty acid productivity in photoautotrophic fresh water microalgae: Screening studies towards biodiesel production. *J. Appl. Phycol., 25*, 931–936.

Adda, M., Merchuk, J. C., & Arad, S. M., (1986). Effect of nitrate on growth and production of cell-wall polysaccharide by the unicellular red alga *Porphyridium*. *Biomass, 10*, 131–140.

Aigner, S., Remias, D., Karsten, U., & Holzinger, A., (2013). Unusual phenolic compounds contribute to ecophysiological performance in the purple colored green alga *Zygogonium ericetorum* (*Zygnematophyceae, Streptophyta*) from a high-alpine habitat. *J. Phycol., 49*(4), 648–660.

Alaguprathana, M., & Poonkothai, M., (2017). *Spirogyra gracilis* – a potent algae for the remediation of textile dyeing effluent. *J. Env. Bio-Sci., 31*, 345–355.

Allakhverdiev, S. I., Los, D. A., & Murata, N., (2009). Regulatory roles in photosynthesis of unsaturated fatty acids in membrane lipids. In: Wada, H., & Murata, N., (eds.), *Lipids in Photosynthesis: Essential and Regulatory Functions* (pp. 373–388). Springer, Dordrecht.

Aminul, I. M., Magnusson, M., Brown, R. J., Ayoko, G. A., Nabi, M. N., & Heimann, K., (2013). Microalgal species selection for biodiesel production based on fuel properties derived from fatty acid profiles. *Energies, 6*(11), 5676–5702.

Ando, H., Ryu, A., Hashimoto, A., Oka, M., & Ichihashi, M., (1998). Linoleic acid and α-linolenic acid lightens ultraviolet-induced hyperpigmentation of the skin. *Arch. Dermatol. Res., 290*, 375–381.

Andosch, A., Affenzeller, M. J., Lütz, C., & Lütz-Meindl, U., (2012). A freshwater green alga under cadmium stress: Ameliorating calcium effects on ultrastructure and photosynthesis in the unicellular model *Micrasterias*. *J. Plant Physiol., 169*(15), 1489–1500.

Asada, K. S., Kanematsu, S., & Uchida, K., (1977). Superoxide dismutases in photosynthetic organisms: Absence of the cuprozinc enzyme in eukaryotic algae. *Arch. Biochem. Biophys., 179*(1), 243–256.

ASTM D6751-08, (2008). *International Standard Specification for Biodiesel Fuel Blend Stock (B100) for Middle Distillate Fuels*. ASTM D6751-08, ASTM International, West Conshohocken.

Bafana, A., Dutt, S., Kumar, S., & Ahuja, P. S., (2011). Superoxide dismutase: An industrial perspective. *Crit. Rev. Biotechnol., 31*(1), 65–76.

Barka, A., & Blecker, C., (2016). Microalgae as a potential source of single-cell proteins: A review. *Biotechnol. Agron. Soc. Environ., 20*(3), 427–436.

Barnard, M. A., Porter, J. W., & Wilde, S. B., (2017). Utilizing *Spirogyra grevilleana* as a phytoremediatory agent for reduction of limnetic nutrients and *Escherichia coli* concentrations. *Am. J. Plant Sci., 8*(5), 1148–1158.

Benalla, W., Bellahcen, S., & Bnouham, M., (2010). Antidiabetic medicinal plants as a source of alpha glucosidase inhibitors. *Curr. Diabetes Rep., 6*(4), 247–254.

Berliner, M. D., & Wenc, K. A., (1976a). Osmotic pressure effects and protoplast formation in *Cosmarium turpinii*. *Microbios Letters, 2*, 39–45.

Berliner, M. D., & Wenc, K. A., (1976b). Protoplast induction in *Micrasterias* and *Cosmarium*. *Protoplasma, 89*, 389–393.

Binder, R. J., (2008). Heat-shock protein-based vaccines for cancer and infectious disease. *Expert Rev. Vaccines, 7*(3), 383–393.

Bishnoi, N. R., Kumar, R., Kumar, S., & Rani, S., (2007). Biosorption of Cr (III) from aqueous solution using algal biomass *Spirogyra* spp. *J. Hazard. Mater., 145*(1, 2), 142–147.

Bleakley, S., & Hayes, M., (2017). Algal proteins: Extraction, application, and challenges concerning production. *Foods, 6*(5), E33.

Brook, A. J., (1965). Planktonic algae as indicators of lake types with special reference to the *Desmidiaceae*. *Limnol. Oceanogr., 10*(3), 403–411.

Brook, A. J., (1981). *The Biology of Desmids* (pp. 15–186). Blackwell Scientific Publications, Oxford.

Brook, A. J., (1982). Desmids of the *Staurastrum tetracerum*-group from a eutrophic lake in mid-Wales. *Brit. Phycol. J., 17*(3), 259–274.

Browse, J., (2009). Oxidation of membrane lipids and functions of oxylipins. In: Wada, H., & Murata, N., (eds.), *Lipids in Photosynthesis: Essential and Regulatory Functions* (pp. 389–405). Springer, Dordrecht.

Campana, F., Zervoudis, S., Perdereau, B., et al., (2004). Topical superoxide dismutase reduces post-irradiation breast cancer fibrosis. *J. Cell. Mol. Med., 8*(1), 109–116.

Cannell, R. J. P., Farmer, P., & Walker, J. M., (1988a). Purification and characterization of pentagalloylglucose, an alpha-glucosidase inhibitor/antibiotic from the freshwater green alga *Spirogyra varians*. *Biochem. J., 255*(3), 937–941.

Cannell, R. J. P., Kellam, S. J., Owsianka, A. M., & Walker, J. M., (1987). Microalgae and cyanobacteria as a source of glucosidase inhibitors. *J. Gen. Microbiol., 133*(7), 1701–1705.

Cannell, R. J. P., Owsianka, A. M., & Walker, J. M., (1988b). Results of a large-scale screening program to detect antibacterial activity from freshwater algae. *Eur. J. Phycol., 23*(1), 41–44.

Caporgno, M. P., & Mathys, A., (2018). Trends in microalgae incorporation into innovative food products with potential health benefits. *Front Nutr., 5*, 58.

Casper-Lindley, C., & Björkman, O., (1998). Fluorescence quenching in four unicellular algae with different light-harvesting and xanthophyll-cycle pigments. *Photosynth. Res., 56*(3), 277–289.

Coesel, P. F. M., & Meesters, K. J., (2007). *Desmids of the Lowlands* (pp. 11–258). KNNV Publishing, Zeist.

Coesel, P. F. M., (1991). Ammonium dependency in *Closterium aciculare* T. West, a planktonic desmid from alkaline, eutrophic waters. *J. Plankton Res., 13*(5), 913–922.

Coesel, P. F. M., (1994). On the ecological significance of a cellular mucilaginous envelope in planktic desmids. *Alg. Studies, 73*, 65–74.

Coesel, P. F. M., (2001). A method for quantifying conservation value in lentic freshwater habitats using desmids as indicator organisms. *Biodivers. Convers., 10*, 177–187.

Coesel, P. F. M., (2003). Desmid flora data as a tool in conservation management of Dutch freshwater wetlands. *Biol. Brat., 58*(4), 717–722.

Cranwell, P. A., Jaworshi, G. H. M., & Bickley, H. M., (1990). Hydrocarbons, sterols, esters and fatty acids in six freshwater chlorophytes. *Phytochemistry, 29*(1), 145–151.

Da Silva, F. K. L., Fonseca, B. M., & Felisberto, S. A., (2018). Community structure of periphytic *Zygnematophyceae* (*Streptophyta*) in urban eutrophic ponds from central Brazil (Goiânia, GO). *Acta Limn. Bras., 30*, e206.

Da Silva, M. L., De Oliveira Da, R. F. A., Odebrecht, C., Giroldo, D., & Abreu, P. C., (2016). Carbohydrates produced in batch cultures of the surf zone diatom *Asterionellopsis glacialis* sensu lato: Influence in vertical migration of the microalga and in bacterial abundance. *J. Exp. Mar. Biol. Ecol., 474*, 126–132.

Darmstadt, G. L., Mao-Qiang, M., Chi, E., et al., (2002). Impact of topical oils on the skin barrier: Possible implications for neonatal health in developing countries. *Acta Paediatr., 91*(5), 546–554.

Dash, P., Tripathy, N. K., & Padhi, S. B., (2014). Novel antioxidant production by *Cladophora* sp. and *Spirogyra* sp. *Med Sci., 7*(25), 74–78.

De Busk, T. A., Grace, K. A., Dierberg, F. E., Jackson, S. D., Chimney, M. J., & Gu, B., (2004). An investigation of the limits of phosphorus removal in wetlands: A mesocosm study of a shallow periphyton-dominated treatment system. *Ecol. Eng., 23*(1), 1–14.

De Jesus, R. F. M., Bernado De, M. A. M., & Santos, C. D. M. R. M., (2015). Marine polysaccharides from algae with potential biomedical applications. *Mar. Drugs, 13*(5), 2967–3028.

De Vries, J., Stanton, A., Archibald, J. M., & Gould, S. B., (2016). Streptophyte terrestrialization in light of plastid evolution. *Trends Plant Sci., 21*(6), 467–476.

Del Campo, J. A., Rodríguez, H., Moreno, J., Vargas, M. A., Rivas, J., & Guerrero, M. G., (2001). Lutein production by *Muriellopsis* sp. in an outdoor tubular photobioreactor. *J. Biotechnol., 85*(3), 289–295.

Demmig-Adams, B., & Adams, W. W., (1992a). Carotenoid composition in sun and shade leaves of plants with different life forms. *Plant Cell Environ., 15*(4), 411–419.

Demmig-Adams, B., & Adams, W. W., (1992b). Photoprotection and other responses of plants to high light stress. *Annu. Rev. Plant Physiol. Plant Mol. Biol., 43*, 599–626.

Demmig-Adams, B., & Adams, W. W., (2006). Photoprotection in an ecological context: The remarkable complexity of thermal energy dissipation. *New Phytol., 172*(1), 11–21.

Demmig-Adams, B., (1998). Survey of thermal energy dissipation and pigment composition in sun and shade leaves. *Plant Cell Physiol., 39*(5), 474–482.

Domozych, D. S., & Domozych, C. E., (2014). Multicellularity in green algae: Upsizing in a walled complex. *Front. Plant. Sci., 5*, 649.

Domozych, D. S., & Rogers-Domozych, C., (1993). Mucilage processing and secretions in the green alga *Closterium* II. Ultrastructure and immunocytochemistry. *J. Phycol., 29*(5), 659–667.

Domozych, D. S., (2007). Exopolymer production by the green alga *Penium margaritanceum*: Implications for biofilm residency. *Int. J. Plant. Sci., 168*(6), 763–774.

Domozych, D. S., Kort, S., Benton, S., & Yu, T., (2005). The extracellular polymeric substance of the green alga *Penium margaritaceum* and its role in biofilm formation. *Biofilms, 2*(2), 129–144.

Domozych, D. S., Popper, Z. A., & Sørensen, I., (2016). Charophytes: Evolutionary giants and emerging model organisms. *Front. Plant Sci., 7*, 1470.

Domozych, D. S., Serfis, A., Kiemle, S. N., & Gretz, M. R., (2007). The structure and biochemistry of charophycean cell walls. I. Pectins of *Penium margaritaceum*. *Protoplasma, 230*(1, 2), 99–115.

Eder, M., & Lütz-Meindl, U., (2010). Analyses and localization of pectin-like carbohydrates in cell wall and mucilage of the green alga *Netrium digitus*. *Protoplasma, 243*(1–4), 25–38.

Ekelhof, A., & Melkonian, M., (2017a). Microalgal cultivation in porous substrate bioreactor for extracellular polysaccharide production. *J. Appl. Phycol., 29*, 1115–1122.

Ekelhof, A., & Melkonian, M., (2017b). Enhanced extracellular polysaccharide production and growth by microalga *Netrium digitus* in a porous substrate bioreactor. *Algal Res., 28*, 184–191.

Elgavish, A., & Elgavish, G. A., (1980). ^{31}P-NMR differentiation between intracellular phosphate pools in *Cosmarium* (*Chlorophyta*). *J. Phycol., 16*(3), 368–374.

Elgavish, A., Halmann, M., & Berman, T. A., (1982). A comparative study of phosphorus utilization and storage in batch cultures of *Peridinium cinctum*, *Pediastrum duplex* and *Cosmarium* sp., from Lake Kinneret (Israel). *Phycologia, 21*(1), 47–54.

Elsharnouby, O., Hafez, H., Nakhla, G., & El-Naggar, M. H., (2013). A critical literature review on biohydrogen production by pure cultures. *Int. J. Hydrogen Energy, 38*(12), 4945–4966.

El-Sheekh, M. M., El-Gamal, A., Bastawess, A. E., & El-Bokhomy, A., (2017). Production and characterization of biodiesel from the unicellular green alga *Scenedesmus obliquus*. *Energ. Source A, 39*(8), 783–792.

European Committee for Standardization, (2008). *Automotive Fuels – Fatty acid Methyl Esters (FAME) for Diesel Engines – Requirements and Test Methods*. EN14214, European Committee for Standardization.

Eshaq, F. S., Ali, M. N., & Mohd, M. K., (2010). *Spirogyra* biomass a renewable source for biofuel (bioethanol) production. *Int. J. Eng. Sci. Technol., 2*(12), 7045–7054.

Fawley, M. W., (1991). Disjunct distribution of the xanthophyll loroxanthin in the green algae (*Chlorophyta*). *J. Phycol., 27*(4), 544–548.

Fehér, G., (2003). The desmid flora of some alkaline lakes and wetlands in Southern Hungary. *Biol. Brat., 58*(4), 671–683.

Ferragut, C., & Bicudo, D. C., (2009). Efeito de diferentes níveis de enriquecimento por fósforo sobre a estrutura da comunidade perifítica em represa oligotrófica tropical (São Paulo, Brasil). *Braz. J. Bot., 32*, 571–585.

Ferragut, C., & Bicudo, D. C., (2012). Effect of N and P enrichment on periphytic algal community succession in a tropical oligotrophic reservoir. *Limnology, 13*, 131–141.

Firdaus, M., & Prihanto, A. A., (2014). α-amylase and α-glucosidase inhibition by brown seaweed (*Sargassum* sp.) extracts. *Res. J. Life Sci., 1*(1), 6–11.

Freire-Nordi, C. S., Vieira, A. A. H., & Nascimento, O. R., (1998). Selective permeability of the extracellular envelope of the microalga *Spondylosium panduriforme* (*Chlorophyceae*) as revealed by electron paramagnetic resonance. *J. Phycol., 34*(4), 631–637.

Freire-Nordi, C. S., Vieira, A. A. H., Nakaie, C. R., & Nascimento, O. R., (2006). Effect of polysaccharide capsule of the microalgae *Staurastrum iversenii* var. *americanum* on diffusion of charged and uncharged molecules, using EPR technique. *Braz. J. Phys., 36*(1), 75–82.

Fuentes-Grünewald, C., Bayliss, C., Zanain, M., Pooley, C., Scolamacchia, M., & Silkina, A., (2015). Evaluation of batch and semi-continuous culture of *Porphyridium purpureum* in a photobioreactor in high latitudes using Fourier transform infrared spectroscopy for monitoring biomass composition and metabolites production. *Bioresour. Technol., 189*, 357–363.

Fužinato, S., Cvijan, M., & Stamenković, M., (2011). A checklist of desmids (Conjugatophyceae, *Chlorophyta*) of Serbia. II. genus *Cosmarium. Cryptogamie Algol., 32*(1), 77–95.

Gadd, G. M., (1988). Accumulation of metals by microorganisms and algae. In: Rehm, H. J., (ed.), *Biotechnology* (pp. 401–434). VCH, Weinheim.

Garcia-Rodríguez, A., Matamoros, V., Fontàs, C., & Salvado, V., (2015). The influence of *Lemna* sp. and *Spirogyra* sp. on the removal of pharmaceuticals and endocrine disruptors in treated wastewaters. *Int. J. Environ. Sci. Technol., 12*, 2327–2338.

Ge, S., Madill, M., & Champagne, P., (2018). Use of freshwater macroalgae *Spirogyra* sp. for the treatment of municipal wastewaters and biomass production for biofuel applications. *Biomass Bioenerg., 111*, 213–223.

Gerrath, J. F., (1993). The biology of desmids: A decade of progress. In: Round, F. E., & Chapman, D. J., (eds.), *Progress in Phycological Research* (Vol. 9, pp. 79–192). Biopress Ltd., Bristol.

Gitzendanner, M. A., Soltis, P. S., Wong, G. K. S., Ruhfel, B. R., & Soltis, D. E., (2018). Plastid phylogenomic analysis of green plants: A billion years of evolutionary history. *Am. J. Bot., 105*(3), 291–301.

Griffiths, M. J., & Harrison, S. T., (2009). Lipid productivity as a key characteristic for choosing algal species for biodiesel production. *J. Appl. Phycol., 21*(5), 493–507.

Griffiths, M. J., Van, H. R. P., & Harrison, S. T. L., (2009). Lipid productivity, settling potential and fatty acid profile of 11 microalgal species grown under nitrogen replete and limited conditions. *J. Appl. Phycol., 24*, 989–1001.

Guerin, M., Huntley, M. E., & Olaizola, M., (2003). *Haematococcus* astaxanthin: Applications for human health and nutrition. *Trends Biotechnol., 21*(5), 210–216.

Guiry, M. D., (2013). Taxonomy and nomenclature of the *Conjugatophyceae* (=*Zygnematophyceae*). *Algae, 28*(1), 1–29.

Gunstone, F. D., (2011). Production and trade of vegetable oils. In: Gunstone, F. D., (ed.), *Vegetable Oils in Food Technology – Composition, Properties and Uses* (pp. 1–25). Blackwell Publishing Ltd., Oxford.

Gupta, V. K., & Rastogi, A., (2008). Biosorption of lead from aqueous solutions by green algae *Spirogyra* species: Kinetics and equilibrium studies. *J. Hazard. Mater., 152*(1), 407–414.

Gupta, V. K., Rastogi, A., Saini, V. K., & Jain, N., (2006). Biosorption of copper (II) from aqueous solutions by *Spirogyra* species. *J. Colloid Interface Sci., 296*(1), 59–63.

Gupta, V. K., Shrivastava, A. K., & Jain, N., (2001). Biosorption of chromium (VI) from aqueous solutions by green algae *Spirogyra* species. *Water Res., 35*(17), 4079–4085.

Ha, Y. W., Dyck, L. A., & Thomas, R. L., (1988). Hydrocolloids from the freshwater microalgae *Palmella texensis* and *Cosmarium turpinii*. *J. Food Sci., 53*(3), 841–844.

Hainz, R., Wober, C., & Schagerl, M., (2009). The relationship between *Spirogyra* (*Zygnematophyceae, Streptophyta*) filament type groups and environmental conditions in Central Europe. *Aquat. Bot., 91*(3), 173–180.

Hall, J. D., & McCourt, R. M., (2015). Conjugating green algae including desmids. In: Wehr, J. D., Sheath, R. G., & Kociolek, J. P., (eds.), *Freshwater Algae of North America: Ecology and Classification* (pp. 429–457). Academic Press, Cambridge.

Han, D., Li, Y., & Hu, Q., (2013). Astaxanthin in microalgae: Pathways, functions and biotechnological implications. *Algae, 28*(2), 131–147.

Han, J. W., Lee, K. P., Yoon, M. K., Sung, H., & Kim, G. H., (2009). Cold stress regulation of a bi-functional 3-dehydroquinate dehydratase/shikimate dehydrogenase (DHQ/SDH)-like gene in the freshwater green alga *Spirogyra varians*. *Bot. Marina, 52*(2), 178–185.

Han, J. W., Yoon, M., Lee, K. P., & Kim, G. H., (2007). Isolation of total RNA from a freshwater green alga *Zygnema cruciatum*, containing high levels of pigments. *Algae, 22*(2), 125–129.

Hasegawa, S., Baba, T., & Hori, Y., (1980). Suppression of allergic contact dermatitis by α-L-fucose. *J. Invest. Dermatol., 75*(3), 284–287.

Healey, F. P., (1979). Short-term responses of nutrient-deficient algae to nutrient addition. *J. Phycol., 15*(3), 289–299.

Hempel, N., Petrick, I., & Behrendt, F., (2012). Biomass productivity and productivity of fatty acids and amino acids of microalgae strains as key characteristics of suitability for biodiesel production. *J. Appl. Phycol., 24*, 1407–1418.

Hendrick, J. P., & Hartl, F. U., (1993). Molecular chaperone functions of heat-shock proteins. *Annu. Rev. Biochem., 62*, 349–384.

Herrmann, R. G., (1968). Die plastidenpigmente einiger desmidiaceen. *Protoplasma, 66*, 357–368.

Holzinger, A., & Karsten, U., (2013). Desiccation stress and tolerance in green algae: Consequences for ultrastructure, physiological, and molecular mechanisms. *Front. Plant Sci., 4*, 327.

Holzinger, A., & Pichrová, M., (2016). Abiotic stress tolerance of charophyte green algae: New challenges for omics techniques. *Front. Plant Sci., 7*, 678.

Holzinger, A., Albert, A., Aigner, S., et al., (2018). Arctic, Antarctic, and temperate green algae *Zygnema* spp. under UV-B stress: Vegetative cells perform better than pre-akinetes. *Protoplasma, 255*, 1239–1252.

Holzinger, A., Roleda, M., & Lütz, C., (2009). The vegetative arctic green alga *Zygnema* is insensitive to experimental UV exposure. *Micron, 40*(8), 831–838.

Hoshaw, R. W., & McCourt, R. M., (1988). The *Zygnemataceae* (*Chlorophyta*): A twenty-year update of research. *Phycologia, 27*(4), 511–548.

Hossain, A., Salleh, A., Boyce, A. N., Chowdhury, P., & Naqiuddin, M., (2008). Biodiesel fuel production from algae as renewable energy. *Am. J. Biotechnol., 4*(3), 250–254.

Hu, Q., Sommerfeld, M., Jarvis, E., et al., (2008). Microalgal triacylglycerols as feedstocks for biofuel production: Perspectives and advances. *Plant J., 54*(4), 621–639.

Ince, M., & Ince, O. K., (2017). An overview of adsorption technique for heavy metal removal from water/wastewater: A critical review. *Int. J. Pure Appl. Sci., 3*(2), 10–19.

Jansen, M. A., Spiering, R., Broere, F., et al., (2018). Targeting of tolerogenic dendritic cells towards heat-shock proteins: A novel therapeutic strategy for autoimmune diseases? *Immunology, 153*(1), 51–59.

Jayaraman, S., & Sarma, Y. S. R. K., (1985). Effects of three metabolic inhibitors on *Cosmarium* and *Closterium*. *J. Indian. Bot. Soc., 64*, 51–54.

Jiao, G., Yu, G., Zhang, J., & Ewart, H. S., (2011). Chemical structures and bioactivities of sulfated polysaccharides from marine algae. *Mar. Drugs, 9*(2), 196–223.

Jones, L. A., & King, C. C., (1996). Cottonseed oil. In: Hui, Y. H., (ed.), *Bailey's Industrial Oil and Fat Products, Edible oil and Fat Products: Oils and Oilseeds* (pp. 159–240). John Wiley and Sons, New York.

Jyothi, B., Sudhakaanrd, G., & Venkateswarlu, V., (1989). Ecological studies on a desmid bloom. *J. Indian Inst. Sci., 69*(4), 285–290.

Kasai, F., & Ichimura, T., (1986). Morphological variabilities of three closely related mating groups of *Closterium ehrenbergii* Meneghinii (Conjugatophyceae). *J. Phycol., 22*(2), 158–168.

Kasai, F., & Ichimura, T., (1990). Temperature optima of three closely related mating groups of the *Closterium ehrenbergii* (*Chlorophyta*) species complex. *Phycologia, 29*(4), 396–402.

Kattner, E., Lorch, D., & Weber, A., (1977). Die bausteine der zellwand und der gallerte eines stammes von *Netrium digitus* (Ehrbg.) Itzigs. & Rothe. *Mitt. Inst. Allg. Bot. Hamburg, 15*, 33–39.

Khalaf, M. A., (2008). Biosorption of reactive dye from textile wastewater by nonviable biomass of *Aspergillus niger* and *Spirogyra* sp. *Bioresour. Technol., 99*(14), 6631–6634.

Khoramabadi, G. S., Jafari, A., & Jamshidi, J. H., (2008). Biosorption of mercury (II) from aqueous solutions by *Zygnema fanicum* algae. *J. Appl. Sci., 8*(11), 2168–2172.

Kiemle, S. N., Domozych, D. S., & Gretz, M. R., (2007). The extracellular polymeric substances of desmids (Conjugatophyceae, *Streptophyta*): Chemistry, structural analyses and implications in wetland biofilms. *Phycologia, 46*(16), 617–627.

Kong, S. K., Bae, Y. S., Park, C. H., & Nam, D. H., (2009). Biosorption of nutrients by *Zygnema sterile* and *Lepocinclism textra* biomass in high rate algae culture system. *Desalin. Water Treat., 2*(1–3), 103–109.

Kovask, V., (1973). On the ecology of desmids. 2. Desmids and the mineral content. *Eesti NSV TA Toimet. Biol., 22*, 334–342.

Kratochvil, D., & Volesky, B., (1998). Advances in the biosorption of heavy metals. *Trends Biotech., 16*(7), 291–300.

Kregel, K. C., (2002). Heat shock proteins: Modifying factors in physiological stress responses and acquired thermotolerance. *J. Appl. Physiol., 92*(5), 2177–2186.

Krejci, M. R., Finney, L., Vogt, S., & Joester, D., (2011a). Selective sequestration of strontium in desmid green algae by biogenic co-precipitation with barite. *Chem. Sus. Chem., 4*(4), 470–473.

Krejci, M. R., Wasserman, B., Finney, L., et al., (2011b). Selectivity in biomineralization of barium and strontium. *J. Struct. Biol., 176*(2), 192–202.

Kroen, W., (1984). Growth and polysaccharide production by the green alga *Chlamydomonas mexicana* (*Chlorophyceae*) on soil. *J. Phycol., 20*(4), 616–618.

Kumar, J., Dhar, P., Tayade, A. B., et al., (2015). Chemical composition and biological activities of trans-Himalayan alga *Spirogyra porticalis* (Muell.) Cleve. *PLoS One, 10*(2), e0118255.

Kumar, V., Gautam, P., Singh, J., & Thakur, R. K., (2016). Assessment of phycoremediation efficiency of *Spirogyra* sp. using sugar mill effluent. *Int. J. Environ. Agr. Biotechn., 1*(1), 54–62.

Laliberté, G., & Hellebust, J. A., (1991). The phylogenetic significance of the distribution of arginine deiminase and arginase in the *Chlorophyta*. *Phycologia, 30*(2), 145–150.

Lang, I., Hodac, L., Friedl, T., & Feussner, I., (2011). Fatty acid profiles and their distribution patterns in microalgae: A comprehensive analysis of more than 2000 strains from the SAG culture collection. *BMC Plant. Biol., 11*, 124.

Lawlor, D. W., (2000). *Photosynthesis* (pp. 34–89). BIOS Scientific Publishers Ltd, Oxford.

Lee, Y. C., & Chang, S. P., (2011). The biosorption of heavy metals from aqueous solution by *Spirogyra* and *Cladophora* filamentous macroalgae. *Bioresour. Technol., 102*(9), 5297–5304.

Lemieux, C., Otis, C., & Turmel, M., (2016). Comparative chloroplast genome analyses of streptophyte green algae uncover major structural alterations in the *Klebsormidiophyceae, Coleochaetophyceae* and *Zygnematophyceae*. *Front. Plant. Sci., 7*, 697.

Lenzenweger, R., (1980). *Staurastrum pingue* teiling im prameter badesee. *Linzer Biol. Beitr., 12*, 389–391.

Lenzenweger, R., (1996). *Desmidiaceenflora von Östereich* (Vol 1, pp. 12–101). Bibliotheca Phycologica, J. Cramer in der Gebrüder Borntraeger Verlagsbuchhandlung, Stuttgart.

Lenzenweger, R., (1997). *Desmidiaceenflora von Östereich* (Vol. 2, pp. 19–81). Bibliotheca Phycologica, J. Cramer in der Gebrüder Borntraeger Verlagsbuchhandlung, Stuttgart.

Lenzenweger, R., (1999). *Desmidiaceenflora von Östereich* (Vol. 3, pp. 70–154). Bibliotheca Phycologica, J. Cramer in der Gebrüder Borntraeger Verlagsbuchhandlung, Stuttgart.

Lenzenweger, R., (2003). *Desmidiaceenflora von Östereich* (Vol. 4, pp. 11–25). Bibliotheca Phycologica, J. Cramer in der Gebrüder Borntraeger Verlagsbuchhandlung, Stuttgart.

Lorch, D. W., (1978). Desmids and heavy metals. II. Manganese: Uptake and influence on growth and morphogenesis of selected species. *Arch. Hydrobiol., 84*, 166–179.

Lorch, D. W., (1986). Desmids and heavy metals I. Uptake of lead by cultures and isolated cell walls of selected species. *Beih. Nova Hedwig., 56*, 105–118.

Lorenz, R. T., & Cysewski, G. R., (2000). Commercial potential for *Haematococcus* microalgae as a natural source of astaxanthin. *Trends Biotechnol., 18*(4), 160–167.

Luciana, W. T., Alba, L. G., Brito, R. M. S., et al., (2014). Safflower oil: An integrated assessment of phytochemistry, antiulcerogenic activity, and rodent and environmental toxicity. *Rev. Bras. Farm., 24*(5), 538–544.

Lütz, C., Seidlitz, H. K., & Meindl, U., (1997). Physiological and structural changes in the chloroplast of the green alga *Micrasterias denticulata* induced by UV-B simulation. *Plant Ecol., 128*, 55–64.

Lütz-Meindl, U., (2016). *Micrasterias* as a model system in plant cell biology. *Front. Plant Sci., 7*, 999.

Lynn, R., & Brock, T. D., (1969). Notes on the ecology of a species of *Zygogonium* (Kütz.) in Yellowstone national park. *J. Phycol., 5*(3), 181–185.

Mane, P. C., Bhosle, A. B., Jangam, C. M., & Vishwakarma, C. V., (2011). Bioadsorption of selenium by pretreated algal biomass. *Adv. Appl. Sci. Res., 2*(2), 202–207.

Masojídek, J., Kopecký, J., Koblížek, M., & Torzillo, G., (2004). The xanthophyll cycle in green algae (*Chlorophyta*): Its role in the photosynthetic apparatus. *Plant Biology, 6*(3), 342–349.

Matei, G. M., Kiptoo, J. K., Oyaro, N. K., & Onditi, A. O., (2014). Biosorption of selected heavy metals by the biomass of the green alga *Spirogyra* sp. *Facta Univers. Phys. Chem. Techn., 12*(1), 1–15.

McCourt, R. M., Karol, K. G., Bell, J., et al., (2000). Phylogeny of the conjugating green algae (*Zygnemophyceae*) based on rbcL sequences. *J. Phycol., 36*(4), 747–758.

Mehta, A., Zitzmann, N., Rudd, P. M., Block, T. M., & Dwek, R. A., (1998). α-glucosidase inhibitors as potential broad based anti-viral agents. *FEBS Letters, 430*(1, 2), 17–22.

Mehta, S. K., & Gaur, J. P., (2005). Use of algae for removing heavy metal ions from wastewater: Progress and prospects. *Crit. Rev. Biotechnol., 25*(3), 113–152.

Meindl, U., (1990). Effects of temperature on cytomorphogenesis and ultrastructure of *Micrasterias denticulata* Bréb. *Protoplasma, 157*, 3–18.

Melčáková, I., & Růžovič, T., (2010). Biosorption of zinc from aqueous solution using algae and plant biomass. *Nova Biotechn., 10*(1), 33–43.

Mix, M., (1972). Die feinstruktur der zellwände über *Mesotaeniaceae* und gonatozygaceae mit einer vergleichenden betrachtung der verschiedenen wandtypen der *Conjugatophyceae* und ber deren systematischen wert. *Arch. Microbiol., 81*, 197–220.

Mohan, S. V., Rao, N. C., Prasad, K. K., & Karthikeyan, J., (2002). Treatment of simulated reactive yellow 22 (Azo) dye effluents using *Spirogyra* species. *Waste Manage, 22*(6), 575–582.

Moorhead, K., Capelli, B., & Cysewski, G. R., (2011). *Spirulina: Nature's Superfood* (pp. 12–96). Cyanotech Corporation, Kailua-Kona, Hawaii.

Moss, B., (1972). The influence of environmental factors on the distribution of freshwater algae: An experimental study. I. Introduction and the influence of calcium concentration. *J. Ecol., 60*(3), 917–932.

Mota, R., Guimarães, R., Büttel, Z., et al., (2013). Production and characterization of extracellular carbohydrate polymer from *Cyanothece* sp. CCY 0110. *Carbohydr. Polym., 92*(2), 1408–1415.

Nakabayashi, T., & Hada, N., (1954). Studies on tannins of *Spirogyra arcta*. 1. Isolation of tannins. *J. Agric. Chem. Soc. Jpn., 28*, 387–391.

Nakabayashi, T., (1955). Studies on tannins of *Spirogyra arcta*. 5. On the structure of *Spirogyra* tannin. *J. Agric. Chem. Soc. Jpn., 29*, 897–899.

Nascimento, I. A., Marques, S. S. I., Cabanelas, I. T. D., et al., (2013). Screening microalgae strains for biodiesel production: Lipid productivity and estimation of fuel quality based on fatty acids profiles as selective criteria. *Bioenergy Res., 6*, 1–13.

Nedbalová, L., & Sklenář, P., (2008). New records of snow algae from the Andes of Ecuador. *Arnaldoa, 15*(1), 17–20.

Neumann, D., Scharf, K. D., & Nover, L., (1984). Heat shock induced changes of plant cell ultrastructure and autoradiographic localization of heat shock proteins. *Eur. J. Cell Biol., 34*(2), 254–264.

Newsome, A. G., & Van, B. R. B., (2012). Characterization of the purple vacuolar pigment of *Zygogonium ericetorum* alga. *Planta Medica, 78*, PJ20.

Ngearnpat, N., & Peerapornpisal, Y., (2007). Application of desmid diversity in assessing the water quality of 12 freshwater resources in Thailand. *J. Appl. Phycol., 19*, 667–674.
Nishikawa, M., Takemoto, S., & Takakura, Y., (2008). Heat shock protein derivatives for delivery of antigens to antigen presenting cells. *Int. J. Pharm., 354*(1, 2), 23–27.
Nover, L., & Scharf, K. D., (1984). Synthesis, modification and structural binding heat shock proteins in tomato cell cultures. *Eur. J. Biochem., 139*(2), 303–313.
Ontawong, A., Saowakon, N., Vivithanaporn, P., et al., (2013). Antioxidant and renoprotective effects of *Spirogyra neglecta* (Hassall) Kützing extract in experimental type 2 diabetic rats. *BioMed Res. Int., 2013*, 820786.
Onyancha, D., Mavura, W., Ngila, J. C., Ongoma, P., & Chacha, J., (2008). Studies of chromium removal from tannery wastewaters by algae biosorbents, *Spirogyra condensata* and *Rhizoclonium hieroglyphicum*. *J. Hazard. Mater., 158*(2, 3), 605–614.
Orosa, M., Torres, E., Fidalgo, P., & Abalde, J., (2000). Production and analysis of secondary carotenoids in green algae. *J. Appl. Phycol., 12*, 553–556.
Ortigueira, J., Pinto, T., Gouveia, L., & Moura, P., (2015). Production and storage of biohydrogen during sequential batch fermentation of *Spirogyra* hydrolyzate by *Clostridium butyricum*. *Energy, 88*, 528–536.
Özer, A., Akkaya, G., & Turabik, M., (2006). The removal of acid red 274 from wastewater. Combined biosorption and biocoagulation with *Spirogyra rhizopus*. *Dyes Pigm., 71*(2), 83–89.
Pacheco, R., Ferreira, A. F., Pinto, T., et al., (2015). The production of pigments and hydrogen through a *Spirogyra* sp. biorefinery. *Energy Convers. Manag., 89*, 789–797.
Palacio-López, K., Tinaz, B., Holzinger, A., & Domozych, D. S., (2019). Arabinogalactan proteins and the extracellular matrix of charophytes: A sticky business. *Front. Plant. Sci., 10*, 447.
Palamar-Mordvintseva, G. M., (1982). *Opredelitelj Presnovodnih Vodoroslei SSSR. Zelenye Vodorosli, Klass konjugaty, Porjadok Desmidievye* (pp. 38–132). Nauka Leningradskoe otdelenie, Leningrad.
Pascual, G., Avgustinova, A., Mejetta, S., et al., (2017). Targeting metastasis-initiating cells through the fatty acid receptor CD36. *Nature, 541*(7635), 41–45.
Paulsen, B. S., & Vieira, A. A. H., (1994). Structure of the capsular and extracellular polysaccharides produced by the desmid *Spondylosium panduriforme* (*Chlorophyta*). *J. Phycol., 30*(4), 638–641.
Péterszegi, G., Fodil-Bourahla, I., Robert, A., & Robert, L., (2003). Pharmacological properties of fucose. Applications in age-related modifications of connective tissues. *Biomed. Pharmacother., 57*, 240–245.
Pichrtová, M., Hájek, T., & Elster, J., (2014a). Osmotic stress and recovery in field populations of *Zygnema* sp. (*Zygnematophyceae, Streptophyta*) on Svalbard (high arctic) subjected to natural desiccation. *FEMS Microb. Ecol., 89*(2), 270–280.
Pichrtová, M., Kulichová, J., & Holzinger, A., (2014b). Nitrogen limitation and slow drying induce desiccation tolerance in conjugating green algae (*Zygnematophyceae, Streptophyta*) from polar habitats. *PLoS One, 9*(11), e113137.
Pichrtová, M., Remias, D., Lewis, L. A., & Holzinger, A., (2013). Changes in phenolic compounds and cellular ultrastructure of Arctic and Antarctic strains of *Zygnema* (*Zygnematophyceae*, *Streptophyta*) after exposure to experimentally enhanced UV to PAR ratio. *Microb. Ecol., 65*(1), 68–83.

Pillsbury, R. W., & Kingston, J. C., (1990). The pH-independent effect of aluminum on cultures of phytoplankton from an acidic Wisconsin lake. *Hydrobiologia, 194*, 225–233.

Pinto, T., Gouveia, L., Ortigueira, J., Saratale, G. D., & Moura, P., (2018). Enhancement of fermentative hydrogen production from *Spirogyra* sp. by increased carbohydrate accumulation and selection of the biomass pretreatment under a biorefinery model. *J. Biosci. Bioeng., 126*(2), 226–234.

Porter, N. A., Caldwell, S. E., & Mills, K. A., (1995). Mechanisms of free radical oxidation of unsaturated lipids. *Lipids, 30*, 277–290.

Powell, N., Shilton, A., Pratt, S., & Chisti, Y., (2011). Luxury uptake of phosphorus by microalgae in full-scale waste stabilisation ponds. *Water Sci. Technol., 63*(4), 704–709.

Rajalakshmi, N., (1986). Nutritional requirements of *Cosmarium subtriordinatum* west and west: A preliminary study. *Phykos, 25*, 57–61.

Ramaraj, R., Unpaprom, Y., Whangchai, N., & Dussadee, N., (2015). Culture of macroalgae *Spirogyra ellipsospora* for long-term experiments, stock maintenance and biogas production. *Emergent Life Sci. Res., 1*(1), 38–45.

Remias, D., Holzinger, A., & Lütz, C., (2009). Physiology, ultrastructure and habitat of the ice alga *Mesotaenium berggrenii* (*Zygnemaphyceae*, *Chlorophyta*) from glaciers in the European Alps. *Phycologia, 48*(4), 302–312.

Remias, D., Holzinger, A., Aigner, S., & Lütz, C., (2012a). Ecophysiology and ultrastructure of *Ancylonema nordenskiöldii* (*Zygnematales*, *Streptophyta*), causing brown ice on glaciers in Svalbard (high arctic). *Polar Biol., 35*, 899–908.

Remias, D., Schwaiger, S., Aigner, S., Leya, T., Stuppner, H., & Lütz, C., (2012b). Characterization of an UV- and VIS-absorbing, purpurogallin-derived secondary pigment new to algae and highly abundant in *Mesotaenium berggrenii* (*Zygnematophyceae*, *Chlorophyta*), an extremophyte living on glaciers. *FEMS Microbiol. Ecol., 79*(3), 638–648.

Rezaee, A., Ramavandi, B., Ganati, F., Ansari, M., & Solimanian, A., (2006). Biosorption of mercury by biomass of filamentous algae *Spirogyra* species. *J. Biol. Sci., 6*(4), 695–700.

Romera, E., González, F., Ballester, A., Blázquez, M. L., & Muñoz, J. A., (2007). Comparative study of biosorption of heavy metals using different types of algae. *Bioresour. Technol., 98*(17), 3344–3353.

Rott, E., Hotzy, R., Cantonati, M., & Sanders, D., (2012). Calcification types of *Oocardium stratum* Nägeli and microhabitat conditions in springs of the Alps. *Freshw. Sci., 31*(2), 610–624.

Růžička, J., (1977). *Die Desmidiaceen Mitteleuropas* (Vol. 1, pp. 20–232). Schweizerbartsche Verlagsbuchhandlung, Stuttgart.

Sajilata, M. G., Singhal, R. S., & Kamat, M. Y., (2008). The carotenoid pigment zeaxanthin – a review. *Compr. Rev. Food Sci. F., 7*(1), 29–49.

Sastry, P. S., & Chaudhary, B. R., (1989). Fixation of CO_2 and incorporation of thymidine under heavy metal stress in *Closterium monliferum*. *Folia Microbiol., 34*, 106–111.

Sathaiah, G., Reddy, Y. R., Reddy, K. L., & Vidyavati, (1984). Effect of quinazolin-4-one on *Cosmarium botrytis*. II. *Indian J. Bot., 7*, 172–175.

Schneider, S., & Lindstrøm, E. A., (2009). Bioindication in Norwegian rivers using non-diatomaceous benthic algae: The acidification index periphyton (AIP). *Ecol. Indic., 9*(6), 1206–1211.

Sen, B., Alp, M. T., Sonmez, F., Turan, K. M. A., & Canpolat, O., (2013). Relationship of algae to water pollution and waste water treatment. In: Elshorbagy, W., & Chowdhury, R. K., (eds.), *Water Treatment* (pp. 335–354). InTech Open, London.

Sibi, G., (2014). Biosorption of arsenic by living and dried biomass of fresh water microalgae – potentials and equilibrium studies. *J. Bioremed. Biodeg., 5*, 6.

Singh, S., Pradhan, S., & Rai, L. C., (2000). Metal removal from single and multimetallic systems by different biosorbent materials as evaluated by differential pulse anodic stripping voltammetry. *Process Biochem., 36*(1, 2), 175–182.

Sivaprakash, K., Blessi, T. L. A., & Madhavan, J., (2015). Biosorption of nickel from industrial wastewater using *Zygnema* sp. *J. Inst. Eng. India Ser. A., 96*, 319–326.

Smýkal, P., Hrdý, I., & Pechan, P. M., (2000). High-molecular-mass complexes formed *in vivo* contain smHSPs and HSP70 and display chaperone-like activity. *Eur. J. Biochem., 267*(8), 2195–2207.

Solit, D. B., & Rosen, N., (2006). Hsp90: A novel target for cancer therapy. *Curr. Top. Med. Chem., 6*(11), 1205–1214.

Song, M., Pei, H., Hua, W., & Maa, G., (2013). Evaluation of the potential of 10 microalgal strains for biodiesel production. *Bioresour. Technol., 141*, 245–251.

Soni, R., & Gupta, A., (2011). Batch biosorption studies of Cr (VI) by using *Zygnema* (green algae). *J. Chem. Pharm. Res., 3*(6), 950–960.

Spijkerman, E., & Coesel, P. F. M., (1996). Phosphorus uptake and growth kinetics of two planktonic desmid species. *Eur. J. Phycol., 31*(1), 53–60.

Stamenković, M., & Cvijan, M., (2008a). High tolerance to water pollution in *Cosmarium boitierense* Kouwets and *Staurastrum bloklandiae* Coesel et Joosten, taxa recorded for the first time from the Balkan Peninsula. *Alg. Studies, 127*(1), 83–94.

Stamenković, M., & Cvijan, M., (2008b). Some new and interesting ecological notes on several desmid taxa from the province of Vojvodina (North Serbia). *Biol. Brat., 63*(6), 921–927.

Stamenković, M., & Cvijan, M., (2008c). Desmid flora (*Chlorophyta, Zygnematophyceae*) of the Danube in the province of Vojvodina (Northern Serbia). *Arch. Biol. Sci., 60*(2), 181–199.

Stamenković, M., & Cvijan, M., (2009). Desmid flora (*Chlorophyta, Zygnematophyceae*) of the river Tisa in the province of Vojvodina (Northern Serbia). *Bot. Serbica, 33*(1), 89–99.

Stamenković, M., & Hanelt, D., (2011). Growth and photosynthetic characteristics of several *Cosmarium* strains (*Zygnematophyceae, Streptophyta*) isolated from various geographic regions under a constant light-temperature regime. *Aquat. Ecol., 45*(4), 455–472.

Stamenković, M., & Hanelt, D., (2013a). Adaptation of growth and photosynthesis to certain temperature regimes is an indicator for the geographical distribution of several *Cosmarium* strains (*Zygnematophyceae, Streptophyta*). *Eur. J. Phycol., 48*(1), 116–127.

Stamenković, M., & Hanelt, D., (2013b). Protection strategies of several *Cosmarium* strains (*Zygnematophyceae, Streptophyta*) isolated from various geographic regions against excessive photosynthetically active radiation. *Photochem. Photobiol., 89*(4), 900–910.

Stamenković, M., & Hanelt, D., (2014). Sensitivity of photosynthesis to UV radiation in several *Cosmarium* strains (*Zygnematophyceae, Streptophyta*) is related to their geographic distribution. *Photochem. Photobiol. Sci., 13*(7), 1066–1081.

Stamenković, M., & Hanelt, D., (2017). Geographic distribution and ecophysiological adaptations of desmids (*Zygnematophyceae, Streptophyta*) in relation to PAR, UV radiation and temperature: A review. *Hydrobiologia, 787*, 1–26.

Stamenković, M., Bischof, K., & Hanelt, D., (2014a). Xanthophyll cycle pool size and composition in several *Cosmarium* strains (*Zygnematophyceae, Streptophyta*) are related to their geographic distribution pattern. *Protist, 165*(1), 14–30.

Stamenković, M., Cvijan, M., & Fužinato, S. A., (2008d). A checklist of desmids (Conjugatophyceae, *Chlorophyta*) of Serbia. I. Introduction and elongate baculiform taxa. *Cryptogamie Algol., 29*(4), 325–347.

Stamenković, M., Steinwall, E., Nilsson, A. K., & Wulff, A., (2019). Desmids (*Zygnematophyceae, Streptophyta*) as a promising freshwater microalgal group for the fatty acid production: Results of a screening study. *J. Appl. Phycol., 31*, 1021–1034.

Stamenković, M., Steinwall, E., Nilsson, A. K., & Wulff, A., (2020). Fatty acids as chemotaxonomic and ecophysiological traits in green microalgae (desmids, *Zygnematophyceae, Streptophyta*): A discriminant analysis approach. *Phytochem., 170*, 112200.

Stamenković, M., Woelken, E., & Hanelt, D., (2014). Ultrastructure of *Cosmarium* strains (*Zygnematophyceae, Streptophyta*) collected from various geographic locations shows species-specific differences both at optimal and stress temperatures. *Protoplasma, 251*(6), 1491–1509.

Stancheva, R., & Sheath, R. G., (2016). Benthic soft-bodied algae as bioindicators of stream water quality. *Knowl. Manag. Aquat. Ecosyst., 417*, 15.

Stancheva, R., Hall, J. D., & Sheath, R. G., (2012). Systematics of the genus *Zygnema* (*Zygnematophyceae, Charophyta*) from Californian watersheds. *J. Phycol., 48*(2), 409–422.

Stancheva, R., Hall, J. D., Herburger, K., et al., (2014). Phylogenetic position of *Zygogonium ericetorum* (*Zygnemataceae, Charophyta*) from a high alpine habitat and ultrastructural characterization of unusual aplanospores. *J. Phycol., 50*(5), 790–803.

Stancheva, R., Hall, J. D., Mccourt, R. M., & Sheath, R., (2013). Identity and phylogenetic placement of *Spirogyra* species (*Zygnematophyceae, Charophyta*) from California streams and elsewhere. *J. Phycol., 49*(3), 588–607.

Stancheva, R., Herburger, K., Sheath, R. G., & Holzinger, A., (2016). Conjugation morphology of *Zygogonium ericetorum* (*Zygnematophyceae, Charophyta*) from a high alpine habitat. *J. Phycol., 52*(1), 131–134.

Stewart, K. D., & Mattox, K. R., (1978). Structural evolution in the flagellated cells of green algae and land plants. *BioSystems, 10*(1, 2), 145–152.

Surek, B., (1983). Mucilage regeneration in the green alga *Cosmocladium saxonicum* De Bary (*Desmidiaceae*): A light microscopic and quantitative study. *Brit. Phycol. J., 18*(1), 73–81.

Syrett, P. J., & Al-Houty, F. A. A., (1984). The phylogenetic significance of the occurrence of urease/urea amidolyase and glycollate oxidase/glycollate dehydrogenase in green algae. *Brit. Phycol. J., 19*(1), 11–21.

Talebi, A. F., Mohtashami, S. K., Tabatabaei, M., et al., (2013). Fatty acids profiling: A selective criterion for screening microalgae strains for biodiesel production. *Algal Res., 2*(3), 258–267.

Thomas, D. G., Minj, N., Mohan, N., & Rao, P. H., (2016). Cultivation of microalgae in domestic wastewater for biofuel applications – an upstream approach. *J. Algal Biomass Utln., 7*(1), 62–70.

Thumvijit, T., Inboot, W., Peerapornpisal, Y., Amornlerdpison, D., & Wongpoomchai, R., (2013a). The antimutagenic and antioxidant properties of *Spirogyra neglecta* (Hassall) Kützing. *J. Med. Plants Res., 7*(34), 2494–2500.

Thumvijit, T., Thuschana, W., Amornlerdpison, D., Peerapornpisal, Y., & Wongpoomchai, R., (2013b). Evaluation of hepatic antioxidant capacities of *Spirogyra neglecta* (Hassall) Kützing in rats. *Interdiscip. Toxicol., 6*(3), 152–156.

Turner, M. A., Howel, E. T., Summerby, M., Hesselein, R. H., Findlay, D. I., & Jackson, M. B., (1991). Changes in epilithon and epiphyton associated with experimental acidification of a lake to pH 5. *Limnol. Oceanogr., 36*(7), 1390–1405.

Url, W., (1955). Resistenz von desmidiaceen gegen schwermetallsalze. *Sitzungsber. Österr. Akad. Wiss. Math.-Nat. Kl. Abt., 164*, 207–230.

Van, H. L., Lewitus, A. J., Kana, T. M., & Craft, N. E., (1992). High-performance liquid chromatography of phytoplankton pigments using a polymeric reversed-phase C. column. *J. Phycol., 28*(6), 867–872.

Venkata, M. S., Ramanaiah, S. V., Rajkumar, B., & Sarma, P. N., (2007). Removal of fluoride from aqueous phase by biosorption onto algal biosorbent *Spirogyra* sp. IO_2: Sorption mechanism elucidation. *J. Hazard Mater., 141*(3), 465–474.

Venkateshwarlu, V., (1983). Ecology of desmids 1. *Staurastrum tetracerum* ralfs. *Indian J. Bot., 6*(1), 68–73.

Vidyashankar, S., Gopal, K. S. V., Swarnalatha, G. V., et al., (2015). Characterization of fatty acids and hydrocarbons of chlorophycean microalgae towards their use as biofuel source. *Biomass Bioenerg., 77*, 75–91.

Vogel, V., & Bergmann, P., (2018). Culture of *Spirogyra* sp. in a flat-panel airlift photobioreactor. *Biotech., 8*(1), 3.

Volland, S., Andosch, A., Milla, M., Stöger, B., Lütz, C., & Lütz-Meindl, U., (2011). Intracellular metal compartmentalization in the green algal model system *Micrasterias denticulata* (*Streptophyta*) measured by transmission electron microscopy-coupled electron energy loss spectroscopy. *J. Phycol., 47*(3), 565–579.

Volland, S., Bayer, E., Baumgartner, V., et al., (2014). Rescue of heavy metal effects on cell physiology of the algal model system *Micrasterias* by divalent ions. *J. Plant Physiol., 171*(2), 154–163.

Volland, S., Lütz, C., Michalke, B., & Lütz-Meindl, U., (2012). Intracellular chromium localization and cell physiological response in the unicellular alga *Micrasterias*. *Aquat. Toxicol., 109*, 59–69.

Wacker, A., Piepho, M., Harwood, J. L., Guschina, I. A., & Arts, M. T., (2013). Light-induced changes in fatty acid profiles of specific lipid classes in several freshwater phytoplankton species. *Front. Plant Sci., 7*, 264.

Wagner, B., Ochs, D., & Bieler, K., (2011). *Benzotropolone Derivatives as Antimicrobial Agents*. PCT Int. Appl. CODEN: PIXXD2 WO 2011048011 A2 20110428.

Wagner, B., Oehrlein, R., Herzog, B., Eichin, K., Baisch, G., & Portmann, S., (2009). *Preparation and Use of Benzotropolone Derivatives as UV Absorbers and Antioxidants and Their use in Sunscreens and/or Cosmetic Compositions*. PCT Int. Appl. CODEN: PIXXD2 WO 2009156324 A2 20091230.

Wang, H. M. D., Chen, C. C., Huynh, P., & Chang, J. S., (2015). Exploring the potential of using algae in cosmetics. *Bioresour. Technol., 184*, 355–362.

Weiss, D., & Lütz-Meindl, U., (1999). Heat response in the green alga *Micrasterias denticulata* (*Desmidiaceae*): Immunodetection and localization of BiP and heat shock protein 70. *Nova Hedwig., 69*(1, 2), 217–228.

Wells, M. L., Potin, P., Craigie, J. S., et al., (2017). Algae as nutritional and functional food sources: Revisiting our understanding. *J. Appl. Phycol., 29*, 949–982.

Wibowo, A. H., Mubarokah, L., & Suratman, A., (2013). The fermentation of green algae (*Spirogyra majuscule* Kuetz) using immobilization technique of Ca-alginate for *Saccharomyces cerevisiae* entrapment. *Indones. J. Chem., 13*(1), 7–13.

Wickett, N. J., Mirarab, S., Nguyen, N., et al., (2014). Phylotranscriptomic analysis of the origin and early diversification of land plants. *Proc. Natl. Acad. Sci. USA., 111*(45), E4859–E4868.

Wilcock, J., Perry, C., Williams, R., & Brook, A., (1989). Biological minerals formed from strontium and barium sulphates. II. Crystallography and control of mineral morphology in desmids. *Proc. R. Soc. Lond. B Biol. Sci., 238*(1292), 203–221.

Woertz, I., Feffer, A., Lundquist, T., & Nelson, Y., (2009). Algae grown on dairy and municipal wastewater for simultaneous nutrient removal and lipid production for biofuel feedstock. *J. Environ. Eng., 135*, 1115–1122.

Wolfstein, K., & Stal, L. J., (2002). Production of extracellular polymeric substances (EPS) by benthic diatoms: Effect of irradiance and temperature. *Mar. Ecol. Prog. Ser., 236*, 13–22.

Xiao, R., & Zheng, Y., (2016). Overview of microalgal extracellular polymeric substances (EPS) and their applications. *Biotechnol. Adv., 34*(7), 1225–1244.

Yeum, K., & Russell, R. M., (2002). Carotenoid bioavailability and bioconversion. *Ann. Rev. Nutr., 22*, 483–504.

Züllig, H., (1982). Untersuchungen über die Stratigraphie von Carotinoiden im geschichteten Sediment von 10 Schweizer Seen zur Erkundung früherer Phytoplankton-entfaltungen. *Schweiz. Z. Hydrol., 44*, 1–98.

CHAPTER 2

Diatoms: Commercial Applications and Prospective Trends

MANJITA MISHRA and SHANTHY SUNDARAM

Advanced Laboratory for Phycological Assessment (ALPHA Plus), Center of Biotechnology, University of Allahabad, Prayagraj, Uttar Pradesh – 211002, India

ABSTRACT

Diatoms are single-cell aquatic microalgae exhibiting remarkable self-assembled patterns of micro and nanoporous three-dimensional silica cell walls called frustules. The structure is quasi-periodically and extremely fine porous network on the surface of these diatoms makes them an attractive tool for various applications based on photonic and optic properties. This has been used in various other areas also like in nanotechnology which includes drug delivery, biosensing, biomimetics, molecular biology, food industries and bioremediations of contaminated water. All these features of diatoms lead us to move ahead in the research where diatoms could be an economical, lucrative source of many novel substances that will be beneficial in the field of medical and health sector in our near future.

2.1 INTRODUCTION

Almost 3 billion years ago, photosynthetic bacteria created the synthesis of oxygen which has been sustained by some of its progeny which gave rise to the aerobic life on to the earth. In an early human history, some of

Microalgal Biotechnology: Bioprospecting Microalgae for Functional Metabolites towards Commercial and Sustainable Applications. Jeyabalan Sangeetha, PhD, Svetlana Codreanu, PhD, & Devarajan Thangadurai, PhD (Eds.)

the photoautotrophs which were related with land-dwelling were mainly cultivated for agriculture and economic use. In recent times, global interest is towards the marine algae of the ocean where novel biomolecules and biomasses are comparatively untouched and unexplored. Diatoms are the major players for biomass production and lower the greenhouse gases (GHGs) from the atmosphere. Studies indicate that these photosynthetic organisms generated all primary production up to 20–23% (Field et al., 1998) and total level of marine production is approx. 40% (Falkowski et al., 1998), which make them the most prominent group of organisms that help as a sequestering agent for carbon from the environment. Diatoms are extremely robust which inhabit the photic zones from equator to hostile sea ice where they can be used as an eminent indicator for changes in environmental condition and provide the rapid response by reacting with frozen sea by their ice-binding proteins (IBPs) molecules "natural antifreeze" (Janech et al., 2006). Therefore, in all types of climate zones, diatoms are showing an extensively high degree of flexible activities, which could be utilized in different aspects of biotechnological applications even though in challenging conditions. Diatoms are getting more attention as it is having huge economic potential.

In the field of solar energy, diatoms can compete with other biological components solar factories as it needs light, water, some diatoms flourish on the waste materials as nutrients like sewage and carbon dioxide (CO_2). Recently, advanced techniques have been used to manipulate the diatoms and can create genetically modified organisms having new capacities.

2.2 EVOLUTION OF DIATOMS

In the later 19th century, German Naturalist Ernst Haeckel proposed the Protista kingdom, which consist the eukaryotic organisms but they were not belong to animals, plants, and fungi, but he proposed the connection between phylogeny and ontogeny. Among the classes of Protista Kingdom, he explained that diatoms are phytoplanktonic organisms mainly, unicellular under Stramenopila group, of supergroup chromalveolates, with completely a separate evolutionary lineage coming from land plants (Katz, 2012). Diatoms are single cellular algae (approx. length 1–500 mm) and it comes under class Bacillariophyceae of division Bacillariophyta and order of Pennales or Centrales that depend on their morphology and habitat. The phytoplanktons are further differentiated into pennate diatoms without raphe, i.e., fragilariophyceae, and with raphe, i.e., bacillariophyceae, and coscinodiscophyceae (centric diatoms), appear in either unicellular or different colonies form, some are of different shapes like filamentous, fans shape (*Meridion*), ribbons-like (*Fragilaria*), zigzag-like (*Tabellaria*), and stellate type (*Asterionella*). The diatoms are the primary producers of the food chain; which can contribute up to 25% of primary productivity globally (Scala and Bowler, 2001). Around 100 million years ago, in Cretaceous, diatoms are separated widespread and developed with vast diversification (Gross, 2012). They are absolutely one of the most diversified categories of eukaryotic organisms (Kooistra and Medlin, 1996) and the estimates suggest that there are more than 250 genera and approx. 10^5 species are of marine and freshwater habitat (Mann and Vanormelingen, 2013).

Diatoms are the single-celled phototrophs which evolved 180 million years before and having more than 10^5 species in total (Drum and Gordon, 2003; Kroth, 2007), and playing an ecologically important role in the process of biogeochemical cycling of elements like carbon, silicon, and phosphate (Bidle et al., 2002; Lopez et al., 2005) their mass reproduction in the form of blooms where pollution in coastal waters are increasing (Allen et al., 2005). Identification of most of the diatoms are by species-specific morphology where cell wall consisting of amorphous silica, and their sizes are ranges in nanoscale meter (Lopez et al., 2005). Their cell is typical eukaryotic cell with different compartments like nucleus, some secretary machines and mitochondria in the first glance, complex plastid ultrastructure and cytoplasm are defined by four layers membranes not like in red and green alga (Gibbs, 1979). These extracellular membrane barriers are like endurable keep shakers which derived from cellular evolution (Cavalier-Smith, 2000). The evolution of diatoms are via secondary endosymbiosis where the heterotrophic host cell engulfed the many unicellular phototrophs whose plastids are covered by two-layered membranes (Maier et al., 2000).

2.3 MORPHOGENESIS OF DIATOM FRUSTULE

These microalgae with the ability of generating a highly fanciful, ornamented, and intricate porous silica present in the cell wall called frustules. These cell walls displaying an amazing diversity in species specific shapes and the pattern of pores, which made the diatoms more popular microbes in the microscopist community in the 19th century. The intricate ultrastructure of frustule was used to check the quality of basic optical microscope (Round et al., 1990). Even today's, high magnification power images of diatom cell walls are continuing to amaze viewers with their huge array of micro and nano-structures (Volcani, 1981) (Figure 2.1). Moreover, its silica cell wall production is by diatoms which characterize the single-celled algae with distinct and influential role in the field of ecology and biogeochemistry of many oceans.

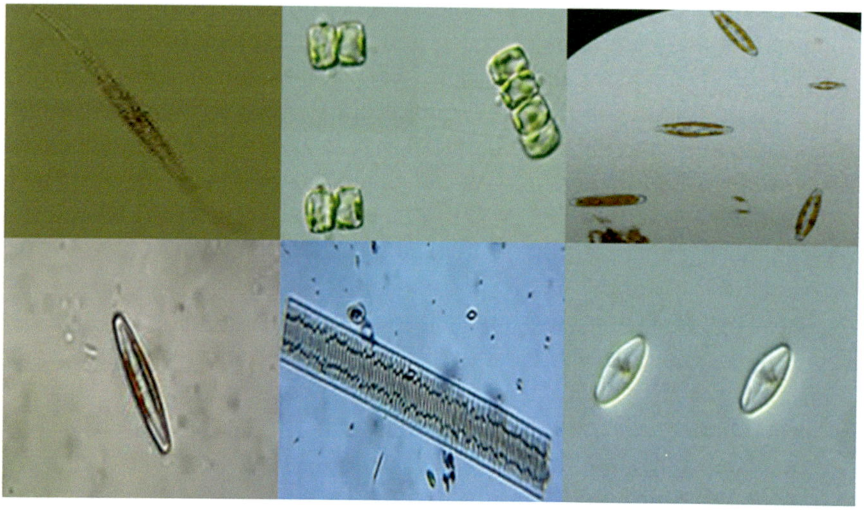

FIGURE 2.1 Diversity in the shape of diatoms. Scanning electron microscopy (SEM) images of diatom biosilica frustules. Scale bar: 10 μm (100 X magnification).

The reproduction that occurs in the diatoms are asexual where cell divides into two daughter cells with inherited one parental valve and another new smaller valve grows within. The features of diatoms are that they can easily procure from aquatic sources and proliferated in the desirable amount. From there they uptake silicon and deposit it in the cell walls, which results in the forming of frustules that are homogenous, intricate, mesoporous, regular patterned, silicious nanostructure, and its feasibility permits for genetic modification to modify the frustule shape and its pore size as per the requirement. The silicification of diatoms which links the carbon and silicon cycles in water: they belong to the very productive organisms on our planet, which are responsible for the lead role in total primary production and 240 Tmol Si/year of total biogenic silica precipitation (Falkowski et al., 1998).

The very first distinction in diatoms is the symmetry of the frustules: Centric or centrales diatoms which are specified by intricate frustules with radial symmetry and they are typical planktonic. Pennates diatoms with bilateral symmetry frustules present in epipelic and benthic regions (Round et al., 1990). In both types, diatom frustules that resembles like a petri-dish of silica box, where hypotheca is inserted in bigger epitheca, which encloses the living cell. The frustule size is species dependent and ranges from microns to millimeters. Both hypotheca and epitheca seems to be a valve which is surrounded by one lateral girdle. These valves are composed of distinct layers like foramen, cribellum, and cribrum each one of them are provided with regular patterns of pores, where dimension (varying from micron to nanometer scale) and structural distribution is depended on the species- and layer-specific.

2.4 BIO-SILICIFICATION OF DIATOMS

Nature has wretched the diatom microbes with a unique feature of uptaking the silicon from its environment and deposit in cell walls; hence generating the biosilica shells which possess the nanomaterials having multifaceted applications.

Silicon element is absorbed by the diatoms from its surrounding with a minimum concentration of 1<1 μM which is then actively transported across the membranes, in the form of silicic acid by using silicic acid transporters (SITs), that lead to an internal silicon pool and subsequently this insoluble silicon incorporated into the cell walls (Martin-Jézéquel et al., 2000; Knight et al., 2016). The formation of biogenic silica in frustules is intracellular polymerization of monomer of silicic acid. The formation of silica structure in diatoms are categorized into three different scales ranges from nano to meso and lastly to the scale of micro-level (Hildebrand et al., 2006, 2007). These microscale structures are combination of shape of valve and girdle bands in the silica deposition vesicle (SDV) via active and passive molding process, this involves the actin, cytoskeleton, and microtubules (Van De Meene and Pickett-Heaps, 2002; Tesson and Hildebrand, 2010a; Knight et al., 2016). Some organic components are needed for the polymerization of biosilica (Kröger and Wetherbee, 2000), i.e., long-chain polyamines (LCPAs, biosilica) and some are silaffins (Kröger et al., 2002; Poulsen and Kröger, 2004; Tesson and Hildebrand, 2010b).

Huge variability has observed in shape of shell which varies from sparse skeleton with crisscross bars to long barrels, stars, pods, triangles, and descriptive disks like flying saucer showed in Figure 2.1. In the process of replication, the diatom divides into halves (called epitheca and hypotheca) and the girdle bands separate out and its new part synthesizes within the cell in SDVs. These girdle bands split into rings or in circulating form which encircle the cell (Round et al., 1990; Hildebrand et al., 2009). In centric diatoms, initially the formation of valve occurs by linear ribs deposition that radiates to the outer part from its center (Taylor et al., 2007; Hildebrand et al., 2009). In spite of this, simple ribbed structure in centric diatoms are seems to be conserved, radiating outside with variations in its structure at nanoscale level.

2.5 CHEMICAL STRUCTURE OF BIOSILICA FRUSTULES

Diatom frustules are mainly composed of amorphous, porous, and hydrated silica structure that are provided with many surface defects such as silanol (Si-OH) and other Si-H groups (Qin et al., 2008). After the accurate analysis done by FTIR and Raman spectroscopies on frustules by subsequent removing of organic content that showed the allowing of signals coming from the organic residuals embedded in the porous matrix (C-H bonds) and sporadic, some localized signals from sulfur composites (C-S and S-H bondings) (Kammer et al., 2010; De Tommasi, 2016). These residuals of sulfur are related with global sulfur cycle, where the phytoplankton playing an important role by releasing the dimethylsulfide (DMS) compounds in the atmosphere (Simó, 2001). Moreover, these frustules are surrounded by an extracellular polymer that is mainly polysaccharides which help in gliding, sessile adhesion, formation of biofilms and colonies and protection against drying (Svetličić et al., 2013).

Throughout the years, the use of diatom frustules are going beyond the microscope quality testing. In addition, the use of diatom biosilica in the form of fossil (called diatomite or diatomaceous earth (DE)) in facial scrubs, toothpaste, and water filtration. The application of physical properties of frustules in various fields of nano, micro, and biotechnology. For example, frustules located in centric diatoms were used efficiently in microlenses form (De Stefano et al., 2007; De Tommasi et al., 2010), able to squeeze the light under its diffraction-limit level (De Tommasi et al., 2014); ability of collecting the light with high efficiency which lead to develop the new generations bio-based and bio-inspired solar cell structure (Toster et al., 2013; Wang et al., 2013); photoluminescence of frustules which is exploited for the realization of biosensors and optical sensors (De Stefano et al., 2005; Gale et al., 2009); the metalized frustules have been used successfully as a nano-structured substrates in field of plasmonics (Payne et al., 2005; Kwon et al., 2014; Ren et al., 2014) functionalized diatomite nanoparticles used as vectors in area of drug-delivery (Delalat et al., 2015; Terracciano et al., 2015); many diatomite and frustules can be used as scatterers in the random lasers equipment (Lamastra et al., 2014).

Other fascinating application can be done after proper modification of frustules: metabolic insertion step in germanium element (Jeffryes et al., 2008a) or titanium (Jeffryes et al., 2008b) which allowed to obtain the efficient nanostructured semiconducting devices in optoelectronics, enhancement in the light trappers of dye-sensitized solar cells and in structured photocatalysts of many toxic chemicals like polymer (Losic et al., 2007b), silicon structure (Bao et al., 2007), and metallic form (Fang et al., 2012). Replicas of frustules have been used as masters in nanofabrications, in electronic and sensing devices and called an enhanced optical transmission (EOT) of plasmonic elements. Therefore, the potential of diatom frustules, lead to complete control of all discussed applications, which even more can be relies on the capability to modify the genes responsible for frustule morphogenesis.

The possibilities to modify the frustule morphology, shape, geometry, and distribution of pores in order to standardize the specific application after the mutation of selective genes (Kröger and Poulsen, 2008). Perhaps, some genes are involved in the process of silica precipitation, spatial structure rearrangement, its aggregation, but relative proteins are not properly characterized. The genome of diatom is basically a melange of many genes, some belong to plants, animals or prokaryotes, which originated through horizontal gene transfers and successive endosymbiosis from bacterial cells (Bowler et al., 2008). In recent years the molecular and genetic details of frustule morphogenesis was elucidated partially. The accurate knowledge of genetics and genomic structure of diatoms will help to understand the processes of construction the decisive structure of frustules.

2.6 COLLECTION OF DIATOM FRUSTULES

To collect the diatoms, firstly it should be isolated from various water bodies and locations. It could be flat surface, covered with water and present at a shallow part of the water. In 2009, Mishra et al. has done the study on the diversity of diatom in freshwater of Prayagraj, Uttar Pradesh in India, where they have isolated them from slimy brown orange green tinge layers on rocks, which were scrubbed using the clean toothbrush and orangish water consisting of diatoms were carefully collected into clean tray. The shape diversity was listed in Table 2.1.

TABLE 2.1 List of Commonly Occurring Diatoms in Different River Sources

Name of Isolated Different Diatom Species	Order of Diatom	Average Number of Diatoms from Different River Sites of Prayagraj, Uttar Pradesh in India (in Triplicates)			
		Saraswati Ghat	Sangam Ghat	Daraganj Ghat	Yamuna Ghat
Cyclotella sp.	Centrales	3	4	3	3
Nitzschia sp.	Pennales	4	3	6	2
Fragilaria sp.	Pennales	2	5	4	4
Navicula sp.	Pennales	5	4	2	4

To collect the diatom frustules, initially it was centrifuged and rinsed thoroughly with distilled water to remove all salt contents. Diatom samples were then treated with 50:50 (v/v) water followed by 30% hydrogen peroxide, and kept for incubation at 90°C for 3–4 h, followed by slowly addition of conc. hydrochloric acid for the removal of organic matter and clean the collected frustules. Finally, samples were collected by sieving to prevent the damage of frustules, rinsed copiously with distilled water and stored in 70% ethanol (Mishra et al., 2009; Yu et al., 2010).

2.7 FUNCTIONALITIES OF DIATOM FRUSTULES

2.7.1 PHYSICAL PROPERTIES OF BIOSILICA FRUSTULES

In diatom biology, the important functionalities of the frustules are to protect the cytoplasm from the external harmful agents, nutrients diffusion and uptake by its porous matrix. In the last decade the new evolutionary advantages were hypothesized (Townley, 2011), which includes harvesting and focusing of light, assimilation, and sinking of carbon (Finkel and Kotrc, 2010). As the ultrastructure anatomy of diatom frustules are mainly related to the physical properties and its functionalities, is not clearly visible by the use of light microscope, but its structure is revealed by other techniques like scanning electron microscope (SEM) images of *Navicula* sp. showed in Figure 2.2, transmission electron microscopy (TEM), atomic force microscopy (AFM) technique (Losic et al., 2007a, b) and other SEM stereo-imaging technique (Chen et al., 2010). In some typical cases, there is merging of two different types of microscopy techniques, like photogrammetric surfaces obtained from SEM images plus confocal microscopy reported by Friedrichs et al. (2012); combination of SEM plus AFM and SEM plus digital holography (DH) seen in Ferrara et al. (2016), and this will guarantee the high level of accuracy and resolution in spatial dimensions. The precise, accurate, and highly detailed structure representation of frustule morphology is important not only per se, but for the retrieval of CAD models used in numerical simulation and for the systematic study and analysis of its physical properties, including mechanical, optical, and fluid dynamics.

Pan introduced a recent technique which allowed the detailed investigation of the internal part of diatom frustules but without thin sectioning of silica walls of its shell (Pan et al., 2014). Indeed, they have obtained the graphene replicas of diatom frustules of *Aulacoseira* genus by deposition of chemical vapor of methane. Since graphene is one of the most transparent materials to the electron beams, that allows the easy visualization of internal structure and morphology of valves and girdles, unveil the intricate dimensions, identification of the interconnected small nanotubes that are connected with different layers. This is the fundamental important to understand the interaction of the living cell with its external environment and the relative exchange of matter.

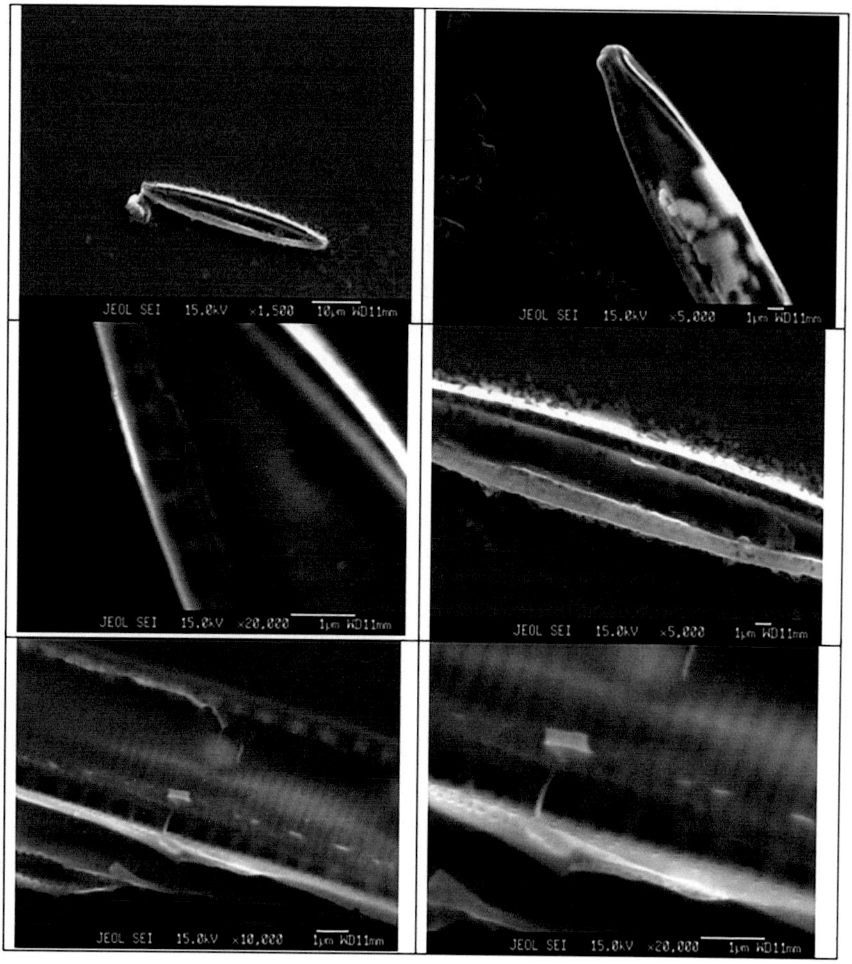

FIGURE 2.2 Depiction of SEM images of diatoms at different magnifications power.

2.7.2 MECHANICAL PROPERTIES OF BIOSILICA FRUSTULES

This specific study is in great demand to understand the functionalities of frustules, but also to design the framework of biomimetics which applied in architecture, nanotechnology (Gordon et al., 2009) and light weight porous constructions (Hamm, 2005). Although, in 2003, Hamm has performed the experiment to measure and numerically simulate the forces required to break the frustules of three different species of diatoms: *Fragilariopsis kerguelensis* (pennate), *Thalassiosira punctigera* (centric), and *Coscinodiscus granii* (centric and larger than *T. punctigera*) by using the calibrated glass having microneedles for loading and breaking these frustules with definitive forces. They found that these diatoms could able to resist the pressures ranges from 1 to 7 N/mm^2 which is equivalent to 100–700 T/m^2. The measurements of an isolated pleura (an open and hoop-shaped part of the girdle) of *T. punctigera* that helped to estimate the Young's modulus E (about 22.4 GPa), when compared with cortical bone and medical dental composites. To understand the distribution of stresses on the ultrastructure of frustules were done after mechanical solicitation by using the technique finite element method (FEM). Here, in this analysis, the objects under the study are sectioned into some finite number of parts (elements) that is equal to the domain which is reduced to the limited amount of degree of freedom.

2.7.3 FLUID DYNAMICAL PROPERTY OF DIATOM FRUSTULE

One of the fundamental functionalities associated with diatom frustules is its ability to filter and sort the nutrients (NH_4^+, NO_3^- and HCO_3^-) from noxious agents (bacteria and viruses). The concentrations differ for each like for bacteria ranges from ~10^6 ml^{-1} (0.2–1 μm in dimension), ~10^7 ml^{-1} for viruses (20–200 nm) up to ~10^9 ml^{-1} for colloids (5 nm–2 μm) and ~10^{14} ml^{-1} for nutrient molecules. The first interaction occurs between these living and the non-living Brownian substances with the external surface of diatom frustules. In 2001 and 2002, Hale, and Mitchell performed the experiments, with *Thalassiosira eccentrica* and *Coscinodiscus* sp. diatoms, where both the advection and diffusion of the added sub-micrometric beads were altered by the presence of micro-topographies in frustule.

2.7.4 OPTICAL PROPERTIES OF THE DIATOM FRUSTULE

There are impressive similarities of diatom frustules that can be comparable with its artificial photonic crystalline structure (Fuhrmann et al., 2004; De Stefano et al., 2009) which induced to hypothesize the ability beyond the mechanical defense and filtration of nutrients from harmful agents. These valves and girdles showed the fundamental role in manipulation and harvesting of light, and therefore enhancing the efficiency of photosynthesis in the extreme environment where sunlight is not accessible. A photonic crystal comprising the periodic and spatial distribution of specific refractive index which is specifically dimensioned and can block the light propagation in specific ranges (also called photonic band-gaps in diatoms) (Joannopoulos et al., 2008).

Several organisms like plants and animals, mainly birds and insects develop through many years of evolution, which is consecutive part of the organisms with sub-micron, periodic, or quasi-periodic artistic architecture and act as role of photonic crystal. This gives rise by means of selective reflectance and transmittance at various wavelength and so-called structural colors (Parker and Townley, 2007; Greanya, 2016). The structural colors are not because of any pigments but due to the geometrical characteristics and its refractive index of enormous micro- and nanostructured scales, plumage, and cuticles. When diatom frustules are interacted with light is become more articulate because of three important phenomena: confinement of light, selective transmission and photoluminescence.

Light confinement was first observed by single valve of *C. wailesii*, of size 150 μm in diameter. This experiment was performed by red coherent radiation, where the incoming beam was adjusted by the valve present in a tiny spot of 10 μm wide and located at the distance of approx. 100 μm from it (De Stefano et al. (2007). At this point, *C. wailesii* valves were assimilated into sort of microlenses, this phenomenon is related with diffraction but not refraction and numerical simulations methods (De Tommasi et al., 2010; Ferrara et al., 2014). The light confinement is because of the contribution of coherent superposition of diffracted lights coming from single pores.

A similar effect was observed in the making of DH images (Di Caprio et al., 2014; Ferrara et al., 2014), that act as a retriever of both phase and intensity of radiation transmitted by small valve. Acquisition of hologram in single diatom valve used in typical Mach-Zender interferometer and its application (Di Caprio et al., 2014). In diatoms *Arachnoidiscus* sp. UV-B transmitted radiation travel in air, the 1st peak of intensity is 500 μm away from its valve while the opposite side the red light has 1st maximum of 90 μm from valve and the value of intensity enhancement was three times higher than incident rays (Ferrara et al., 2014). A similar type of behavior is also observed in the mechanism of diatom frustules while screening the cells from harmful radiations of UV; the other process is adsorption by hydrates amorphous silica of diatom frustules. Indeed, the photonic property of frustules which showed the efficient conversion of destructive UV rays into useful photosynthetically active radiation (PAR). Absolute structure of diatom frustules that are composed by nanoporous silica with surface defects and embedded organic compounds, their characterization is done by intense photoluminescence at specific wavelength (Qin et al., 2008). Specifically, under the irradiation with UV-B light they mostly emit blue radiation, which suggests that harmful radiation can promote the accumulation of pyrimidine dimers in both DNA strands. This is converted into noxious radiation which is located in correspondence of maxima intensity of photosynthetic radiation spectrum (De Tommasi, 2016). Although, fucoxanthine (important pigments present in diatom chloroplasts) of around 450 nm was a significant absorption band (Table 2.2).

TABLE 2.2 Some Research Studies Describing the Use of Silica-based Diatom Biosensors

Diatomic Species as Biosensors	Fabricated Molecules	Detective Methods	Detected Biomolecules	References
Cyclotella sp.	Immunoglobulin G of Rabbit	Biosilica quenching photoluminescence	Antigen	Gale et al. (2009)
Thalassiosira rotula	–	Biosilica quenching photoluminescence	NO_2	Bismuto et al. (2008)
Coscinodiscus wailesii	Antibody	Electrochemical detection with no label tag	Biomarker proteins of Cardiovascular region	Lin et al. (2010)
Coscinodiscus concinnus	Antibody	Biosilica quenching photoluminescence	Antigen	De Stefano et al. (2009)

Note: From "Applications of diatoms and silica nanotechnology in biosensing, drug, and gene delivery, and formation of complex metal nanostructures," by Dolatabadi and De la Guardia (2011).

2.8 APPLICATIONS OF DIATOM FRUSTULES

Diatoms are minute microalgae, mainly found in the water habitat. Their occurrence and wide distribution make them an ideal tool for a broad range of its applications in fossilized and living form both (Atazadeh and Sharififi, 2010). An ecological perspective towards diatom is that it produces 40% of annual organic carbon and 20% of fresh oxygen (Mann, 1999). Diatoms are the natural producers of the substances like antibiotics, foodstuffs, and pharmaceuticals. These characteristics make them a valuable bioresources for the procurement of food supplements and can substitutes the synthetic substances which ranges from cosmetic to fuel. The commercial use of diatoms in industrial sectors like nitrogen-fixing biofertilizers, fuel production, renewable energy, and waste detoxification which uses biological waste as its substrate.

In natural environment, diatoms have potent ability to remove the carbon-dioxide from affected atmosphere and widely utilizes for reconstruction of environment and audit, investigation of forensic matters like drowning of victims and monitoring of water quality. These several properties lead to the use of diatoms and their frustules in several areas of technology.

2.8.1 POTENTIAL SOURCE OF NANOMATERIALS

Diatoms are fast replicable organisms and after the genetic engineering, it could be used as a cost-effective and desirable industrialized model system. Many efforts have been done to substitute the silicon with oxides of different metals like titanium, germanium, and zinc with their established properties of optical, thermal, electrical, biological, and chemical areas and paid off fabulously (Rorrer et al., 2005; Jeffryes et al., 2008a; Jaccard et al., 2009). In 2005, Rorrer et al. demonstrated that how diatom can fabricate the semiconductor nanostructured titanium dioxide by bottom-up assembly course on hugely parallel scale. They have done the insertion of nanostructured TiO_2 by metabolic process, and formed the nanocomposite of two elements, i.e., titanium, and silicon in diatom *Pinnularia* sp., by cultivating them into two stage-controlled bioreactor process. Significantly, useful for designing of solar cells which is dye-sensitized to improve the efficiency of light trapping and structured photocatalyst for the breakdown of toxic compounds. Lang et al. (2013) used the live cells of diatom for the formulation of organo-silica assembly with no defects in the intricate frustule pattern. Addition of different metals in the existing silica frustules improves the usability and durability in several nanotechnological applications.

The specialty of diatoms are to incorporate the desired material into its frustules and enhance its application for the making of biosensors, bioreactors, and in the field of biotechnology, photonic devices, nanomedicines, and microfluidics. To obtain the intact frustules live diatoms can be used with limited abrasive chemical treatments, this biosilica structure can be further processed depending on the desire of the final goal. Since many years they have been using successfully as a template to synthesize the advanced bio-hybrid nanostructures (Nassif and Livage, 2011).

In this chapter, attempt has been done to conscientiously gather the information regarding multidisciplinary applicability of miniature diatoms frustules in several field of biotechnology, nanotechnology, mainly in drug delivery, biosensor design, immunodiagnostics, and metabolite production were done. Table 2.3 shows the properties of diatom and its application in various fields.

TABLE 2.3 Tabulation of the Different Properties of Diatoms for its Suitable Uses

Study Area	Properties of Materials Used	References
Nanotechnological field	• Reproducibility of 3D structures • Self-replicative • Genetic engineered and economical • Complex pore sizes but modifiable • Used as abrasives due to hard surface	Hildebrand et al. (2006, 2007); Losic et al. (2006); Jeffryes et al. (2008a); Lang et al. (2013); Rorrer and Wang (2016)
Use of biosensor in area of forensic science	• Creation of number of channels on surface of single silicon chip • Low-cost and easily available • Shape of frustules vary depending upon the environment useful in identification in different types of crime scenes	Dempsey et al. (1997); De Stefano et al. (2009); Gordon et al. (2009); Verma (2013)
Immunosensors and immunodiagnostics	• Highly sensitive and its surface can be chemically modified by attaching the bioactive molecules • Property of Encapsulation and filtration of diatom frustules • Pore size is controllable	Desai et al. (1998); Townley et al. (2008); Rorrer and Wang (2016)
Filtration for water purification	• Act as filters to remove the micro-organisms • Permeability and accurate pore size • Transportable • Low cost	Lobo et al. (1991)

TABLE 2.3 *(Continued)*

Study Area	Properties of Materials Used	References
Aquaculture and fishing	• Rich source of amino acids and lipid • Presence of blue-green pigment • Easily available in nature	Lebeau et al. (2002)
In solar panel and biofuel production	• Storage of reserve foods in form of oil and volutin • Secretion of anti-microbial and anti-tumor peptides • Formation of lipid-fuel precursors • Produces more oil in nutrient-deprived condition • Consisting of photosynthetic pigments and able to genetic engineering	Lincoln et al. (1990); Pesando (1990); Alonso et al. (1996); Dunahay et al. (1996); Ramachandra et al. (2009)
Bioremediation	• Resistance against heavy metal due to phytochelatin production • Efficient withdrawing of cadmium and phosphorous metals	Pistocchi et al. (2000); Schmitt et al. (2001)
Drug delivery	• Chemically stable and biocompatible • Sustained release • Uniform pore size • Non-toxic	Zhang et al. (2013); Milovic et al. (2014); Rea et al. (2014); Vasani et al. (2015)

Note: From "All new faces of diatoms: potential source of nanomaterials and beyond" by Mishra et al. (2017).

2.8.2 AS FILTRATE IN WATER PURIFICATION

One of the natural forms of diatom is DE which consists of a heterogeneous concoction of dead diatom fossil residues with filtration ability. The use of diatom frustules instead of DE is more useful as single culture which provides them the homogeneous permeability having unique pore size (Hildebrand, 2008). They are cost-effectively transportable in small amount and grow in desired confluence that will be ideal in industrial applications (Lobo et al., 1991).

2.8.3 AS BIODEVICES

Cells of diatoms was cultured on self-assembled monolayers, where glass surface was activated by adding the trifluoromethyl, carboxyl, methyl, and amino groups. After rinsing the post adhesion, diatoms formed the 2D array which help to use in the development of bio-devices (Umemura et al., 2001). Diatoms of freshwater can be utilized as a biosensor for the assessment of water quality by using alternate current di-electrophoresis where live cells of diatom are chained in one line to create the 2D array (Siebman et al., 2017).

2.8.4 INDUSTRIAL APPLICATIONS

The algal biomass consists of three important components for fuel production: use of carbohydrates substances for ethanol production via fermentation technology, proteins for the methane production by anaerobic gasification and biodiesel production from natural oils.

2.8.4.1 METABOLITE PRODUCTION

Some diatoms are synthetically cultivated for the production of intracellular metabolites such as eicosapentaenoic acid (EPA), amino acids and essential lipids for cosmetics and pharmaceutical utilities (Lebeau and Robert, 2003; Hemaiswarya et al., 2011).

The live species of diatoms *Thalassiosira* and *Chaetoceros* are used for feeding the larvae (Spolaore et al., 2006), and *Thalassiosira pseudonana*, *Tetraselmis suecica*, *Isochrysis galbana*, *Pavlova lutheri*, and *Skeletonema costatum* are used as a feed of bivalve mollusks species (Hemaiswarya et al., 2011). Its extracellular metabolites are used to feed chicken and fish. The diatom species of *Nitzschia laevis* and *P. tricornutum* have been growing in photoreactors like helical tubular photobioreactor, perfusion cell bleeding, glass tube outdoors and glass tank photobioreactors for the production of EPA (Lebeau et al., 2002). This EPA is used to treat heart diseases, blood platelet aggregation, hyper-triglyceridemia as well as to reduce the level of cholesterol in blood, preventing the risk of inflammation and arteriosclerosis. The EPA obtained from fish oil products are unstable, poor taste and high purification cost which reduces its demand (Abedi and Sahari, 2014). Mainly *Nitzschia laevis* produces (2.50–2.76% dw of EPA) *Nitzschia inconspicua* with (1.9–4.7% dw of EPA), *Phaeodactylum tricornutum* and *Navicula saprophila* (2.2–3.9% dw of EPA) are cultivated for EPA (Lebeau and Robert, 2003; Abedi and Sahari, 2014; Wah et al., 2015).

Diatom *Nitzschia inconspicia* species has been used for the production of arachidonic acid (ARA) approx. 0.6–4.7% of total fatty acids (FAs) components (Chu et al., 1994; Lebeau and Robert, 2003). Isoleucine and aspartic acid are produced by diatomic species of *Chaetoceros calcitrans* and *S. costatum.* Amino acid leucine is prepared by *C. calcitrans, S. costatum* produces ornithine, serine, tyrosine, and glutamic acid are synthesized by *Thalassiosira* sp. (Derrien et al., 1998; Hildebrand et al., 2012). Domic acid is produced by *Nitzschia navis-varingica* (Kotaki et al., 2000; Martin-Jézéquel et al., 2015) as an insecticidal and anti-helminthic (Lincoln et al., 1990; Lebeau and Robert, 2003). The diatoms are also utilized for antibacterial and antifungal properties by having a complex of different FAs (Thillairajasekar et al., 2009).

Diatom species *S. costatum* inhibits the growth of bacteria *Vibrio* sp. in water culture (Naviner et al., 1999). The combination of organic extracts from *S. costatum* species (Bergé et al., 1996) and aqua extract from diatom *Haslea ostrearia* (Rowland et al., 2001) also showed the antitumoral property which is effective against the lung cancer and deadly HIV (Hildebrand et al., 2012).

2.8.4.2 BIOFUELS

In diatom oil is produced as a food reservoir during vegetative stage which help to keep them in a floating condition while waiting for favorable situations. These oils can produce the neutral lipids that act as a lipid fuel precursor: yield more oil than soyabean, palm, and oilseeds. Ramachandra et al. (2009) reported that under stress conditions diatom releases more oil as if silica content or nitrogen content are lesser in the culture. The comparative analysis done by micro-spectrometer to show that diatom oil consists the 60–70% high saturated fatty acids (FAs) compared to well-known crude oil.

Diatoms can imbibe the CO_2 and if sink on the floor of the ocean, where it preserved to produce petroleum chemicals (Ramachandra et al., 2009; Vinayak et al., 2015). Ramachandra et al. (2009) demonstrated the time-saving method for diatom oil production by reducing the production time. Additionally, they have genetically tailored the diatoms for direct secretion of gasoline which averts the additional benefits. Diatom's fuel can be substituted with fossil fuel hence substantially reducing the burden of greenhouse. The diatom *Cyclotella cryptica* has been altered genetically for the production of biodiesel (Dunahay et al., 1996). Another species is *Phaeodactylum tricornutum* Bohlin UTEX 640 has been mutated for 44% or higher EPA production (Lebeau and Robert, 2003).

Diatoms can restore carbon as a natural oil. The algae cultivation, e.g., diatoms with 70% dry weight of lipid content require at least a surface of 4 MHa, which is 2.2% of the fertile land in the USA where high sun exposure zone replace the petroleum consumption (Chisti, 2007). Diatoms flourishing in aqueous suspension and easy access to water, CO_2 and nutrients and convert these molecules into lipid stocks of approx. 85% of its weight (Lebeau and Robert, 2003) and it can produce 30% more amount of oil per area of oilseed crops in land. It is calculated that 210 kHa of diatoms in open ponds can generate liquid fuel energy into 10 BTUs when no gene modification or optimization is there in photobioreactors (Sheehan et al., 1998).

The presence of a high degree of complexity and intricate hierarchical structure showed by diatom cell wall is acquired under mild physiological environmental condition. The biological processes which generate the patterned frustule biosilica are of great interest to understand the emerging area of nanotechnology.

2.8.4.3 PHYTOREMEDIATION OF HEAVY METALS

Exposure of heavy metals, some plants, algae, and fungi can synthesize the glutathione oligomers called phytochelatins (PCs) (Grill et al., 1985). These phytochelatins are both intracellular and extracellular chelators with general structure of (γ-Glu-Cys) n-Gly, whereas it varies from 2 to 11 in number (Toppi and Gabbrielli, 1999). In aggregation plants, algae, and fungi PCs are activated by some heavy metals such as Ag, Al, Co, Cd, Cs, Cr, Cu, Mo, Mn, Mg, Hg, Ni, Pb, Se, and Zn which are present in soil and water (Salt et al., 1995; Cobbett and Goldsbrough, 2002; Prasad and Freitas, 2003). PCs are ecologically important components for the detoxification of heavy metal both in the cells and tissues as well as the external chelation of metals will be done by organic chelations. To understand the mechanism behind the chelator substrate-binding system where their combination varies from one another and it depend upon the substrate specificity and that make them more suitable for industrial use by overexpression in the process of phytoremediation of affected sites in diatoms (Salt et al., 1998).

Intracellular PCs when combined with metal-binding peptides are represented by a high content of cysteine in peptides and it has been analyzed in marine diatom *Phaeodactylum tricornutum* in laboratory and exposed to Pb, Cd or Zn. The anticipation of discovering the novel natural and artificial phytochelatins, facilitators for cation diffusion and compound metallothioneins which help to broaden the spectrum of the available metals (Prasad and Freitas, 2003). From peptides to proteins, biomolecules of plants, fungal, and algae all express in diatoms which could help to bind and collect the metal pollutants from contaminated waters when compared with transgenic plants that cannot withstand without soil. Therefore, to optimize the metals, nitrogen, and phosphate recovery could be done by using the raw materials of wastewater and that in addition can reduce the cost of wastewater purification (Lebeau and Robert, 2003).

2.8.5 NANOTECHNOLOGICAL APPLICATIONS

In the nanotechnology field, diatoms are able to biomineralize the combination of various compounds like silica, carbohydrates, and proteins and form the matrix of an intricate pattern of inorganic silica shell which surpasses the modern engineering technologies. The diatom frustules act as a biomimetic model of silica structures from micro to nanoscale range (Lopez et al., 2005; Hildebrand et al., 2008) and now called diatomaceous filters which used in filtration applications ranges from liquid filtration to DNA purification techniques (Gilmore et al., 1993) that can adsorb the heavy metals (Al-Degs et al., 2001). Perhaps, the most beneficial characteristic of the nano-structural silica is to synthesize themselves under different ambient conditions, i.e., amorphous silica which is transported into the vesicles and precipitated on peptide and carbohydrate scaffolding (Parkinson et al., 1999). Silica aggregates smoothened in the vesicles and this process affected by temperature and pH *in vivo* condition (Parkinson et al., 1999). The sequencing of diatom genome was done (Armbrust et al., 2004;

Kroth et al., 2008), which enlighten the nature and genetic base of the interspecies variability that will be no doubt forthcoming and boost its future manipulation and designing of the frustule nanostructures, that apparently will be vary significantly within the organisms (De Martino et al., 2007). The capability to genetically tailor the diatoms for the synthesis of designer frustules which is anticipated for the wider range of applications such as microelectronic devices, biological, and chemical sensing and disease diagnosis (Bismuto et al., 2008), for drug delivery scheme (Wee et al., 2005), as economical nanofiltration system (Parkinson et al., 1999) in catalysis (Jia et al., 2007), for storage of energy (Pérez-Caberoa et al., 2008) and capacitors (Weatherspoon et al., 2005). All these ultrastructures having the potential of light-emitting display as well as optical storage step (Parkinson et al., 1999). Besides, the frustule silicate can easily be replaced at the atomic level by magnesium oxide and implicates the use of frustules silica in nano-metallurgical area for casting many metals nanostructure form (Drum and Gordon, 2003).

2.8.5.1 USE IN PHOTONIC APPLICATIONS

Diatoms are magnificent distributed microalgae possessing silica shells known as frustules with three-dimensional porous architecture which provides larger surface areas. The biophotonic area has been more considerable in recent years (Prasad, 2003). Currently, biologists, chemists, and physicists worked in this area by using various approaches to investigate the biological growth in algae and physical properties like microfluidity and optical characteristics achieving the interests (Wang et al., 2013). These attempts providing many opportunities for the use of biosilica structure in the microsystems and commercial products, specifically interesting possibilities are open to manipulate and techniques to exploit the frustules in photonic field for that high surface area and less pores are required, for example, use as a photoanodes in dye-sensitized solar cells (Toster et al., 2013), as a probes for holographic optical tweezing (Olof et al., 2012) and as a photonic crystals for gas sensing (Lettieri et al., 2008). The RL has been using for different materials such as powders (Wiersma, 2008), polymeric composite materials doped with dye with fixed scattering particles (Long Wu and Deng, 2012; Yadav et al., 2013) and liquid solutions of particles and dyes (dielectric and metallic) (Lawandy et al., 1994).

These RL composites showing various advantages like low fabrication cost, adjustment of wavelength by altering the chromophore spread of broad angular emission that makes this an ideal component for display application. Furthermore, they also have the features of shape flexibility and compatibility when combine with various substrates materials which holding the strong promise in various applications like laser print (Lawandy et al., 1994) and deliver as a rugged or low-cost method for the identification of downed ships, satellites, and aircrafts (Laine et al., 2003). Since in early 19th RL emission in different biomaterials have been observed with many applications in biomaterials like cellulose fibers and bacterial cellulose (Dos Santos et al., 2014) and demonstrated the use of single-cell in biological laser (Gather and Seok, 2011). Recently reported the first experimental demonstration of this RL effect in commercial diatomite and by using multispecies frustules of living diatoms (Lamastra et al., 2014).

2.8.5.2 USE AS BIOSENSORS

In current years, ultrasensitive detection of many biological compounds with the help of nanomaterials are receiving much attention due to its unique chemical, optical, electronic, and some mechanical properties. Example of nanomaterials like gold and silver NPs, other are quantum dots (QDs), nanowires, biosilca, polymers, and carbon nanotubes (CNTs) to detect the macromolecules (antibodies, enzymes, and nucleic acids) (Dolatabadi et al., 2011).

The probability of cheaper formation of such arrays in channel which lead to the Lab on a chip (diverse channels on one silicon chip) and its filtration ability is favorable for designing biosensors (Gordon et al., 2009; Siebman et al., 2017). Although, these biosensitive devices have biological molecular recognition with transducer, inducer which induces the signal obtained from the molecules that have been sensed (Collings and Caruso, 1997). In biosensors, frustules can filter, and its pore size is well controllable to incorporate the particular frustules molecules into a well-designed sensing chamber in biosensors and selective trafficking of biomolecules. As it is of extensive refractive nature, the signal will amplify and can be utilized as fluorescent probe.

2.8.5.2.1 *Diatoms in Biosensing*

Diatoms are explored as a biosensor platform due to its large surface area and specific optical properties (De Stefano et al., 2009; Losic et al., 2009). De Stefano et al. presented the fluorescence image of diatom seen by a microscope under the high-pressure mercury lamp irradiation of diatoms frustules being photoluminescent (De Stefano et al., 2009). However, these biosilica-based templates are provided with the advantage of simple integration with the conventional well-tested processing methods in the field of the semiconductor industry (Lin et al., 2010).

The silica beads are able to conjugate with the dyes (organic and inorganic), lanthanide chelates and some biomolecules via surface chemistry. The incorporation of fluorophores in the glassy biomaterials can improve the stability and protecting it from photobleaching due to its porous nature (De Stefano et al., 2009). Some studies have been done where the effect of its surrounding environment was seen on the photoluminescent property of diatom species (*Thalassiosira rotula* Meunier) (De Stefano et al., 2005). Bismuto et al. reported that *Thalassiosira rotula* could check the nitrogen dioxide (NO_2). They have used the continuous-wave photoluminescence to detect the luminescence quenching by NO_2 (Bismuto et al., 2008). De Stefano et al. also checked the NO_2 effect on photoluminescence of marine species of diatom *Coscinodiscus wailesii* represented in Table 2.2. They have shown the quenching of photoluminescence signal because of electrophilic nature of NO_2 which attract the electrons from biosilica skeleton and also quench the photoluminescence (De Stefano et al., 2009).

The valve present on the diatom surface could be chemically altered by using the alkoxysilane compounds as the reactive Si–OH groups cover it completely. For the construction of Si–O–Si bridge, the –OH group have undergone the condensation reaction with alkoxysilanes (Stewart and Buriak, 2000). Moreover, because of intricate geometrical nanostructure of diatom frustules, they could act as a template for patterning the biomolecules at micro and nanoscale level (Sumper and Brunner, 2006).

The functionalization of diatom frustules by antibody has been reported by many researchers which provide the microscale biosensor platform as label-free, photoluminescence-based identification of the immunocomplex compounds (Bismuto et al., 2008; De Stefano et al., 2009; Gale et al., 2009). In the current study, De Stefano has done the chemically modification of biosilica frustules of marine diatom species *Coscinodiscus concinnus* and there attached the highly selective bioprobe, i.e., antibody. Also reported the event of molecular recognition between antibody and complementary antigen by showing the variation in the photoluminescence spectrum of diatoms (De Stefano et al., 2008; De Stefano et al., 2009). Gale et al. also functionalized the biosilica frustules of 10^{-1} m *Cyclotella* sp. with the help of rabbit-immunoglobulin G (IgG) that is covalently linked through the process of silanol amination. This demonstrated that the IgG antibody of nucleophilic nature enhanced the intrinsic blue color photoluminescence in diatom biosilica by 6 times. However, the formation of immunocomplex with complimentary antigen, the antibody anti-rabbit IgG enhanced the intensity of peak photoluminescence by a factor of 3, but this effect was not seen in the non-complimentary antigen (Gale et al., 2009). Mishra et al. (2020a) also reported the fabrication of diatom-based biosensor for the detection of *Tilletia indica* of Karnal Bunt in wheat crops (Figure 2.3).

First time, Lin et al. also used the diatoms as a sensor pattern which is analyzed by changes in the electrical properties for the selective and quantitative detection of antibody-antigen interactions. They have performed this assay to detect the two inflammatory proteins (myeloperoxidase and C-reactive protein) (Lin et al., 2010). Other examples of biomolecules that have also immobilized on the silica is enzymes (Zamora et al., 2009). DNA oligonucleotides also conjugated to diatom biosilica, and served as a template where sequence-specific multilayer conjugation of gold NPs which functionalized the complementary DNA strands (Losic et al., 2009).

In a recent study, Mishra has done the best work to simulate the diatom-based immuno-biosensor microcantilever for the early estimation of Karnal bunt disease in wheat (Mishra et al., 2020b). In Figure 2.4, the deflections can be seen as the molecules of antigen were bound on the antibody-coated microcantilever.

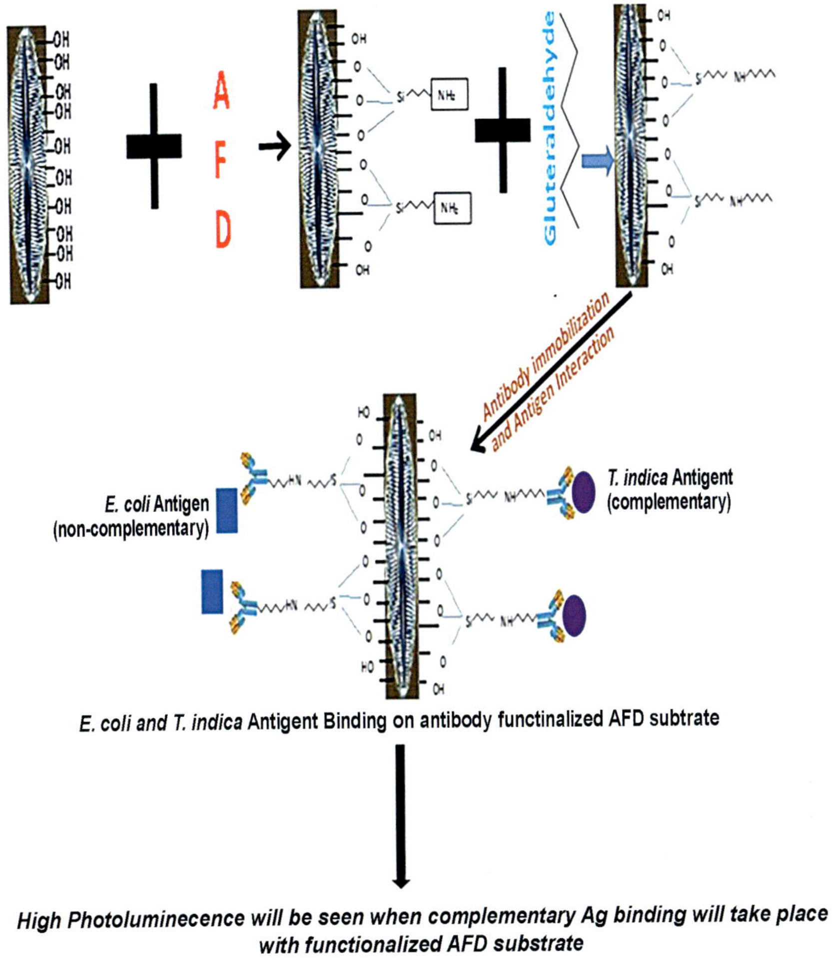

FIGURE 2.3 Detailed diagram of different fabrication steps of diatom-based sensor.

2.8.5.2.2 Silica in Biosensing

The exclusive characteristics of thin silica films of mesoporous type are getting attention because of its high prospective in catalysts, molecular sieving, optical, and sensor devices and adsorbents (Yuliarto et al., 2009). Its application is interesting due to its large surface area, patterned porous channels, high pore volumes and high mechanical and thermal stability (Yuliarto et al., 2009; Li et al., 2011; Wei et al., 2011). The characteristics of surface morphology of mesoporous silica NPs (MSNs) was modified flexibility by the addition of functional groups which can be used for encapsulation or internal immobilization of many biomolecules like enzymes, mediators, antibodies, and DNA (Wei et al., 2010, 2011).

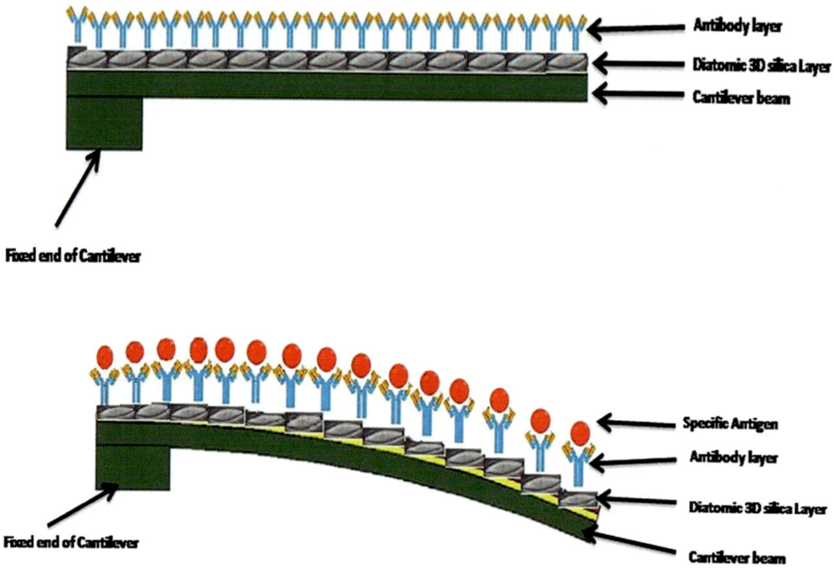

FIGURE 2.4 The schematic representation of microcantilever-based 3D biosensor at different states: resting state (A); and sensing state (B).

2.8.5.2.3 Use in Drug and Gene Delivery

The concept of designing a new, efficient vehicles to deliver the molecular probes and drugs into the targeted cells is important because the cellular penetration of small molecules is poor, easily degradation of small molecules via different pathways. The biosilica of diatom has been suggested as potential promising substitutes of other available popular vehicles because of its special structural, chemical, optical, and mechanical properties that would overcome the hurdles observing while usual delivery of some therapeutic agents and its advantageous over the existing system of microparticle delivery (Gordon et al., 2009). The larger surface area, highly porous at nanoscale, biodegradability, and biocompatibility of amorphous diatom biosilica are the salient features which help to make these frustules appropriate for its use in drug delivery system. Moreover, they can be functionalized, designed, and protected for the controlled release of drugs through pores of nanosized or by embedding in the pores of silica (Poulsen et al., 2007; Gordon et al., 2009).

Drug delivery vehicles should be designed practically to overcome the various physical barriers (Dolatabadi and de la Guardia, 2011). In a recent study, the poorly permeating drugs have been attached covalently to the transporter and form the cell-penetrating conjugate and solve the problem of penetration (Kam et al., 2004; Dolatabadi et al., 2011). The field of drug delivery is focusing the target drugs and genes to its destination of desired cells while minimizing the side effects of drugs (Pagona and Tagmatarchis, 2006; Tran et al., 2009).

Some are reported the functionalization of diatom frustules with enzymes and antibodies (Poulsen et al., 2007; Townley and Parker, 2008; Gale et al., 2009). The antibody-functionalized biosilica frustules could be utilized in antibody techniques and arrays (e.g., immunoprecipitation) (Townley and Parker, 2008). Losic et al. has introduced the diatoms biosilica with magnetic properties, where he functionalized the diatom-biosilica microcapsules with conjugated compound dopamine-modified FeO-NPs and which showed the tremendous potential when using as a magnetically-routed drug-delivery microcarriers (Losic et al., 2010). More examples are tabulated in Table 2.4.

Hard form of biosilica, homogenous pore size, designed spacing, genetically transformable, biocompatibility, and chemically inert are the crucial features which facilitate the utilization of frustules as a drug delivery vehicle. The pore size and the rate of release of drug from diatom biosilica is species-specific that helps the investigators for ample choice of designing the vehicles. Drug laden biosilica is directed towards the specific site of release by incorporating the ferromagnetic elements in its structure and using the magnet also.

High resolution imaging recent techniques form the baseline to investigate the biomineralization process in diatoms that will ultimately affect the manufacturing capacities of the device used. Zhang et al. (2013) have used the diatoms efficiently to deliver the drugs orally in gastrointestinal diseases. Usage of biosilica microparticles are best as it is having no toxicity in fact it enhances the permeability of drugs like mesalamine and prednisone while enabling its sustained release. In one report, it has been shown the diatom is used as a solid carrier for the class II BCS drugs which is notorious for low water solubility via self-emulsifying drug delivery system. Two approaches were used for diverse self-emulsifying suspension mainly phospholipid of carbamazepine (CBZ), in first method, direct mixing was used with diatoms, and in second the diatoms are dispersed into ethanolic preparations (Milovic et al., 2014). Diatom biosilica has also been reported to transport the siRNA into desired tumor cellular system (Rea et al., 2014).

TABLE 2.4 Representation of Diverse Application of Diatom Biosilica and its Mesoporous Silica Nanoparticles in the Area of Drug and Gene Delivery

Different Forms of Silica	Functionalized Materials Added	Area of Applications	References
Microcapsules of diatom silica	• Iron-oxide • Nanoparticles modified by dopamine	Drug delivery	Losic et al. (2010)
MSN matrix templated by surfactant	Cetyl trimethyl ammonium bromide, Triton X-100	Delivery of anticancer drug	He et al. (2010)
Mesoporous silica nanoparticles	Carboplatin	Cancer treatment in lungs and ovaries	Di Pasqua et al. (2009)
Silica nanoparticles (SNP)	Aminosilanes	Delivery of DNA	Kneuer et al. (2000)
Silica nanoparticles (SNP)	–	Plasmid DNA incorporation in COS-7 and 293T cells under *in vitro*	Bhakta et al. (2011)
Silica nanotubes (SNT)	Quantum dots, cationic polymers and iron oxides	Plasmid DNA incorporation in HeLa cells by endocytosis	Namgung et al. (2011)

Note: From "Applications of diatoms and silica nanotechnology in biosensing, drug, and gene delivery, and formation of complex metal nanostructures," by J. E. N. Dolatabadi and M. de la Guardia (2011). *Trends Analyt. Chem., 30*(9), 1538–1548.

2.8.5.2.4 *Immunodiagnostics*

Immunoisolation or bioencapsulation are characteristic use from filtration features of frustules. Later on, biocapsule competent of immunoisolating transplants has been used. The scientists have used the methods of UV lithography, deposition of thin film of silicon and specific etching process (Desai et al., 1998). These immunocapsules are capable to shield the enclosure part from defensive material of immune system simultaneously permits the flow of vital nutrients and oxygen to desired tissue. Hence, frustules are mesoporous in nature and this became the ideal vehicle for the transportation of nutrients into girdle cells.

The diatom biosilica can be tailored chemically to artificially tether the antibodies and other biomolecules to it and these attached molecules retain the inherent bioactivity. This optimum structure is essential in antibody arrays and are responsible for the immunodiagnostics. As culture of diatom requires energy in the form of light and minimal nutrients hence can be cultivable at low price and renewable starter material (Townley et al., 2008).

2.8.5.2.5 Use in the Formation of Complex Metal Nanostructures

Some biological materials and methods are recommended as an inspirational source for the design and fabrication of latest nanostructured materials. The advanced microfabricated technologies includes the template synthesis with frustule biosilica provides as a platform with spatially patterned, micro, and nanosized structures (Losic et al., 2005). Many studies have done which confirmed the fabrication of various metal nanostructures with a broad range of geometries (such as wires, particles, and pores) by using biological substances as templates (Losic et al., 2006). Assemblies of porous-nanostructured noble metals has been used as interesting materials in various field like biochemical, gas sensing, thermal, and electronic applications (Payne et al., 2005; Bao et al., 2009).

Different techniques used to synthesize the porous noble metal nanostructure are: (i) selective etching of noble-metals which bear alloys (dealloying); (ii) combustion synthesis; and (iii) method of physical and chemical-vapor, for the deposition on porous inorganic or organic templates (Bao et al., 2009).

Gold and silver nanoparticle deposition onto diatom biosilica via thermal deposition (Payne et al., 2005; Losic et al., 2006). Losic et al. has demonstrated the gold replicas of frustule templates of two distinct diatom species, for example, *Thalassiosira eccentrica* and *Coscinodiscus* sp. They suggested that their replicas had patterned nanoscale designs with comparable dimensions to that of the visible light (Losic et al., 2006). Yu et al. reported that self-supporting microstructure of gold having 3D morphologies that can be prepared by using no support electrodeless deposition of gold on the substrate of diatom biosilica (Yu et al., 2010). They were reported for the formation of gold diatom replicas with a wider range of micro to nanoscales structure by the feasible scalable process (Yu et al., 2010). In a recent study, Bao et al. used the scalable combination of solid/gas and chemical processes to generate the free-standing 3D porous microscale assemblies of many NPs like silver, gold, and palladium with controlled selectable morphologies of biosilica frustules (Bao et al., 2009). To explore the genetic and molecular parts as well as their mechanisms which help in the synthesis of diatom silica frustules (Mock et al., 2007).

2.9 GENETIC MODIFICATION OF DIATOM AND ITS FUTURE ASPECTS

Apart from harvesting the natural occurring compounds from diatoms which grown in the bioreactors at high optical densities, the next mandatory step for achieving the lucrative production in diatoms is to gain the knowledge as much as possible, and that will be continuing for long years ahead (Round et al., 2000). Many milestones are achieved when the complete genomic study of diatoms was done in *Thalassiosira pseudonana* and *P. tricornutum* to reconstruct the cellular biochemistry and study the compartmentalization of diatoms (Mock et al., 2007; Kroth et al., 2008). Many examples of manipulation are compiled in Table 2.5. The localization assays with green florescent proteins led to realize the targets in various compartments in the organelles that are highly differentiated (Kilian and Kroth, 2005; Gould et al., 2006 a, b; Sommer et al., 2007). The confirmation was done by using *in silico* analysis which showed that the enzymes required for biochemical pathways are not localized to one compartment only in the diatoms (Kroth et al., 2008).

These knowledges are important for genetically manipulating the diatoms for selective compartmentalization of the novel biochemical capacitors at specific cellular locations where key precursors should present before and can be used to direct the desired or targeted biochemical pathway.

TABLE 2.5 Listing the Name of Transformable Diatoms by Different Molecular Tools Involved in Genetic Recombination

Diatom Organisms	Marker Gene	Promoter Gene	Antibiotic Gene	Expressed Gene
Navicula saprophila	nptII	acc	G418	–
Cyclotella cryptica	nptII	acc	G418	–
Cylindrotheca fusiformis	sh-ble	fcp nr frustulin	Zeocin	eGFP
Thalassiosira pseudonana	nat1 sh-ble	fcp nr	Nourseothricin Zeocin	eGFP
Phaeodactylum tricornutum	sh-ble nat1	fcp nr ca PCMV	Zeocin nourseothricin G418	eGFP uidA Cat Luciferase

acc: acetyl-CoA carboxylase; fcp: fucoxanthin chlorophyll; nr: nitrate reductase; ca: carbonic anhydrase; PCMV: promoter sequences of the cytomegalovirus; Sh-ble: zeocin binding protein; nat1: nourseothricin acetyltransferase; nptII: neomycin phosphotransferase II; cat: chloramphenicol acetyltransferase.

Note: From "Diatoms in biotechnology: modern tools and applications," by A. Bozarth, U. G. Maier and S. Zauner, (2009). *Appl. Microbiol. Biotechnol.*, *82*, 195–201.

2.10 SUMMARY

The most probable problem in diatoms for the application of nutritional, commercial, and industrial level, formerly the cost of raw substance of fossil fuels and feasibility of easy processing steps of various plant crops to produce fuels which is much cheaper via fermentation. As the oil price is increasing over the last 10 years and enhanced the environmental awareness in public, interest to increase the carbon-neutral and recovery of renewable resources has been increased worldwide. The land area for the production of fuel crops is limited and arable space for agriculture is decreasing and demand of algal application is enticing.

2.11 CONCLUSION

In recent years it has been investigated the natural materials with unique properties (such as aerodynamics, hydrodynamics, adhesive, and wetting).

One such microbes are diatoms and the use of its mesoporous silica nanomaterials for biological purposes. These biological systems are intelligent enough to respond against specific stimuli and display the highly selective detection of various biomolecules. When in compared with other available solid nanoparticle biosensors, this mesoporous and microporous silica structure provides the high porosity structure and optical transparency. The highly porous property providing the large surface areas and high pore volumes that encapsulate or immobilize the large number of sensing molecules, hence providing the faster response time and lower detection limits. The characteristic of optical transparency that allows the proper optical detection by its biosilica layers. To consider these advantages, highly porous silica-based materials used in biosensor. While several reported results in exciting and exhibiting the advanced potential for its future applications, still some new breakthroughs are required for the purpose of drug and gene-delivery systems. In the field of nanotechnology, many new nanodevices are yet to be investigated under *in vivo* conditions.

KEYWORDS

- **biosensing**
- **diatoms**
- **drug delivery**
- **frustules**
- **gene delivery**
- **nanomaterials**
- **nanotechnology**
- **silica**

REFERENCES

Abedi, E., & Sahari, M. A., (2014). Long-chain polyunsaturated fatty acid sources and evaluation of their nutritional and functional properties. *Food Sci. Nutr., 2*, 443–463.

Al-Degs, Y., Khraisheh, M. A., & Tutunji, M. F., (2001). Sorption of lead ions on diatomite and manganese oxides modified diatomite. *Water Res., 35*, 3724–3728.

Allen, J. T., Brown, L., Sanders, R., et al., (2005). Diatom carbon export enhanced by silicate upwelling in the northeast Atlantic. *Nature, 437*, 728–732.

Alonso, D. L., Segura, D. C. C. I., Grima, E. M., et al., (1996). First insights into improvement of eicosapentaenoic acid content in *Phaeodactylum tricornutum* (*Bacillariophyceae*) by induced mutagenesis. *J. Phycol., 32*, 339–345.

Atazadeh, I., & Sharififi, M., (2010). *Algae as Bioindicators*. Lambert Academic Publishing, Saarbrucken, Germany.

Bao, Z., Ernst, E. M., Yoo, S., et al., (2009). Syntheses of porous self-supporting metal-nanoparticle assemblies with 3D morphologies inherited from biosilica templates (diatom frustules). *Adv. Mater., 21*, 474–478.

Bao, Z., Weatherspoon, M. R., Shian, S., et al., (2007). Chemical reduction of three-dimensional silica micro-assemblies into microporous silicon replicas. *Nature, 446*, 172–175.

Bergé, J., Bourgougnon, N., Carbonnelle, D., et al., (1996). Antiproliferative effects of an organic extract from the marine diatom *Skeletonema costatum* (Grev.) Cleve. against a non-small-cell bronchopulmonary carcinoma line (NSCLC-N6). *Anticancer Res., 17*, 2115–2120.

Bhakta, G., Sharma, R. K., Gupta, N., et al., (2011). Multifunctional silica nanoparticles with potentials of imaging and gene delivery. *Nanomed., 7*, 472–479.

Bidle, K. D., Manganelli, M., & Azam, F., (2002). Regulation of oceanic silicon and carbon preservation by temperature control on bacteria. *Science, 298*, 1980–1984.

Bismuto, A., Setaro, A., Maddalena, P., et al., (2008). Marine diatoms as optical chemical sensors: A time-resolved study. *Sens. Actuators B Chem., 130*, 396–399.

Bowler, C., Allen, A. E., Badger, J. H., et al., (2008). The *Phaeodactylum* genome reveals the evolutionary history of diatom genomes. *Nature, 456*, 239–244.

Bozarth, A., Maier, UG. & Zauner, S. Diatoms in biotechnology: modern tools and applications. *Appl Microbiol Biotechnol 82*, 195–201 (2009). https://doi.org/10.1007/s00253-008-1804-8

Cavalier-Smith, T., (2000). Membrane heredity and early chloroplast evolution. *Trends Plant Sci., 5*, 174–182.

Chen, X., Ostadi, H., & Jiang, K., (2010). Three-dimensional surface reconstruction of diatomaceous frustules. *Anal. Biochem., 403*, 63–66.

Chisti, Y., (2007). Biodiesel from microalgae. *Biotechnol. Adv., 25*, 294–306.

Chu, W. L., Phang, S. M., & Goh, S. H., (1994). Studies on the production of useful chemicals, especially fatty acids in the marine diatom *Nitzschia conspicua* Grunow. In: Sasekumar, A., Marshall, N., & Macintosh, D. J., (eds), *Ecology and Conservation of Southeast Asian Marine and Freshwater Environments including Wetlands* (pp. 33–40). Berlin: Springer.

Cobbett, C., & Goldsbrough, P., (2002). Phytochelatins and metallothioneins: Roles in heavy metal detoxification and homeostasis. *Annu. Rev. Plant. Biol., 53*, 159–182.

Collings, A., & Caruso, F., (1997). Biosensors: Recent advances. *Rep. Prog. Phys., 60*, 1397–1445.

De Martino, A., Meichenin, A., Shi, J., et al., (2007). Genetic and phenotypic characterization of *Phaeodactylum tricornutum* (*Bacillariophyceae*) accessions. *J. Phycol., 43*, 992–1009.

De Stefano, L., De Stefano, M., Bismuto, A., et al., (2005). Marine diatoms as optical chemical sensors. *Appl. Phys. Lett., 87*, 233902.

De Stefano, L., Lamberti, A., Rotiroti, L., et al., (2008). Interfacing the nanostructured biosilica microshells of the marine diatom *Coscinodiscus wailesii* with biological matter. *Acta Biomater., 4*, 126–130.
De Stefano, L., Maddalena, P., Moretti, L., et al., (2009). Nano-biosilica from marine diatoms: A brand-new material for photonic applications. *Superlattice. Microst., 46*, 84–89.
De Stefano, L., Rea, I., Rendina, I., et al., (2007). Lensless light focusing with the centric marine diatom *Coscinodiscus wailesii*. *Opt. Express., 15*, 18082–18088.
De Tommasi, E., (2016). Light manipulation by single cells: The case of diatoms. *J. Spectrosc.* 1–13.
De Tommasi, E., De Luca, A. C., Lavanga, L., et al., (2014). Biologically enabled subdiffractive focusing. *Opt. Express., 22*, 27214–27227.
De Tommasi, E., Rea, I., Mocella, V., et al., (2010). Multi-wavelength study of light transmitted through a single marine centric diatom. *Opt. Express., 18*, 12203–12212.
Delalat, B., Sheppard, V. C., Rasi, G. S., et al., (2015). Targeted drug delivery using genetically engineered diatom biosilica. *Nat. Commun., 6*, 8791.
Dempsey, E., Diamond, D., Smyth, M. R., et al., (1997). Design and development of a miniaturized total chemical analysis system for on-line lactate and glucose monitoring in biological samples. *Anal. Chim. Acta., 346*, 341–349.
Derrien, A., Coiffard, L. J., Coiffard, C., et al., (1998). Free amino acid analysis of five microalgae. *J. Appl. Phycol., 10*, 131–134.
Desai, T. A., Chu, W. H., Tu, J. K., et al., (1998). Microfabricated immunoisolating biocapsules. *Biotechnol. Bioeng., 57*, 118–120.
Di Caprio, G., Coppola, G., De Stefano, L., et al., (2014). Shedding light on diatom photonics by means of digital holography. *J. Biophotonics., 7*, 341–350.
Di Pasqua, A. J., Wallner, S., Kerwood, D. J., et al., (2009). Adsorption of the Pt^{II} anticancer drug carboplatin by mesoporous silica. *Chem. Biodivers., 6*, 1343–1349.
Dolatabadi, J. E. N., & De La Guardia, M., (2011). Applications of diatoms and silica nanotechnology in biosensing, drug and gene delivery, and formation of complex metal nanostructures. *Trends Anal. Chem., 30*, 1538–1548.
Dolatabadi, J. E. N., Mashinchian, O., Ayoubi, B., et al., (2011). Optical and electrochemical DNA nanobiosensors. *Trends Anal. Chem., 30*, 459–472.
Dos Santos, M. V., Dominguez, C. T., Schiavon, J. V., et al., (2014). Random laser action from flexible biocellulose-based device. *J. Appl. Phys., 115*, 083108.
Drum, R. W., & Gordon, R., (2003). Star trek replicators and diatom nanotechnology. *Trends Biotechnol., 21*, 325–328.
Dunahay, T. G., Jarvis, E. E., Dais, S. S., et al., (1996). Manipulation of microalgal lipid production using genetic engineering. In: *Proceedings of the Seventeenth Symposium on Biotechnology for Fuels and Chemicals* (pp. 223–231). Berlin: Springer.
Falkowski, P. G., Barber, R. T., & Smetacek, V. V., (1998). Biogeochemical controls and feedbacks on ocean primary production. *Science, 281*, 200–207.
Fang, Y., Chen, V. W., Cai, Y., et al., (2012). Biologically enabled synthesis of freestanding metallic structures possessing subwavelength pore arrays for extraordinary (surface Plasmon-mediated) infrared transmission. *Adv. Funct. Mater., 22*, 2550–2559.
Ferrara, M. A., Dardano, P., De Stefano, L., et al., (2014). Optical properties of diatom nanostructured biosilica *Arachnoidiscus* sp: Micro-optics from mother nature. *PLoS One, 9*, e103750.

Ferrara, M. A., De Tommasi, E., Coppola, G., et al., (2016). Diatom valve three-dimensional representation: A new imaging method based on combined microscopies. *Int. J. Mol. Sci., 17*, 1645.

Field, C. B., Behrenfeld, M. J., Randerson, J. T., et al., (1998). Primary production of the biosphere: Integrating terrestrial and oceanic components. *Science, 281*, 237–240.

Finkel, Z. V., & Kotrc, B., (2010). Silica use through time: Macroevolutionary change in the morphology of the diatom frustule. *Geomicrobiol. J., 27*, 596–608.

Friedrichs, L., Maier, M., & Hamm, C., (2012). A new method for exact three-dimensional reconstructions of diatom frustules. *J. Microsc., 248*, 208–217.

Fuhrmann, T., Landwehr, S., El Rharbi-Kucki, M., et al., (2004). Diatom as living photonic crystals. *Appl. Phys. B Lasers Opt., 78*, 257–260.

Gale, D. K., Gutu, T., Jiao, J., et al., (2009). Photoluminescence detection of biomolecules by antibody-functionalized diatom. *Biosilica. Ad. Funct. Mater., 19*, 926–933.

Gather, M. C., & Seok, H. Y., (2011). Single-cell biological lasers. *Nature Photon, 5*, 406.

Gibbs, S. P., (1979). The route of entry of cytoplasmically synthesized proteins into chloroplasts of algae possessing chloroplast ER. *J. Cell Sci., 35*, 253–266.

Gilmore, S., Weston, P., & Thomson, J., (1993). A simple, rapid, inexpensive and widely applicable technique for purifying plant DNA. *Aust. Syst. Bot., 6*, 139–142.

Gordon, R., Losic, D., Tiffany, M. A., et al., (2009). The glass menagerie: Diatoms for novel applications in nanotechnology. *Trends Biotechnol., 27*, 116–127.

Gould, S. B., Sommer, M. S., Hadfi, K., et al., (2006a). Protein targeting into the complex plastid of cryptophytes. *J. Mol. Evol., 62*, 674–681.

Greanya, V., (2016). *Bioinspired Photonics: Optical Structures and Systems Inspired by Nature*. Boco Raton: CRC Press.

Grill, E., Winnacker, E. L., & Zenk, M. H., (1985). Phytochelatins: The principal heavy-metal complexing peptides of higher plants. *Science, 230*, 674–676.

Gross, M., (2012). The mysteries of the diatoms. *Curr. Biol., 22*, 15.

Hale, M. S., & Mitchell, J. G., (2001). Functional morphology of diatom frustule microstructures: Hydrodynamic control of Brownian particle diffusion and advection. *Aquat. Microb. Ecol., 24*, 287–295.

Hale, M. S., & Mitchell, J. G., (2002). Effects of particle size, flow velocity, and cell surface microtopography on the motion of submicrometer particles over diatoms. *Nano Lett., 2*, 657–663.

Hamm, C., (2005). The evolution of advanced mechanical defenses and potential technological applications of diatom shells. *J. Nanosci. Nanotechnol., 5*, 108–119.

He, Q., Shi, J., Chen, F., et al., (2010). An anticancer drug delivery system based on surfactant-templated mesoporous silica nanoparticles. *Biomaterials, 31*, 3335–3346.

Hemaiswarya, S., Raja, R., Kumar, R. R., et al., (2011). Microalgae: A sustainable feed source for aquaculture. *World J. Microbiol. Biotechnol., 27*, 1737–1746.

Hildebrand, M., Davis, A. K., Smith, S. R., et al., (2012). The place of diatoms in the biofuels industry. *Biofuels, 3*, 221–240.

Hildebrand, M., Doktycz, M. J., & Allison, D. P., (2008). Application of AFM in understanding biomineral formation in diatoms. *Pflugers Arch., 456*, 127–137.

Hildebrand, M., Frigeri, L. G., & Davis, A. K., (2007). Synchronized growth of *Thalassiosira pseudonana* (*Bacillariophyceae*) provides novel insights into cell wall synthesis processes in relation to the cell cycle. *J Phycol., 43*, 730–740.

Hildebrand, M., Kim, S., Shi, D., et al., (2009). 3D imaging of diatoms with ion-abrasion scanning electron microscopy. *J. Struct. Biol., 166*, 316–328.

Hildebrand, M., York, E., Kelz, J. I., et al., (2006). Nanoscale control of silica morphology and three-dimensional structure during diatom cell wall formation. *J. Mater. Res., 21*, 2689–2698.

Jaccard, T., Ariztegui, D., & Wilkinson, K. J., (2009). Incorporation of zinc into the frustule of the freshwater diatom *Stephanodiscus hantzschii*. *Chem. Geol., 265*, 381–386.

Janech, M. G., Krell, A., Mock, T., et al., (2006). Ice binding proteins from sea ice diatoms (*Bacillariophyceae*). *J. Phycol., 42*, 410–416.

Jeffryes, C., Gutu, T., Jiao, J., et al., (2008a). Two-stage photobioreactor process for the metabolic insertion of nanostructured germanium into the silica microstructure of the diatom *Pinnularia* sp. *Mater. Sci. Eng. C., 28*, 107–118.

Jeffryes, C., Gutu, T., Jiao, J., et al., (2008b). Metabolic insertion of nanostructured TiO_2 into patterned biosilica of the diatom *Pinnularia* sp. by a two-stage bioreactor cultivation process. *ACS Nano, 2*, 2103–2112.

Jia, Y., Hana, W., Xionga, G., et al., (2007). Diatomite as high performance and environmentally friendly catalysts for phenol hydroxylation with H_2O_2. *Sci. Technol. Adv. Mater., 8*, 106–109.

Joannopoulos, J. D., Johnson, S. G., Winn, J. N., et al., (2008). *Photonic Crystals: Molding the Flow of Light*. Princeton University Press.

Kam, N. W. S., Jessop, T. C., Wender, P. A., et al., (2004). Nanotube molecular transporters: Internalization of carbon nanotube–protein conjugates into mammalian cells. *J. Am. Chem. Soc., 126*, 6850, 6851.

Kammer, M., Hedric, R., Ehrlich, H., et al., (2010). Spatially resolved determination of the structure and composition of diatom cell walls by Raman and FTIR imaging. *Anal. Bioanal. Chem., 398*, 509–517.

Katz, L. A., (2012). Origin and diversification of eukaryotes. *Annu. Rev. Microbiol., 66*, 411–427.

Kilian, O., & Kroth, P. G., (2005). Identification and characterization of a new conserved motif within the presequence of proteins targeted into complex diatom plastids. *Plant J., 41*, 175–183.

Kneuer, C., Sameti, M., Bakowsky, U., et al., (2000). A nonviral DNA delivery system based on surface-modified silica nanoparticles can efficiently transfect cells *in vivo*. *Bioconjug. Chem., 11*, 926–932.

Knight, M. J., Senior, L., Nancolas, B., et al., (2016). Direct evidence of the molecular basis for biological silicon transport. *Nat. Commun., 7*, 11926.

Kooistra, W. H. C. F., & Medlin, L. K., (1996). Evolution of the diatoms (Bacillariophyta): IV. A reconstruction of their age from small subunit rRNA coding regions and the fossil record. *Mol. Phylogenet. Evol., 6*, 391–407.

Kotaki, Y., Koike, K., Yoshida, M., et al., (2000). Domoic acid production in *Nitzschia* sp. (*Bacillariophyceae*) isolated from a shrimp-culture pond in Do Son, Vietnam. *J. Phycol., 36*, 1057–1060.

Kröger, N., & Poulsen, N., (2008). Diatoms – from cell wall biogenesis to nanotechnology. *Annu. Rev. Genet., 42*, 83–107.

Kröger, N., & Wetherbee, R., (2000). Pleuralins are involved in theca differentiation in the diatom *Cylindrotheca fusiformis*. *Protist, 151*, 263–273.

Kröger, N., Lorenz, S., Brunner, E., et al., (2002). Self-assembly of highly phosphorylated silaffins and their function in biosilica morphogenesis. *Science, 298*, 584–586.

Kroth, P. G., Chiovitti, A., Gruber, A., et al., (2008). A model for carbohydrate metabolism in the diatom *Phaeodactylum tricornutum* deduced from comparative whole-genome analysis. *PLoS One, 3*, e1426.

Kroth, P., (2007). *Molecular Biology and the Biotechnological Potential of Diatoms*. Springer, Berlin.

Kwon, S. Y., Park, S., & Nichols, W. T., (2014). Self-assembled diatom substrates with plasmonic functionality. *J. Korean Phys. Soc., 64*, 1179–1184.

Laine, R. M., Rand, S., Hinklin, T., et al., (2003). *Ultrafine Powders and Their Use as Lasing Media*. US Patent 6,656,588.

Lamastra, F. R., De Angelis, R., Antonucci, A., et al., (2014). Polymer composite random lasers based on diatom frustules as scatterers. *RSC Adv., 4*(106), 61809–61816.

Lang, Y., Del Monte, F., Collins, L., et al., (2013). Functionalization of the living diatom *Thalassiosira weissflogii* with thiol moieties. *Nat. Commun., 4*, 2683.

Lawandy, N. M., Balachandran, R. M., Gomes, A. S. L., et al., (1994). Laser action in strongly scattering media. *Nature, 368*, 436.

Lebeau, T., & Robert, J. M., (2003). Diatom cultivation and biotechnologically relevant products. Part I: Cultivation at various scales. *Appl. Microbiol. Biotechnol., 60*, 612–623.

Lebeau, T., Gaudin, P., Moan, R., et al., (2002). A new photobioreactor for continuous marennin production with a marine diatom: Influence of the light intensity and the immobilized-cell matrix (alginate beads or agar layer). *Appl. Microbiol. Biotechnol., 59*, 153–159.

Lettieri, S., De Stefano, L., De Stefano, M., et al., (2008). The gas-detection properties of light-emitting diatoms. *Adv. Funct. Mater., 18*(8), 1257–1264.

Li, A., Zhang, W., Ghaffarivardavagh, R., et al., (2016). Towards uniformly oriented diatom frustule monolayers: Experimental and theoretical analyses. *Microsyst. Nanoeng., 2*, 16064.

Li, H., He, J., Zhao, Y., et al., (2011). Immobilization of glucose oxidase and platinum on mesoporous silica nanoparticles for the fabrication of glucose biosensor. *Electrochim. Acta, 56*, 2960–2965.

Lin, K. C., Kunduru, V., Bothara, M., et al., (2010). Biogenic nanoporous silica-based sensor for enhanced electrochemical detection of cardiovascular biomarkers proteins. *Biosens. Bioelectron., 25*, 2336–2342.

Lincoln, R. A., Strupinski, K., & Walker, J. M., (1990). Biologically active compounds from diatoms. *Diatom Res., 5*, 337–349.

Lobo, E. A., Oliveira, M. A., Neves, M., et al., (1991). Characterization of wetland environments in the State of Rio Grande do Sul, where species of Anatidae with game value occur. *Acta Biol. Leopoldensia, 13*(1), 19–60.

Long, W. L., & Deng, L., (2012). Random lasers in dye-doped polymer-dispersed liquid crystals containing silver nanoparticles. *Phys. B: Cond. Mat., 407*(24), 4826–4830.

Lopez, P. J., Descles, J., Allen, A. E., et al., (2005). Prospects in diatom research. *Curr. Opin. Biotechnol., 16*, 180–186.

Losic, D., Mitchell, J. G., & Voelcker, N. H., (2005). Complex gold nanostructures derived by templating from diatom frustules. *J. Chem. Commun., 39*, 4905–4907.

Losic, D., Mitchell, J. G., & Voelcker, N. H., (2006). Fabrication of gold nanostructures by templating from porous diatom frustules. *New J. Chem., 30*, 908–914.
Losic, D., Mitchell, J. G., Lal, R., et al., (2007a). Rapid fabrication of micro- and nanoscale patterns by replica molding from diatom biosilica. *Adv. Funct. Mater., 17*, 2439–2446.
Losic, D., Pillar, R. J., Dilger, T., et al., (2007b). Atomic force microscopy (AFM) characterization of the porous silica nanostructure of two centric diatoms. *J. Porous. Mater., 14*, 61–69.
Losic, D., Yu, Y., & Aw, M. S., et al., (2010). Surface functionalization of diatoms with dopamine modified iron oxide nanoparticles: Toward magnetically guided drug microcarriers with biologically derived morphologies. *J. Chem. Commun., 46*, 6323–6325.
Maier, U. G., Douglas, S. E., & Cavalier-Smith, T., (2000). The nucleomorph genomes of cryptophytes and chlorarachniophytes. *Protist, 151*, 103–109.
Mann, D. G., & Vanormelingen, P., (2013). An inordinate fondness? The number, distributions, and origins of diatom species. *J. Eukaryot. Microbiol., 60*(4), 414–420.
Mann, D. G., (1999). The species concept in diatoms. *Phycologia, 38*, 437–495.
Martin-Jézéquel, V., Calu, G., Candela, L., et al., (2015). Effects of organic and inorganic nitrogen on the growth and production of domoic acid by *Pseudo-nitzschia* multiseries and *P. australis* (*Bacillariophyceae*) in culture. *Mar. Drugs, 13*, 7067–7086.
Martin-Jézéquel, V., Hildebrand, M., & Brzezinski, M. A., (2000). Silicon metabolism in diatoms: Implications for growth. *J. Phycol., 36*, 821–840.
Milovic, M., Simovic, S., Losic, D., et al., (2014). Solid self-emulsifying phospholipid suspension (SSEPS) with diatom as a drug carrier. *Eur. J. Pharm. Sci., 63*, 226–232.
Mishra, M., Arukha, A. P., Bashir, T., et al., (2017). All new faces of diatoms: Potential source of nanomaterials and beyond. *Front. Microbiol., 8*, 1239. doi: 10.3389/fmicb.2017.01239.
Mishra, M., Singh, S. K., Bhardwaj, A., et al., (2019). Growth analysis and frustule characterization of benthic diatomic isolates for nanotechnological applications. *Inter. J. Biotech. Res. Dev., 2*(1), 43–52.
Mishra, M., Singh, S. K., Bhardwaj, A., et al., (2020a). Development of diatom based photoluminescent immunosensor for the early detection of Karnal bunt disease of wheat crop. *ACS Omega, 5*, 8251–8257.
Mishra, M., Singh, S. K., Shanker, R., et al., (2020b). Design and simulation of diatom based microcantilever immuno-biosensor for the early detection of Karnal bunt. *3Biotech, 10*, 201.
Mock, T., Samanta, P., Iverson, V., et al., (2007). Whole-genome expression profiling of the marine diatom *Thalassiosira pseudonana* identifies genes involved in silicon bioprocesses. *Proc. Natl. Acad. Sci. USA, 105*, 1579–1584.
Namgung, R., Zhang, Y., Fang, Q. L., et al., (2011). Multifunctional silica nanotubes for dual-modality gene delivery and MR imaging. *Biomaterials, 32*, 3042–3052.
Nassif, N., & Livage, J., (2011). From diatoms to silica-based biohybrids. *Chem. Soc. Rev., 40*, 849–859.
Naviner, M., Bergé, J. P., Durand, P., et al., (1999). Antibacterial activity of the marine diatom *Skeletonema costatum* against aquacultural pathogens. *Aquaculture, 174*, 15–24.
Olof, S. N., Grieve, J. A., Phillips, D. B., et al., (2012). Measuring nanoscale forces with living probes. *Nano Lett., 12*(11), 6018–6023.

Pagona, G., & Tagmatarchis, N., (2006). Carbon nanotubes: Materials for medicinal chemistry and biotechnological applications. *Curr. Med. Chem.,13*(15), 1789–1798.
Pan, Z., Lerch, S. J. L., Xu, L., et al., (2014). Electronically transparent graphene replicas of diatoms: A new technique for the investigation of frustule morphology. *Sci Rep., 4*, 6117.
Parker, A. R., & Townley, H. E., (2007). Biomimetics of photonic nanostructures. *Nat. Nanotech., 2*, 347–353.
Parkinson, J., Brechet, Y., & Gordon, R., (1999). Centric diatom morphogenesis: A model based on a DLA algorithm investigating the potential role of microtubules. *Biochem. Biophys. Acta, 1452*, 89–102.
Payne, E. K., Rosi, N. L., Xue, C., et al., (2005). Sacrificial biological templates for the formation of nanostructured metallic microshells. *Angew. Chem., Int. Ed. Engl., 44*, 5064.
Pérez-Caberoa, M., Puchola, V., Beltrána, D., et al., 2008.*Thalassiosira pseudonana* diatom as biotemplate to produce a macroporous ordered carbon-rich material. *Carbon, 46*, 297–304.
Pesando, D., (1990). Antibacterial and antifungal activities of marine algae. In: Akatsuka, I., (ed.), *Introduction to Applied Phycology* (pp. 3–26). The Hague: SPB Academic Publishing B.V.
Pistocchi, R., Mormile, M., Guerrini, F., et al., (2000). Increased production of extra- and intracellular metal-ligands in phytoplankton exposed to copper and cadmium. *J. Appl. Phycol., 12*, 469–477.
Poulsen, N., & Kröger, N., (2004). Silica morphogenesis by alternative processing of silaffins in the diatom *Thalassiosira pseudonana. J. Biol. Chem., 279*, 42993–42999.
Poulsen, N., Berne, C., Spain, J., et al., (2007). Silica immobilization of an enzyme through genetic engineering of the diatom *Thalassiosira pseudonana. Angew. Chem. Int. Ed. Engl., 46*, 1843–1846.
Prasad, M., & Freitas, H., (2003). Metal hyperaccumulation in plants – biodiversity prospecting for phytoremediation technology. *Electron J. Biotechnol., 93*, 285–321.
Prasad, P. N., (2003). *Introduction to Biophotonics*. New York: Wiley-Interscience.
Qin, T., Gutu, T., Jiao, J., et al., (2008). Photoluminescence of silica nanostructures from bioreactor culture of marine diatom *Nitzchia frustulum. J. Nanosci. Nanotechnol., 8*, 2392–2398.
Ramachandra, T. V., Mahapatra, D. M., & Gordon, R., (2009). Milking diatoms for sustainable energy: Biochemical engineering versus gasoline-secreting diatom solar panels. *Ind. Eng. Chem. Res., 48*, 8769–8788.
Rea, I., Martucci, N. M., De Stefano, L., et al., (2014). Diatomite biosilica nanocarriers for siRNA transport inside cancer cells. *Biochim. Biophys. Acta, 1840*, 3393–3403.
Ren, F., Campbell, J., Rorrer, G. L., et al., (2014). Surface-enhanced Raman spectroscopy sensors from nanobiosilica with self-assembled plasmonic nanoparticles. *IEEE J. Sel. Top. Quant., 20*, 6900806.
Rorrer, G. L., & Wang, A. X., (2016). Nanostructured diatom frustule immunosensors. *Front. Nanosci. Nanotechnol., 2*, 128–130.
Rorrer, G. L., Chang, C. H., Liu, S. H., et al., (2005). Biosynthesis of silicon–germanium oxide nanocomposites by the marine diatom *Nitzschia frustulum. J. Nanosci. Nanotechnol., 5*, 41–49.

Round, F. E., Crawford, R. M., & Mann, D. G., (1990). *The Diatoms, Biology and Morphology of the Genera*. Cambridge, England: Cambridge University Press.

Rowland, S., Belt, S., Wraige, E., et al., (2001). Effects of temperature on polyunsaturation in cytostatic lipids of *Haslea ostrearia. Phytochem., 56*, 597–602.

Salt, D. E., Blaylock, M., Kumar, N. P., et al., (1995). Phytoremediation: A novel strategy for the removal of toxic metals from the environment using plants. *Biotechnol., 13*, 468–474.

Salt, D. E., Smith, R. D., & Raskin, I., (1998). Phytoremediation. *Annu. Rev. Plant Physiol. Plant Mol. Biol., 49*, 643–668.

Scala, S., & Bowler, C., (2001). Molecular insights into the novel aspects of diatom biology. *Cell. Mol. Life Sci., 58*, 1666–1673.

Schmitt, D., Müller, A., Csögör, Z., et al., (2001). The adsorption kinetics of metal ions onto different microalgae and siliceous earth. *Water Res., 35*, 779–785.

Sheehan, J., Dunahay, T., Benemann, J., et al., (1998). *A Look Back at the U.S. Department of Energy's Aquatic Species Program: Biodiesel from Algae*. Close-Out Report. National Renewable Energy Lab, Department of Energy, Golden, Report Number NREL/TP-580-24190.

Siebman, C., Velev, O. D., & Slaveykova, V. I., (2017). Alternating current dielectrophoresis collection and chaining of phytoplankton on chip: Comparison of individual species and artificial communities. *Biosens. (Basel), 7*(1), 4.

Simó, R., (2001). Production of atmospheric sulfur by oceanic plankton: Biogeochemical, ecological and evolutionary links. *Trends Ecol. Evol., 16*, 287–294.

Sommer, M. S., Gould, S. B., Lehmann, P., et al., (2007). Der1-mediated preprotein import into the periplastid compartment of chromalveolates? *Mol. Biol. Evol., 24*, 918–928.

Spolaore, P., Joannis-Cassan, C., Duran, E., et al., (2006). Commercial applications of microalgae. *J. Biosci. Bioeng., 101*, 87–96.

Stewart, M. P., & Buriak, J. M., (2000). Chemical and biological applications of porous silicon technology. *Adv. Mater., 12*, 859–869.

Sumper, M., & Brunner, E., (2006). Learning from diatoms: Nature's tools for the production of nanostructured silica. *Adv. Funct. Mater., 16*(1), 17–26.

Svetličić, V., Žutić, V., Pletikapić, G., et al., (2013). Marine polysaccharide networks and diatoms at the nanometric scale. *Int. J. Mol. Sci., 14*, 20064–20078.

Taylor, J. C., Harding, W. R., & Archibald, C., (2007). *An Illustrated Guide to Some Common Diatom Species from South Africa*. Water Research Commission: Gezina.

Terracciano, M., Shahbazi, M. A., Correia, A., et al., (2015). Surface bioengineering of diatomite based nanovectors for efficient intracellular uptake and drug delivery. *Nano, 7*, 20063–20074.

Tesson, B., & Hildebrand, M., (2010a). Dynamics of silica cell wall morphogenesis in the diatom *Cyclotella cryptica*: Substructure formation and the role of microfilaments. *J. Struct. Biol., 169*, 62–74.

Tesson, B., & Hildebrand, M., (2010b). Extensive and intimate association of the cytoskeleton with forming silica in diatoms: Control over patterning on the meso- and micro-scale. *PLoS One, 5*, e14300.

Thillairajasekar, K., Duraipandiyan, V., Perumal, P., et al., (2009). Antimicrobial activity of *Trichodesmium erythraeum* (Ehr) (microalga) from Southeast coast of Tamil Nadu, India. *Int. J. Integr. Biol., 5*, 167–170.

Toppi, L. S. D., & Gabbrielli, R., (1999). Response to cadmium in higher plants. *Environ. Exp. Bot., 41*, 105–130.

Toster, J., Iyer, K. S., Xiang, W., et al., (2013). Diatom frustules as light traps enhance DSSC efficiency. *Nano, 5*, 873–876.

Townley, H. E., (2011). Diatom frustules: Physical, optical, and biotechnological applications. In: Seckbach, J., & Kociolek, P., (eds.), *The Diatom World. Cellular Origin, Life in Extreme Habitats and Astrobiology* (Vol. 19). Springer, Dordrecht.

Townley, H. E., Parker, A. R., & White-Cooper, H., (2008). Exploitation of diatom frustules for nanotechnology: Tethering active biomolecules. *Adv. Funct. Mater., 18*, 369–374.

Tran, P. A., Zhang, L., & Webster, T. J., (2009). Carbon nanofibers and carbon nanotubes in regenerative medicine. *Adv. Drug Deliv. Rev., 61*, 1097–1114.

Umemura, K., Ishikawa, M., & Kuroda, R., (2001). Controlled immobilization of DNA molecules using chemical modification of mica surfaces for atomic force microscopy: Characterization in air. *Anal. Biochem., 290*, 232–237.

Van De, M. A. M., & Pickett-Heaps, J. D., (2002). Valve morphogenesis in the centric diatom *Proboscia alata* Sundstrom. *J. Phycol., 38*, 351–363.

Vasani, R., Losic, D., Cavallaro, A., et al., (2015). Fabrication of stimulus-responsive diatom biosilica microcapsules for antibiotic drug delivery. *J. Mater. Chem. B, 3*, 4325–4329.

Verma, K., (2013). Role of diatoms in the world of forensic science. *J. Forensic Res., 4*, 181.

Vinayak, V., Manoylov, K. M., Gateau, H., et al., (2015). Diatom milking: A review and new approaches. *Mar. Drugs, 13*, 2629–2665.

Volcani, B. E., (1981). Cell wall formation in diatoms: Morphogenesis and biochemistry. In: Simpson, T. L., & Volcani, B. E., (eds.), *Silicon and Siliceous Structures in Biological Systems*. Springer, New York, NY.

Wah, N. B., Ahmad, A. L. B., Chieh, D. C. J., et al., (2015). Changes in lipid profiles of a tropical benthic diatom in different cultivation temperature. *Asian J. Appl. Sci. Eng., 4*, 91–101.

Wang, Y., Cai, J., Jiang, Y., et al., (2013). Preparation of biosilica structures from frustules of diatoms and their applications: Current state and perspectives. *Appl. Microbiol. Biotechnol., 97*, 453–460.

Weatherspoon, M. R., Allan, S. M., Hunt, E., et al., (2005). Sol-gel synthesis on self-replicating single-cell scaffolds: Applying complex chemistries to nature's 3-D nanostructured templates. *Chem. Commun., 5*, 651–653.

Wee, K. M., Rogers, T. N., Altan, B. S., et al., (2005). Engineering and medical applications of diatoms. *J. Nanosci. Nanotechnol., 5*, 88–91.

Wei, Q., Li, R., Du, B., et al., (2011). Multifunctional mesoporous silica nanoparticles as sensitive labels for immunoassay of human chorionic gonadotropin. *Sens. Actuators B Chem., 153*(1), 256–260.

Wei, Q., Xin, X., Du, B., et al., (2010). Electrochemical immunosensor for norethisterone based on signal amplification strategy of graphene sheets and multienzyme functionalized mesoporous silica nanoparticles. *Biosens. Bioelectron., 26*(2), 723–729.

Wiersma, D. S., (2008). The physics and applications of random lasers. *Nature Phys., 4*, 359.

Yadav, A., De Angelis, R., Casalboni, M., et al., (2013). Spectral properties of self-assembled polystyrene nanospheres photonic crystals doped with luminescent dyes. *Opt. Mater., 35*, 1538–1543.

Yu, Y., Addai-Mensah, J., & Losic, D., (2010). Synthesis of self-supporting gold microstructures with three-dimensional morphologies by direct replication of diatom templates. *Langmuir, 26*, 14068–14072.

Yuliarto, B., Kumai, Y., Inagaki, S., et al., (2009). Enhanced benzene selectivity of mesoporous silica SPV sensors by incorporating phenylene groups in the silica framework. *Sens. Actuators B, 138*, 417–421.

Zamora, P., Narvaez, A., & Domınguez, E., (2009). Enzyme-modified nanoparticles using biomimetically synthesized silica. *Bioelectrochem., 76*(1), 100–106.

Zhang, H., Shahbazi, M. A., Makila, E. M., et al., (2013). Diatom silica microparticles for sustained release and permeation enhancement following oral delivery of prednisone and mesalamine. *Biomaterials, 34*, 9210–9219.

CHAPTER 3

Microalgae as a Potential Source of Bioactive Compounds and Functional Ingredients

MD. AKHLAQUR RAHMAN,[1] RUPALI KAUR,[2] and SHANTHY SUNDARAM[2]

[1]*Department of Biotechnology, S.S. Khanna Girls' Degree College, Prayagraj, Uttar Pradesh – 211003, India*

[2]*Center of Biotechnology, Nehru Science Center, University of Allahabad, Prayagraj, Uttar Pradesh – 211002, India*

ABSTRACT

Cyanobacteria (blue-green algae) are the photosynthetic microorganism having tremendous application potential in the field of human wellbeing with lots of biological activities in the form of nutritive complement. Different cyanobacteria (*Nostoc*, *Anabaena*, *Microcystis*, *Lyngbya*, *Oscillatoria*, *Synechocystis*, *Calotherix*, etc.), produce a large number of secondary metabolites (pigments, vitamins, food supplements) containing biological functions. The secondary byproducts obtained from cyanobacteria are great factor of antioxidant and anti-inflammatory response. Cyanobacteria also produce some biochemical diverse compounds belonging to alkaloids, cyclic peptides, lipopolysaccharides (LPSs), fatty acids (FAs), etc. Besides these, this photosynthetic microorganism has novel potency of biosynthesis of UV absorbing/protective compound. Approximately 60% of marine cyanobacteria are reported as biomodulator. They play a role as anticancer, antimicrobial, antiviral, antioxidant.

Microalgal Biotechnology: Bioprospecting Microalgae for Functional Metabolites towards Commercial and Sustainable Applications. Jeyabalan Sangeetha, PhD, Svetlana Codreanu, PhD, & Devarajan Thangadurai, PhD (Eds.)

However, these compounds are also at the clinical trial stage, and few are available in the market. Therefore, we can conclude that microalgae have potential application in the area of biomedical science, cosmetic industries, nutraceuticals, animal feed, etc.

3.1 INTRODUCTION

Algae are gram-negative oxygenic autotrophic organisms that live in marine and freshwater environmental conditions. They range from small unicellular microalgae, for example, from diatoms and cyanobacteria to multicellular macroalgae as giant kelp or giant bladed kelp (Kouzuma and Watanabe, 2015). They are larger than bacteria and are mostly aquatic organisms. Due to their aquatic and photosynthetic nature, they are often called "blue-green algae." The photosynthetic microorganism Cyanobacteria are unicellular organisms capable of growing in both colonies and in filamentous form and mostly it is surrounded by a mucilaginous or gelatinous sheath (Figure 3.1). Cyanobacteria belong to the Monera Kingdom and Cyanophyta division. This Cyanophyta is classified and called cyanobacteria; it is not placed in the group of algae due to their prokaryotic characteristics. Microalgae ranges in size from 0.2 to 2 μm up to filamentous form having sizes of 100 μm or may be more than it (Ravindran et al., 2016). The photosynthetic microorganisms have a great capacity for synthesis of biomolecules such as proteins, different pigments (chlorophyll, carotenoids, phycocyanin (PC)), antioxidants, polyunsaturated fatty acids (PUFA), etc. In addition, cyanobacteria are an important organism grow in marine/freshwater that synthesize different bioactive compounds with the help of primary and secondary metabolism. Some cyanobacteria species such as *Nostoc*, *Anabaena*, *Microcystis*, *Lyngbya*, *Oscillatoria*, *Synechocystis*, *Calotherix*, etc., produces different kind of secondary metabolites. In recent scenario, some microalgae are studied, and it is found that they contain active pharmacological molecules like anti-inflammatory, anticancer, antifungal, some antibiotics, and other pharmaceuticals (Bonotto, 1988; Centella et al., 2017). Blue-green algae (*Spirulina* sp., *Chlorella* sp., and *Aphanizomenon flos-aquae* (AFA)) act as a dietary source. They also contain probiotic compound which is beneficial for health (Singh et al., 2005). AFA is an important microalga which shows hypocholesterolemic effect due to rich source of chlorophyll, this chlorophyll content stimulates the liver function and maintains the cholesterol level in blood (Vlad et al., 1995). It is well known that *Spirulina* sp. is a rich source of single-cell protein. It is also a source of different vitamins (including B_{12}), pigments, and minerals including magnesium, calcium, iron, manganese, selenium, zinc, and potassium. They also have some FAs that activate hair and skin growth.

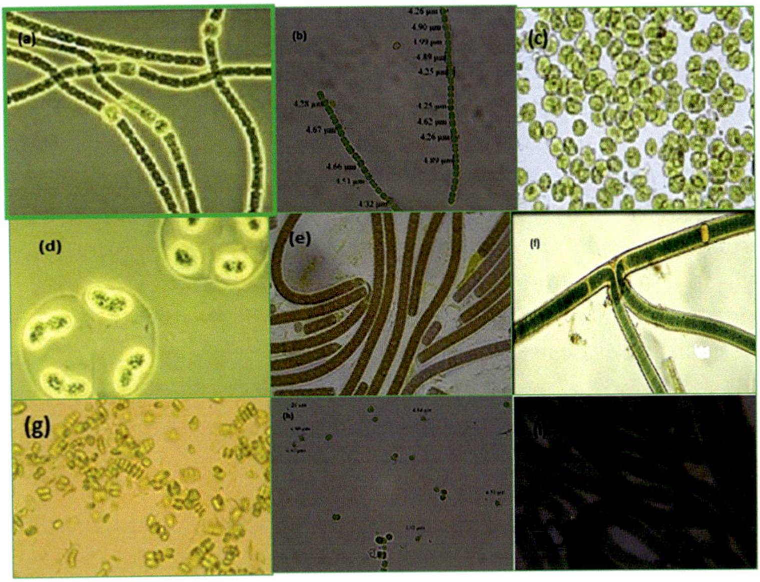

FIGURE 3.1 (a) *Anabaena cylindrica*; (b) *Nostoc muscorum*; (c) *Chlorella*; (d) *Gloeothece*; (e) *Ocillatoria;* (f) *Scytonema* sp.; (g) *Scenedesmus dimorphus*; (h) *Synechocystis* PCC 6803; (i) *Spirulina platensis*.
Source: Photographs taken from laboratory cultures present in Center of Biotechnology, University of Allahabad.

It is important to note that several cyanobacterial species can grow in desert region, i.e., hyper-arid, and produce some secondary metabolites in the form of photo protecting compounds for example Mycosporine like amino acids (MAAs) and Scytonemin, have a tendency to provide defense from UV radiation and other drought conditions (Fleming and Castenholz, 2007). Since some last decades, cyanobacteria have achieved great attention in the form of food supplements, bioactive compounds, biofertilizer, living cell factories for biofuel and high valuable pharmaceutical compounds with biomodulatory effects. So, we can also see that these extremophilic cyanobacteria can act as a rich source of cosmetic products are potentially applicable as sun blocking lotion and moisturizing agents.

3.2 MICROALGAE AS SECONDARY METABOLITES

3.2.1 CAROTENOIDS

Microalgae have the capacity to synthesize a superfluity of high value carotenoids, which are pigments usually of yellow, orange or red color (Chen, 1996). On the basis of structure, carotenoids are classified as lutein, xanthophylls, carotenes (α and β carotenes) and zeaxanthin (Safafar et al., 2015). Till day 400 carotenoids are known, out of them only few such as β carotene, astaxanthin, and in less amount zeaxanthin, lycopene, bixin, lutein, and fucoxanthin are available in market. The carotenoids act as important ingredients in human nutrition and have a tendency to control the risk of diseases caused due to nutrient deficiency (Shao et al., 2013).

3.2.1.1 CAROTENE

Microalgae as rich source of carotene and due to its high utility and commercialization the cost of carotene shoots up to 700€/kg (Mojaat et al., 2008). The production of carotene from microalgae can be done at very less expense. For the production of carotene some microalgal species such as *Spirulina maxima, Hematococcus*, and *Dunaliella* are used (Cardozo et al., 2007). Carotene is mostly used as a food colorant and it is also being added in packaged food and beverage products. Besides this, β carotene is also a rich source of vitamin A and being used in animal feed. Recently β carotene is in demand and being used in preparation of multivitamins as a source of pro-vitamin A (retinol) (Da et al., 2016). This carotene is being used in the medical science for the treatment of some metabolic disorders. It is used to treat lung cancer and breast cancer (Limon et al., 2015) and also helpful to treat eye-related disease (Gong and Bassi, 2016). From research, it was found that during 1970s, under high salt, high temperature and nutrient stress condition thc microalgal species *Dunaliella salina*, accumulates 14% of dry weight in the form of carotene. On the basis of this research, β carotene obtained from *Dunaliella salina* is a blooming industry.

3.2.1.2 ZEAXANTHIN AND LUTEIN

Most of the microalgae are significant source of naturally occurring zeaxanthin as well as lutein. The major source of these compounds, *viz.* zeaxanthin and lutein are *Spirulina* spp. and *Dunaliella salina* (Vo et al., 2015). So, the two metabolic products, i.e., Lutein-3 and zeaxanthin are important in the field of valuable nutraceutical market. Lutein-3 is applicable in animal tissues for egg yolks and chicken skin coloring, cosmetic items, packaging of foods and pharmaceutical products also (Perez-Garcia et al., 2011). From research it is also reported that Lutein obtained from *C. vulgaris* shows significant biomedical especially anti-cancerous activities, it is being applicable for human colon cancer cell line namely HCT-116 and it is also reported that consumption of lutein-rich food reduces the risk of cancer disease (Parveen and Nadumane, 2016).

3.2.1.3 ASTAXANTHIN

Astaxanthin is a secondary metabolite having chemical formula of $C_{40}H_{52}O_4$ and its molecular weight 596.8 g/mol in geometric cis- and trans-isomers. Astaxanthin is mostly obtained from green algae such as *Haematococcus pluvialis*, *Chlorella vulgaris*, *Chlorella zofingiensis* and *Chlorococcum* sp. Astaxanthin isolated from microalgae *Haematococcus pluvialis* is much greater up to 4–5% dry weight than other reported source. Astaxanthin is the sole source of pinkish color of aquatic fish. It has much higher antioxidant activity in comparison to other antioxidants, for example, carotene and vitamin E, so it works as the strongest antioxidant molecule within carotenoids (Gong et al., 2016). Due to the high antioxidant activity the compound astaxanthin enhances the level of catalase, superoxide dismutase (SOD) and peroxidase activity, and thus it prevents lipid peroxidation *in vivo* condition (Ranga Rao et al., 2013). Astaxanthin has immunological function, it stimulates antibody production, anti-inflammatory response, anti-aging, and temperature proofing when it is administrated with aspirin.

3.2.2 PIGMENTS

Pigments obtained from cyanobacteria such as chlorophyll, carotenoids, and phycobiliproteins (PBPs) are very useful tools in biological industries. Cyanobacteria contain collection of accessory light-harvesting complex of pigments having a significant role in photonic energy harvesting for the purpose of carbon storage during photosynthesis (Sinha et al., 1995). This important light-harvesting protein complexes termed as PBPs, contains different types of chromophores for the absorption of light energy (Kannaujiya and Sinha, 2015, 2017). PBPs are water-soluble compounds and classified into three groups such as PC, allophycocyanin (APC) and phycoerythrin (PE) having absorbance between 450 and 660 nm (Grossman et al., 1993). These PC, APC, and PE have antioxidant and anti-inflammatory properties (Bhaskar et al., 2005). It is also reported that cyanobacterial PC comprise of approximately 15% of the dehydrated mass of the blue-green algae which shows anti-inflammatory, antioxidant, neuroprotective, and hepatoprotective activity (Eriksen, 2008). Cyanobacterial PC is pharmacologically used for the treatment of Parkinson's, and Alzheimer's diseases (Rimbau et al., 2001) and also controls constipation, cataract, pancreatitis, some degenerative diseases, skin, and oral cancers. Recently, algal pigments have been most applicable in the medical field in diagnostic as fluorescent tags. Besides these medicinal properties, these algal pigments are used as natural colorants for food, cosmetics, and pharmaceuticals, which is the substitute of synthetic dyes. *Nostoc*, *Anabaena* sp., and *Spirulina platensis*, are blue-green algae which produce PC in large amount. PC isolated from *Anabaena cylindrica* is shown in Figure 3.2.

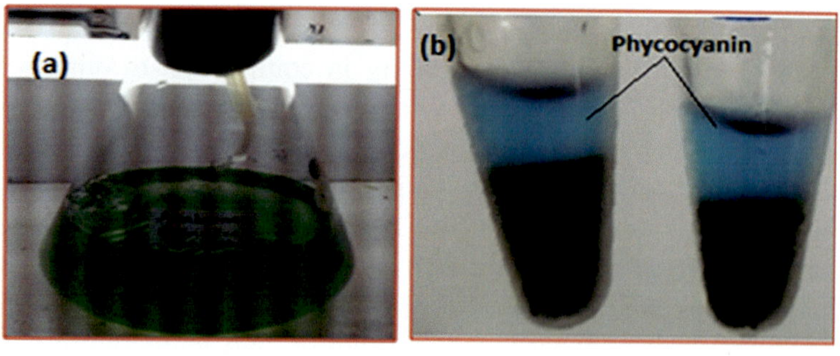

FIGURE 3.2 (a) Culture of cyanobacteria *Anabaena cylindrica*; and (b) phycocyanin isolated from *Anabaena cylindrica* in phosphate buffer.

3.2.3 VITAMINS

Microalgae are rich source of vitamins. *Spirulina* sp. is the richest source of single-cell protein, vitamin B_{12} and vitamin E of whole food source. From previous research, it was found that consumption of 20 gm of *Spirulina* sp. is equal to approximately 70% of vitamin A, 50% of vitamin B_2, 12% of vitamin B3 and also provides the body requirement of vitamin B_{12} (Watanabe et al., 2002). It is reported that *Spirulina* sp. is an affluent resource of vitamin E (tocopherol) content (190 mg kg^{-1}); it is three times greater than pure wheat germ content (Challem et al., 1981). Microalgae such as *Nostoc punctiforme*, *Anabaena hassali*, *Microcystis pulverea*, AFA, and *Phormidium bijugatum* are rich sources of vitamins B-complex, nicotinic acid and pentothene. Microalgae besides a single cell protein source, is also a major constituent in food, aquaculture, and pharmaceuticals due to the presence of various useful biochemical substances (Table 3.1).

TABLE 3.1 Useful Biochemical Substances Present in Microalgae

Vitamins	A, B_1, B_6, B_{12}, C, E, biotin, folic acid, nicotinic acid, pantothenate, riboflavin
Pigments	Chlorophyll, β-carotene, phycoerythrin, astaxanthin, lutein, fucoxanthin, zeaxanthin, canthaxanthin
Antioxidants	Catalases, superoxide dismutase, tocopherols, polyphenols
Polyunsaturated Fatty Acids (PUFAs)	ARA(C20:4), DHA(C22:6), GAL(C18:3), EPA(C20:5)
Others	Antimicrobial, antiviral agents, antifungal, proteins, amino acids, toxins, sterols, MAAs

3.3 MICROALGAE AS NUTRITIONAL SUPPLEMENTS

Since thousands of years, some algal species such as *Spirulina platensis*, *Spirulina maxima*, *Spirulina fusiformis*, *Nostoc commune,* and AFA have been broadly consumed as nutritional supplement by human beings. Out of most blue-green algae, *Spirulina platensis* is the most common algal species which is being used by humans as a supplement. *Spirulina platensis* is richest source of single-cell protein. The algae contain 70% of dry weight of protein and it is also rich in minerals, vitamins, some essential FAs and other important nutrients (Vonshak, 1997). However, protein obtained from *Spirulina* contains less amount of cystein, methionine, and lysine and it is inferior to milk or meat protein (Ciferri, 1983) while it is superior to protein obtained from plant source including legumes.

The history of application of *Nostoc commune* is so overwhelming, it is being used for medicinal point of view for the treatment of night blindness, inflammation, chronic fatigue, and indigestion (Qiu et al., 2002). Besides these medicinal properties, lipid isolated from *Nostoc* sp. is being used in restrain of cholesterol biosynthesis. So *Nostoc* sp. can be accepted as a good candidate for the treatment of hypercholesterolemia.

In addition to these nutritional activities of algae, it is shown that if AFA taken orally by healthy person it shows the reduction in phagocytic activity of polymorphic nucleated cell *in vitro* condition (Jensen et al., 2000). Hence by way it also shows the bio-modulatory effect in humans.

TABLE 3.2 Biomass Composition of Microalgae Expressed on a Dry Matter Basis

Algal Strain	Protein Content (%)	Lipid (%)	Carbohydrate (%)
Anabaena cylindrica	43–56	4–7	25–30
Chlamydomonas reinhardtii	48	21	17
Chlorella vulgaris	41–58	10–22	12–17
Dunaliella tertiolecta	29	11	14
Dunaliella salina	57	6	32
Porphyridium cruentum	28–39	9–14	40–57
Scenedesmus dimorphus	8–18	16–40	21–52
Scenedesmus obliquus	50–56	12–14	10–17
Spirulina maxima	60–61	6–7	13–16
Spirulina platensis	42–63	4–11	8–14
Synechococcus sp.	63	11	15
Spirogyra sp.	6–20	11–21	33–64

3.4 MICROALGAE AS FOOD

As we have earlier shown that microalgae are important source of proteins, different vitamins such as vitamin A, B_1, B_2, B_6, B_{12}, C, E, and minerals such as iodine, potassium, niacin, magnesium, calcium, and iron. Being as a large source of nutrients, it is used as a major part of food. It is being used as a food source mostly in countries China, Korea, and Japan. One of best examples is *Spirulina platensis*, it is a blue-green alga which is world widely used as a nutritious food for humans. Another one example is *Spirulina maxima*, it is commercially available and used mostly as food supplement for human as well as animal feed source. Presently algal species such as *Chlorella* and *Spirulina* is available in market as a fish food. Microalgae contain three most important food components: Protein, Lipid (oil) and Carbohydrate (Um and Kim, 2009). Microalgal biomass composition expressed in dry matter is mentioned in Table 3.2.

3.5 MICROALGAE AS A PHOTO PROTECTING/UV-PROTECTING COMPOUND

A number of motivating properties of microalgae either it is freshwater or marine water has property of biosynthesis of photo protecting compounds which have tendency to protect themselves from UV radiation (Singh et al., 2010). Along with this, cyanobacteria have the capacity to synthesize some other extracellular polysaccharide which provides a template for UV protecting compounds such as scytonemin and mycosporine-like amino acids (MAAs) when it is exposed to UV-B radiation (Ehlin Schulz et al., 1997; Rahman et al., 2016). Cyanobacteria adopt strategies as a defensive barrier that provides protection as well as adaptation to counteract damaging effects of UV radiation (Figure 3.3). Mycosporine-like amino acids (MAAs) are produced in different organisms; it is a kind of secondary metabolic product which has a tendency to absorb solar radiation either directly or indirectly and gives protection to organisms from solar radiation (Häder et al., 2007). MAAs is a water-soluble compound having a small molecular weight (<400 Da). The amount of MAAs synthesized from cyanobacteria has natural properties to absorb 10–26% of photon energy which directly comes from UV-B radiation; it provides defense against deleterious effect from UV-B radiation (Ehling-Shulz and Scherer, 1999). Therefore, due of their UV-absorption capacity, MAAs work as strong photo-protectant derived from microalgae against UV-radiation.

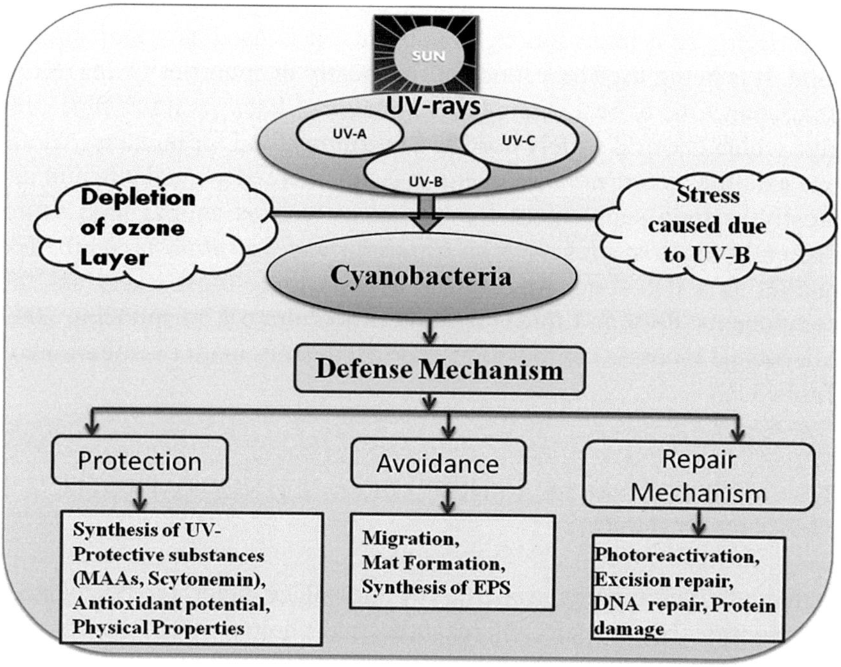

FIGURE 3.3 Strategies opted by cyanobacteria to counteract damage from UV radiation.

It is shown that MAAs is synthesized and accumulated at the epidermis of skin; at this location MAAs provides much sunscreen activities (Oren and Cimerman, 2007). A number of research findings show that MAAs is much stable and effectively works as sunscreen compounds. From previous study it is shown that porphyra-334 gained from *Porphyra vietnamensis* has more photoprotecting effects as against commonly used *Aloe vera* gel (Bhatia et al., 2010). From different research findings it is confirmed that MAAs has less photodynamic reactivity in comparison to other sunscreen compounds commercially available in the market. MAAs as porphyra-334 and shinorine isolated from different microalgae are being used as sunscreen products on commercial level specifically Helioguard 365 and Helionori which is safeguarded against UV-A radiation (Schmid et al., 2006). Besides this another UV protecting compound is scytonemin. Scytonemin is a yellow-brown, lipid-soluble photo protective compound deposit in the exopolysaccharide sheath of several cyanobacteria which is generally present on rock surface, bark of trees and marine intertidal mats (Garcia-Pichel and Belnap, 1996). Scytonemin contains indolic and phenolic subunits having molecular mass 544 Da. This photoprotecting was first of all reported in some terrestrial cyanobacteria (Nägeli and Gattungene, 1849). From research findings, it is reported that Scytonemin is a very stable bioactive substance having potency to screen lack of any metabolic activity, even after better physiological inactivity. Tables 3.3 and 3.4 show the different reported MAAs and other UV-protecting compounds.

These UV protecting compounds have tremendous potential as anti-oxidative substances, natural sunscreen compounds and therapeutic agents beside photo-protecting properties. Hence, it can be summarized that these UV protecting compounds MAAs and Scytonemin have a valuable role in pharmaceutical as well as cosmetic and healthcare industries, and their commercial values can be explored with the help of biotechnology.

TABLE 3.3 Different Reported MAAs/UV-Screening Compounds from Diverse Microalgae

Microalgae	Reported MAAs/UV-Screening Compounds
Ulva lactuca	Porphyra-334
Gymnogongrus griffithsiae	Shinorine, Porphyra-334, Palythine, Asterina
Gymnogongrus antatcticus	Shinorine
Gelidium pusillum	Porphyra-334, Shinorine, Palythine, Asterina, Palythinol
Pilayella littoralis	Porphyra-334
Pophyra columbina	Mycosporine-glycine, Shinorine, Porphyra-334, Palythine, Asterina,
Kallymenia antartica	Shinorine, Palythine, Asterina-330
Porphyra endiviifolium	Shinorine, Porphyra-334
Porphyra leucosticta	Shinorine, Palythine, Porphyra-334, Asterina
Calothrix sp., *Scytonema* sp., *Rivularia* sp., *Nostoc commune, Nostoc punctiforme*	Scytonemin
Scenedesmus sp., *Scotiella chlorelloidea, Scotiellopsis rubescens, Spongiochloris spongiosa, Dunaliella salina, Chlorella fusca*	Sporopollenin

TABLE 3.4 Different Reported MAAs with Their Corresponding Absorption Spectra

Mycosporine-like Amino Acids	λ_{max} (nm)
Asterina-330	330
Mycosporine-taurine	309
Mycosporine–glycine	310
Palythine	320
Palythine-serine-sulfate	320
Palythine-serine	320
Mycosporine-methylamine-serine	327
Mycosporine-methylamine-threonine	327
Mycosporine-glutamic acid-glycine	330
Palythinol	332
Mycosporine-2-glycine	334
Scytonemin	386
Porphyra-334	334

3.6 HIGH VALUABLE METABOLITES OBTAINED FROM CYANOBACTERIA

3.6.1 CYANOVIRIN-N (CN-N)

Cyanovirin-N (CN-N) is an effective virucidal 101 amino acid long, 11 kDa protein. This unique protein irreversibly inactivates varied strain of HIV virus such as HIV-1, HIV-2, M-tropic strains involved in sexually transmission of HIV and also affective against cell to cell and virus to cell broadcast of HIV contagion (Burja et al., 2001). From different research findings, it is confirmed that the efficiency of this potent protein CN-N obtained from *Nostoc* sp., has great antivirus activity against Influenza, Ebola, and HCV (Hepatitis C) (Barrientos et al., 2003; O'Keefe et al., 2003; Helle et al., 2006; Buffa et al., 2009). Different researches related to cell-associated gp120 confirm that this potent protein CN-N interacts with viral envelope and blocks together CD-4 independent and CD-4 dependent binding and HIV-1 with cells (Esser et al., 1999). CN-N is synthesized by genetically modified *Escherichia coli* bacteria and purified monomeric protein (Colleluori, 2005). For large scale production of CN-N a vector having pel-B signal peptide sequence is applicable (Mori et al., 1998). Scytovitin is also an antiviral peptide compound obtained from cyanobacterium *Scytonema varium*. Structurally scytovirin is 95 peptide residues which is dissimilar from CN-N which works against HIV-1. This novel compound scytovirin binds only with viral cover proteins gp120, gp160 and gp41 while it is unable to bind with cellular receptor CD4 (Bokesch et al., 2003).

3.6.2 BOROPHYCIN

Borophycin is an important metabolic product obtained from marine cyanobacteria *Nostoc linckia* and *Nostoc spongiaeforme* var. *tenue* (Figure 3.4), which are boron-containing metabolites. This compound has great cytotoxic activity against human epidermoid carcinoma and also affective against colorectal adenocarcinoma cell lines. Besides, it also shows antimicrobial activity (Burja et al., 2001).

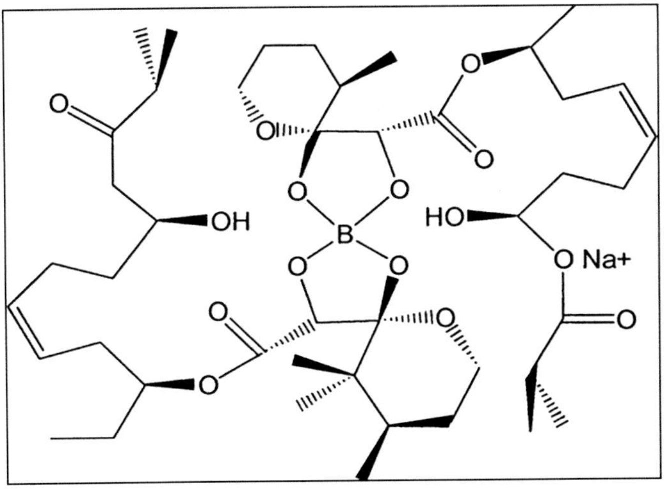

FIGURE 3.4 Chemical structure of borophycin.

3.6.3 CRYPTOPHYCIN

The bioactive compound Cryptophycin (Figure 3.5) first of all isolated from cyanobacteria *Nostoc* sp. ATCC 53789. It displays an exclusive function as fungicidal. From different research sources it was also identified as toxic and also divided as natural product. Cryptophycin has also been retrieved from *Nostoc* sp. GSV 224 and shows a potential cytotoxic effect in opposition to human carcinoma cell lines. Due to this cytotoxic effect, it shows broad-spectrum drug resistant and drug-sensitive human tumors (Burja et al., 2001). From mechanism point of view, the cryptophycin interacts with tubulin, which disrupts tubulin-dynamics and finally, the apoptosis of tumor cells takes place (Panda et al., 1998).

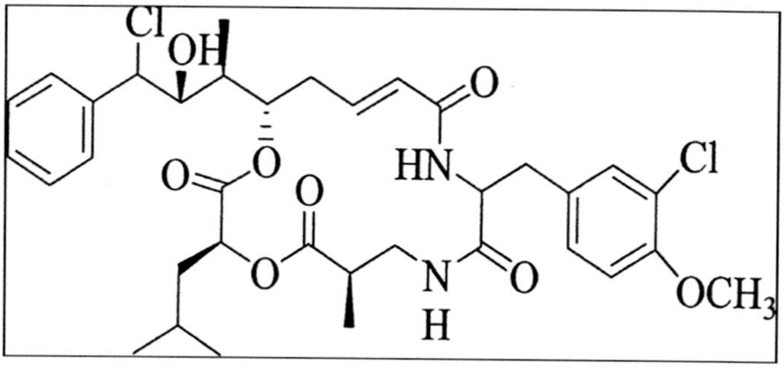

FIGURE 3.5 Chemical structure of cryptophycin.

3.6.4 LIPOPEPTIDES

It is assumed that approximately 65–68% of product (natural) obtained from different species of cyanobacteria having nitrogen. The product obtained from different marine cyanobacteria having amino-acid fragment which is directly linked to FA portion, form a compound known as lipopeptides. Previously natural products obtained from 424 marine cyanobacteria were studied and it was found that this natural product contains 40.2% lipopeptides, 9.4% amides, 5.6% pure amino acid, 4.2% macrolides and 4.2% FAs.

The bioactive compounds lipopeptides are very useful biochemical substance with a great potency with cytotoxic, antimicrobial, anticancer, antiviral, antifungal, and enzyme inhibitor activities (Burja et al., 2001). A number of lipopeptides are reported from cyanobacteria, for example, lyngbyatoxin, barbamide, curacin A, Hassallidin, jamaicamide. One known example of lipopeptides is hapalosin (Figure 3.6), it is a cyclic depsipeptide obtained from cyanobacterium, *Hapalosiphon welwitschii*, having multidrug resistance activity (Kashihara et al., 2000). The important feature of lipopeptides is that they have low molecular weight and it passes through different blood barrier tissues, so it is being used as a drug delivery system (Burja et al., 2001). Anabaenolysins are another example of lipopeptides produced by cyanobacteria *Anabaena* sp. isolated from the Baltic Sea. The anabaenolysins exhibit antifungal activity (Jokela et al., 2012).

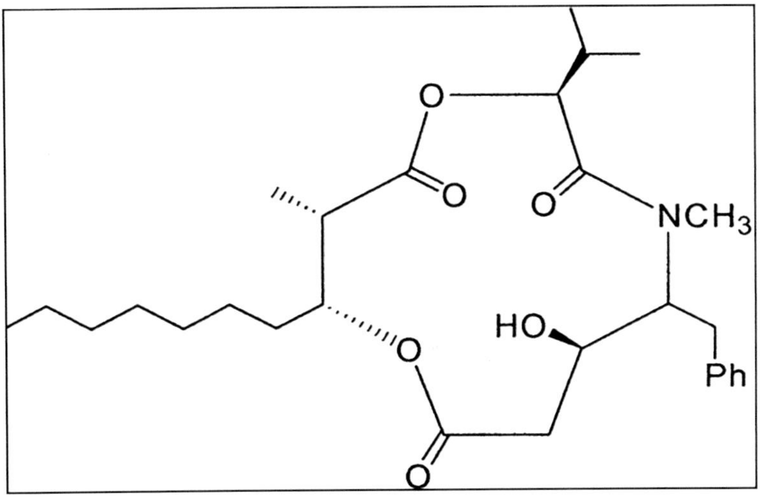

FIGURE 3.6 Chemical structure of hapalosin.

3.7 TOXIC COMPOUNDS FROM CYANOBACTERIA

Cyanobacteria are an important source of toxic secondary compounds which are responsible for intoxication for human and domestic as well as wildlife. It is reported that a lot of bloom-forming cyanobacteria obtained from several habitats have the potency to synthesize diverse cyanotoxins (Rastogi et al., 2014). On the basis of chemical nature cyanotoxins are classified into three categories: (i) cyclic peptides (Nodularins and microcystins (MCs); (ii) alkaloids (anatoxin-a, anatoxin-a(s), lyngbyatoxin-a, cylindrospermopsin (CYN), aplysiatoxin, saxitoxins (STX)); and (iii) lipopolysaccharides (LPSs) (Kaebernick and Neilan, 2001). Table 3.5 shows common cyanotoxins present in several cyanobacteria and its probable harmfulness and activities. Beside this on the basis of biological character cyanotoxins are classified into five groups: hepatotoxins, cytotoxins, neurotoxins, irritant toxins and dermatotoxins (Sivonen and Jones, 1999; Codd et al., 2005).

3.7.1 CYCLIC PEPTIDES

Microcystins (MCs) are the furthermost commonly cyanotoxins present on the surface and drinking water. MCs (Figure 3.7(a)) are cyclic heptapeptides produced from some cyanobacteria such as planktonic *Anabaena*, *Nostoc*, *Oscillatoria*, *Planktothrix*, *Microcystis*, and *Anabaenopsis* species, and beginning terrestrial *Hapalosiphon* genera (Aboal and Puig, 2009; Rastogi et al., 2014). Like microcystine, Nodularin (Figure 3.7(b)) is the second group of hepatotoxins which is reported from *Nodularia spumigena*, and *Nostoc* sp. So, on the basis of delicate toxicity, microcystin-LR (MC-LR) is measured as the strongest hepatotoxin (Funari and Testai, 2008).

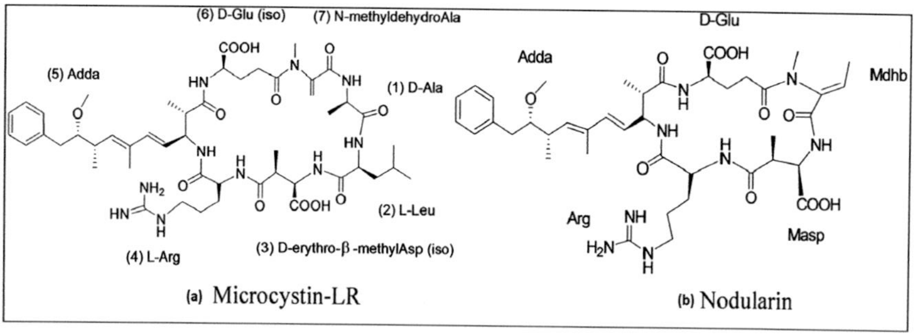

FIGURE 3.7 Chemical structure of (a) microcystin (MC-LR); and (b) nodularin (NOD).

On the basis of computational analysis, the specific gene accountable for microcystin biosynthesis has been reported in numerous cyanobacterial species (Christiansen et al., 2008; Gehringer et al., 2012). From research, it is shown that the microcystin gene of *Microcystis aeruginosa* PCC7806, have 55 kb of DNA and it is composed of 10 (mcyABCDEFGHIJ) bidirectional transcribed open reading frames (ORFs) set in two transcribed operons, namely *mcyA-C* and *mcyD-J* (Tillett et al., 2000). The cyanotoxin

MC-LR are accountable for inhibition of particles serine-threonine protein phosphatases 1 (PP1) and 2A (PP2A) which Ki values is below from 0.1 nM. Furthermore, it is shown that the microcystin and nodularin shows similar biological action regardless of their chemical structures are quite different. Another one toxic species of cyanobacteria is *Lyngbya*, which is filamentous and present abundantly within tropical and subtropical waters. There are several cytotoxic compounds obtained from cyanobacteria, *Lyngbya* such as lyngbyatoxin A, B, and C, aplysiatoxin, antillatoxin, and debromoaplysiatoxin.

3.7.2 ALKALOIDS

A number of alkaloid toxins are reported from different cyanobacteria and have varying chemical stability. The alkaloid anatoxin-a have low molecular weight of 165 Da is fast working neurotoxin. Its homolog toxin is Homoanatoxin-a having molecular weight 179 Da which also a fast-working neurotoxin like anatoxin-a, so they are known as fast death factors (FDFs). The alkaloid Anatoxin-a (Figure 3.8) is isolated from several cyanobacterial species such as *Anabaena flos-aquae*, *A. circinalis*, *Aphanizomenon*, *Cylindrospermum*, *A. spiroides*, *M. aeruginosa*, *A. planctonica* and *Planktothrix* (Edwards et al., 1992; Park et al., 1993). The homolog toxin Homoanatoxin-a (MW = 179) is recovered from *Phormidium formosum* strain. Alkaloid homoanatoxin-a carry methylene group at C-2 position on the place of acetyl group in anatoxin-a. From previous study, it is shown that the fatal dosage of anatoxin-a and its homolog homoanatoxin-a resulting in 50% deaths (LD50) are 200–250 μg kg^{-1} bw (Carmichael et al., 1990; Skulberg et al., 1992).

Another one homolog of anatoxin-a is anatoxin-a(S). The Anatoxin-a(S) is an exclusive phosphate ester of cyclic N-hydroxyguanine which molecular weight is 252 Da. Anatoxin-a(S) is obtained from *Anabaena flos-aquae* and *A. lemmermannii* is a powerful acetylcholinesterase (AChE) inhibitor (Matsunaga et al., 1989). Currently, it is documented in huge amount and obtained from *Anabaena lemmermannii* (Henriksen et al., 1997; Onodera et al., 1997a). It is much toxic when compared to anatoxin-a (Carmichael et al., 1990; Mejean et al., 2014).

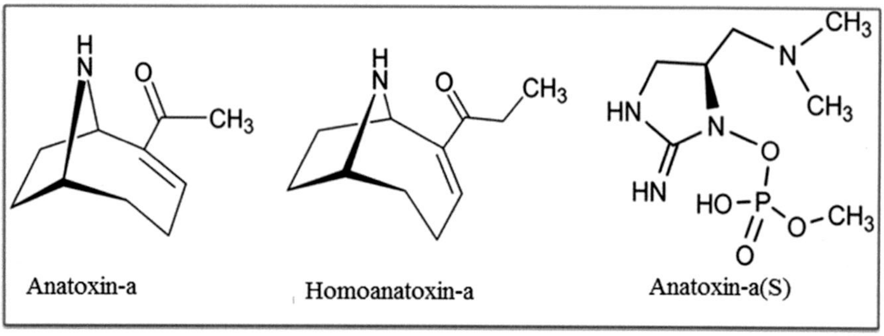

FIGURE 3.8 Chemical structures of anatoxins.

Saxitoxin and its equivalent compounds such as neosaxitoxin (Figure 3.9) are a collection of carbamate alkaloid toxins which may be either non-sulfated (STX), singly sulfated (gonyautoxins (GTX)) or doubly sulfated (C-toxins) on the basis of chemical structure. They all are tremendous neurotoxins. These toxins are tricyclic compounds, having a tetrahydropurine group and two other guanidine subunits, generally known as paralytic shellfish poisons (PSPs). So, STX also called PSPs which have tendency to block the neuronal communication via binding to the voltage-gated Na^+ channels (Strichartz et al., 1986; Su et al., 2004) From current research it is found that approximate 27 kind of STX have been found in various cyanobacteria such as *Anabaena circinalis*, *Lyngbya wollei*, *Anabaena flos-aquae*, *Cylindrospermopsis raciborskii*, and *Aphanizomenon*.

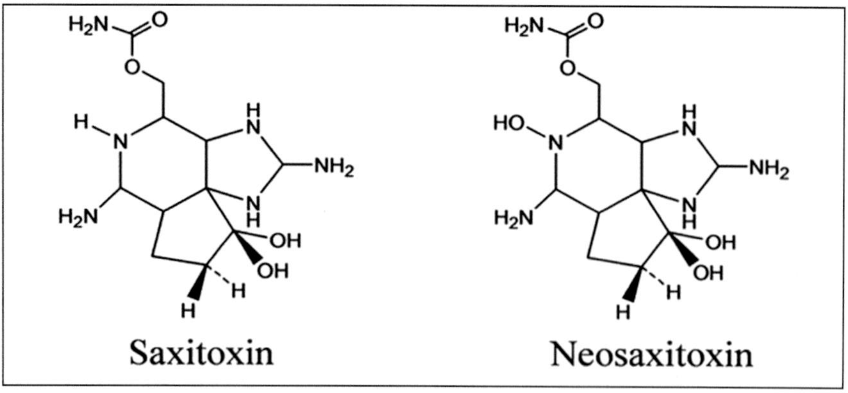

FIGURE 3.9 Chemical structure of saxitoxin and neosaxitoxin.

Structurally the CYN cyanotoxin (Figure 3.10) is a polyketide-alkaloid having a sulfate groups and tricyclic guanidine moiety. Its molecular weight is 415 Da. It is reported that, lots of cyanobacterial species such as *Lyngbya wollei*, AFA, *Anabaena lapponica*, *Cylindrospermopsis raciborskii*, *Aphanizomenon ovalisporum*, *Anabaena bergii*, *Oscillatoria* (*Planktothrix*), *Raphidiopsis curvata*, and *Umezakia natans*, synthesize CYN and their homolog (Spoof et al., 2006; Seifert et al., 2007; Mazmouz et al., 2010). The cyanotoxin CYN has cytotoxic, hepatotoxic, and nephrotoxic effects and also acts as a potential carcinogen having the tendency of reticence of cytochrome P450, glutathione, and protein biosynthesis (Humpage et al., 2000; Froscio et al., 2003; Neumann et al., 2007).

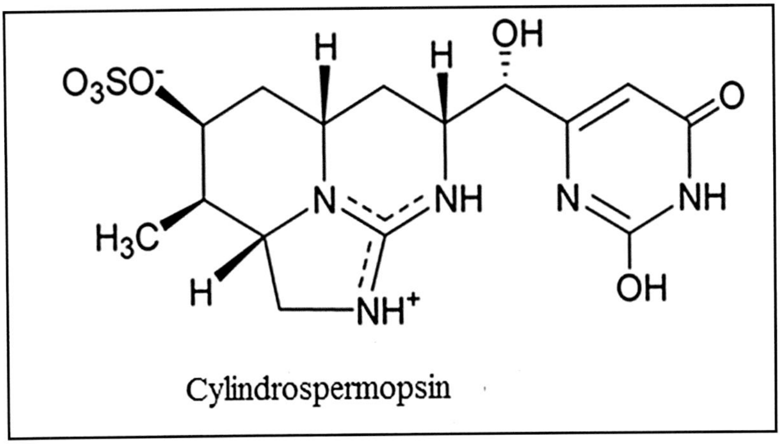

FIGURE 3.10 Chemical structure of cylindrospermopsin.

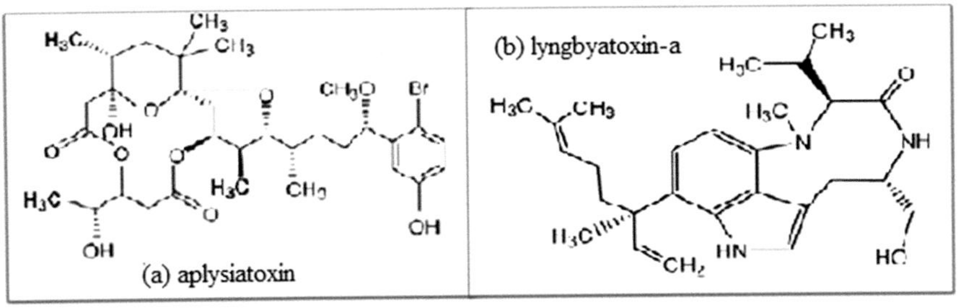

FIGURE 3.11 Chemical structure of (a) aplysiatoxin; and (b) lyngbyatoxin-a.

TABLE 3.5 Different Cyanotoxins Present in Various Cyanobacteria and Their Potential Toxicity and Mode of Actions

Toxins	Cyanobacterial Genera	Biological Toxicity	Targeted Organs in Mammals	Mode of Action
Microcystins	*Anabaena*, *Anabaenopsis*, *Aphanocapsa*, *Aphanizomenon*, *Arthrospira*, *Cyanobium*, *Cylindrospermopsis*, *Fischerella*, *Hapalosiphon*, *Limnothrix*, *Lyngbya*, *Microcystis*, *Nostoc*	Hepatotoxic	Liver	Inhibitors of protein phosphatases 1, 2A and 3, genotoxicity, tumor promoter
Nodularins	*Nodularia*	Hepatotoxic	Liver	Inhibitors of protein phosphatases 1, 2A and 3, tumor promoter
Antillatoxin	*Lyngbya*	Neurotoxic	Nerve synapse	Blocking neuronal communication by binding to the voltage-gated Na^+ channels
Anatoxin-a	*Anabaena*, *Aphanizomenon*, *Cylindrospermum*, *Microcystis*, *Planktothrix*, *Raphidiopsis*	Neurotoxic	Nerve synapse	Depolarizing neuromuscular blocking
Anatoxin a-(s)	*Anabaena*	Neurotoxic	Nerve synapse	Inhibition of Ach-esterase activity, hyper-excitability of nerve
Saxitoxins	*Anabaena*, *Aphanizomenon*, *Cylindrospermopsis*, *Lyngbya*, *Planktothrix*, *Raphidiopsis*, *Scytonema*	Neurotoxic	Nerve axons	Blocking neuronal communication by binding to the voltage-gated Na^+ channels

TABLE 3.5 *(Continued)*

Toxins	**Cyanobacterial Genera**	**Biological Toxicity**	**Targeted Organs in Mammals**	**Mode of Action**
Aplysiatoxins	*Lyngbya*, *Oscillatoria*, *Schizothrix*, *Trichodesmium*	Dermatotoxic	Skin	Potent tumor promoters and protein kinase C activators
Cylindrospermopsin	*Anabaena*, *Aphanizomenon*, *Cylindrospermopsis*, *Lyngbya*, *Oscillatoria* (*Planktothrix*), *Rhaphidiopsis*, *Umezakia*	Hepatotoxic, nephrotoxic, and cytotoxic	Liver	Irreversible inhibition of protein and glutathione synthesis, implicating cytochrome P-450, overexpression of DNA damage repair proteins
Lyngbyatoxin-a	*Lyngbya*, *Schizothrix*, *Oscillatoria*	Cytotoxic, gastroenteritis dermatotoxic	Gastro-intestinal tract, skin	Dermonecrotic, potent tumor promoters and protein kinase C activator
Lipopolysaccharides (LPS)	*Anabaena*, *Spirulina*, *Microcystis*, *Oscillatoria*, and almost all cyanobacteria	Dermatotoxic	Affects any uncovered tissue	Irritation, and allergic property

Some other alkaloid cyanotoxins such as aplysiatoxin and lyngbyatoxins are also reported in some fresh and/or marine water cyanobacteria (Figure 3.11). It is reported that few cyanobacteria such as *Oscillatoria nigroviridis*, *Lyngbya majuscula*, *Trichodesmium erythraeum,* and *Schizothrix calcicola* has tendency to produce alkaloid aplysiatoxin (Mynderse et al., 1977; Gupta et al., 2014). This aplysiatoxin acts as protein kinase C activators, tumor promoters and also generates several mortal effects. Another one reported cyanotoxin lyngbyatoxin, is structurally a cyclic dipeptide compound, which was obtained from cyanobacteria *Lyngbya majuscula* (Taylor et al., 2014) which is similar in toxicity and potent tumor promoter as aplysiatoxin.

3.7.3 LIPOPOLYSACCHARIDES (LPSs)

On the basis of structure, the endotoxin LPSs are made up of an oligosaccharide, an inside acylated glycolipid, and an external polysaccharide chain (Raetz and Whitfield, 2002). Usually, the FA constituent of LPS generates toxicity and negative effects such as allergenic reaction and irritation in humans and animal tissues (Mankiewicz et al., 2003). LPS were first of all isolated from the cyanobacterium *Anacystis nidulans.* LPSs are the building blocks of the cell wall of Gram-negative bacteria and cyanobacteria also, where they form complexes within proteins and phospholipids. Some cyanobacterial species like *Anabaena*, *Microcystis*, *Anacystis nidulans*, *Oscillatoria*, and *Spirulina* are reported which produces lipopolysaccharide toxin (Smith et al., 2008; Blahova et al., 2013). But, still the accurate procedure of LPS toxicity produced by diverse cyanobacteria is unknown. Some reported cyanotoxins are mentioned in Table 3.5.

3.8 BIOMODULATORY EFFECT OF MICROALGAE

3.8.1 ANTICANCER ACTIVITY

In the current scenario of medical science treatment of cancer is done using ionizing radiation, alkylating agents, DNA topoisomerase inhibitors, hyperthermia, and platinum compounds which induce DNA damage erratically destroy both healthy and fast multiplying of tumor cell (Beesoo et al., 2014). While, it is shown that pigments isolated from microalgae are able to protect healthy cells from genetic damages and put forth for cytotoxic, antiproliferative, and pro-apoptotic events in tumor cells, so it may be a suggestive tool for cancer prevention or chemotherapy (Baudelet et al., 2013). *Streptomyces* sp. NPS853 synthesize a new anthramycin-type compound known as Usabamycins-37. This Usabamycins express less destruction of He-La cell growth and selective inhibition of serotonin (5-hydroxytrypamine) 5-HT2B uptake (Manivasagan et al., 2014). It is shown that *Spirulina* and *Dunaliella* extract has a tendency to inhibit the chemically synthesized carcinogenesis in model known as hamster buccal pouches (Schwartz and Shklar, 1987; Schwartz et al., 1988).

Studies show that PC extracted from microalgae AFA destroyed the *in vitro* expansion of one cell line out of four tested tumors, so it indicates that cell lines are sensitive to the PC. Some reported algal toxins such as Antillatoxin-51, debromoaplysiatoxin-53 and aplysiatoxin-52, are cytotoxic compounds having anticancer activity (Burja et al., 2001). Beside these toxins, apratoxins is a lipopeptide cyanobacterial secondary metabolite having potent cancer cell cytotoxicity.

3.8.2 ANTIMICROBIAL ACTIVITIES OF MICROALGAE

Several bacteria and fungi are the causative agents for several pathogenic diseases in animals, humans, and plants. These microorganisms decrease the crop production and also are causative agents for food spoilage. Over the past few decades lifestyle of human being is changed, and antibiotics are being widely used, resulting in microbes becoming resistant to different antibiotics. So nowadays, it is necessary to find out the new antimicrobial agents, which can be potentially useful. If we consider the chemically synthesized antibiotics, there is a drawback of negative impact as well as risk for creation of several resistant pathogenic microbial strains. Scientists are working for isolation of some natural and novel antibiotics from natural resources such as plants and microorganisms having broad mode of action. Microalgae are one of the sources of photosynthetic microorganisms able to produce bioactive compounds which are effective, in crude as well as purified form or any one of them, as antimicrobial agents. The first reported green microalgae, i.e., *Chlorella*, which has the potency to inhibit both Gram-positive as well as Gram-negative bacteria (Washida et al., 2006). It is also reported that *Dunaliella salina* produces some bioactive compounds which is affective against several microorganisms (bacterial and fungus) strains, for example, *Escherichia coli*, *Staphylococcus aureus*, *Pseudomonas aeruginosa*, *Aspergillus niger*, and *Candida albicans* (Cho et al., 2011). The bioactive compound extracted from *D. salina* prevents the growth of bacteria *Klebsiella pneumoniae*. Besides these some microalgae also produce bioactive compounds that have antifungal activities (Volk et al., 2006).

In a report, it was shown that the c-lactone malyngolide-9 an antibiotic isolated from shallow-water habitat of the blue-green algae *Lyngbya majuscula* is effective against bacteria *M. smegmatis* and *S. pyogenes*. It is also found that the crude extract of *Microcystis aeruginosa* (Cyanophycota) displays a high level of antifungal as well as antibacterial activity (Khalid et al., 2010). So, it can be summarized that microalgae may act as potent antimicrobial agent.

3.8.3 ANTIOXIDANT ACTIVITY OF MICROALGAE

Antioxidants are important molecules utilized by the human body to protect themselves from oxidative damage from the negative effects of free radicals. Free radicals, such as ROS (reactive oxygen species) and NOS (nitrogen reactive species) cause oxidative damage and induce rheumatoid arthritis, cataracts, atherosclerosis, neurological damage, muscular dystrophy, cancer, and aging (Thomas et al., 2011). On the basis of health concerns the risks created from artificial antioxidants for example butylated hydroxytoluene (BHT) and propyl gallate (PG), it is necessary to think about the natural sources of antioxidants. It is seen that the photoautotrophic microorganism is extremely exposed to oxidative and radical damages in its natural environments, to protect themselves from oxidative stress (Singh et al., 2010). Antioxidant molecules available in microalgae, such as phenolic compounds, flavonoids, carotenoids, tocopherol, FAs, and alkaloids play an effective role to fight against the oxidative process (Stamenic et al., 2014; Régnier et al., 2015). The photosynthetic pigment chlorophyll a and their metabolic product obtained from microalgal species are reported to have many antioxidant activities (Cho et al., 2011). The microalga *Undaria pinnatifida* produce fucoxanthin pigment and its derivatives for example auroxanthin, which have powerful radical scavenging activity (Sachindra et al., 2007). Another pigment such as phycoerythrobilin, created by some microalgal species, also possesses antioxidant activity (Yabuta et al., 2010). So, the extensive application of natural products produced by microalgae having powerful antioxidant potency enhances the nutritional value as food, pharmaceutical, and nutraceutical point of view.

3.8.4 ANTI-OBESITY ACTIVITY OF MICROALGAE

Nowadays, due to fast life and unbalanced nutrition humans are suffering from obesity. The obesity is caused due to accumulation of adipose tissue, i.e., fat in the human body. From study it is found that the obesity is a type of metabolic disorder associated with several kinds of health-related complications and several diseases for example diabetes mellitus, cardiovascular disease, aging, and cancer (Kopelman, 2000). From several researches, it is found that some medicinal plants synthesize a variety of fat-lowering agents and anti-hyperlipidemic. In the current scenario, another natural source such as microalgae are studied as a budding source of fat-lowering agents and anti-hyperlipidemic. It has been also shown that ROS and NOS are greater responsible for obesity. So, antioxidants agents maybe useful to counteract free radical-induced fats accumulation. As reported, fucoxanthin, and fucoxanthinol compound has a tendency to degrade adipocyte differentiation by the mechanism of down-regulating peroxisome proliferator-activated receptor-c (Hayato et al., 2006).

As reported, it is seen that neoxanthin and fucoxanthin has a tendency to completely inhibit fat accumulation in tissues (Plaza et al., 2008). The microalgal species, *Cylindrotheca closterium* and *Phaeodactylum tricornutum* synthesize compounds named fucoxanthin (Kim et al., 2012). This fucoxanthin compound also works as an anti-oxidant, anti-obesity, anti-diabetic, anti-inflammatory, and anticancer agent (Maeda et al., 2007). Besides all these, they have antituberculosis activity, antiprotozoal activity, and anthelminthic activity.

3.9 CONCLUSIONS

Microalgae are a budding producer of renewable, natural, maintainable, and inexpensive foundations of bioactive compounds and food ingredients. In recent scenario, the demand of valuable compounds obtained from microalgae has increased due to their potential application in the field of pharmaceuticals, nutraceuticals, cosmeceuticals, biological waste treatment, animal feed, etc. Microalgae are promising but still unexplored as a natural source providing a variety of chemical compounds in the form of drugs. Up to 80% drug used as antibacterial and anticancer approved between 1983 and 1994 are isolated from natural products. It is also true that commonly used microbial drug producers such as *Actinomycetes* and *Hyphomycetes* are the focusing area of research from a pharmaceutical point of view. But now the discovery rate of important compounds from these classical sources is decreasing day by day, so it's the time to move to the photosynthetic organism, i.e., microalgae and exploit its potential benefits. From this study, it concluded that microalgae are a prominent source of bioactive compounds such as carotenoids, pigments, vitamins, animal feed, and lipopeptides. Microalgae produce a broad range of toxins which is responsible for intoxication for human and domestic as well as wildlife. Besides this, microalgae also show the biomodulatory effect, which is applicable for the treatment of inflammation, Cancer, Malaria, Alzheimer, HIV, cardiovascular diseases (CDVs), Leishmaniasis, and other diseases. Cyanobacteria are good food supplement. *Spirulina* (*Arthrospira*) are a rich source of single-cell protein and it is being used as heath supplement for human beings and also for animal/aqua feed. This photosynthetic microorganism also has novel potency that they produce UV absorbing/screening compounds as mycosporine-like amino acid and scytonemin.

Therefore, it is concluded that by exploring novel compounds with improved action of lately identified as well as older microalgal classes, we can determine the biotechnological application of cyanobacteria as a natural source of bioactive compounds for the cosmetic and pharmaceuticals industries.

KEYWORDS

- **alkaloids**
- **anticancer activity**
- **bioactive compounds**
- **borophycin**
- **cyanobacteria**

- **lipopolysaccharides**
- **microalgae**
- **microorganism**
- **mycosporine-like amino acids**
- **pigments**
- **scytonemin**
- **vitamin**

REFERENCES

Aboal, M., & Puig, M. A., (2009). Microcystin production in *Rivularia* colonies of calcareous streams from Mediterranean Spanish basins. *Algol. Stud., 130*, 39–52.

Barrientos, L. G., O'Keefe, B. R., & Bray, M., (2003). Cyanovirin-N binds to the viral surface glycoprotein, GP1,2 and inhibits infectivity of Ebola virus. *Antiviral Res., 58*, 47–56.

Baudelet, P. H., Gagez, A. L., Berard, J. B., Juin, C., Bridiau, N., Kaas, R., Thiery, V., Cadoret, J. P., & Picot, L., (2013). Antiproliferative activity of *Cyanophora paradoxa* pigments in melanoma, breast and lung cancer cells. *Mar. Drugs., 11*, 4390–4406.

Beesoo, R., Neergheen-Bhujun, V., Bhagooli, R., & Bahorun, T., (2014). Apoptosis-inducing lead compounds isolated from marine organisms of potential relevance in cancer treatment. *Mutat. Res. Mol. Mech. Mutagen., 768*, 84–97.

Bhaskar, S. U., Gopalswamy, G., & Raghu, R., (2005). A simple method for efficient extraction and purification of C-phycocyanin from *Spirulina platensis* Geitler. *Indian J. Exp. Biol., 43*, 277–279.

Bhatia, S., Sharma, K., Namdeo, A. G., Chaugule, B. B., Kavale, M., & Nanda, S., (2010). Broad-spectrum sun-protective action of Porphyra 334 derived from *Porphyra vietnamensis*. *Pharmacog. Res., 2*, 45–49.

Blahova, L., Adamovsky, O., Kubala, L., Švihalkova, Š. L., Zounkova, R., & Blaha, L., (2013). The isolation and characterization of lipopolysaccharides from *Microcystis aeruginosa*, a prominent toxic water bloom-forming cyanobacteria. *Toxicon., 76*, 187–196.

Bokesch, H. R., Barry, R., & O'Keefe, T. C., (2003). A potent novel anti-HIV protein from the cultured cyanobacterium *Scytonema varium*. *Biochemistry, 42*, 2578–2584.

Bonotto, S., (1988). Food and chemicals from microalgae. *Prog. Oceanogr., 21*, 207–215.

Buffa, V., Stieh, D., & Mamhood, N., (2009). Cyanovirin-N potently inhibits human immunodeficiency virus type 1 infection in cellular and cervical explant models. *J. Gen. Virol., 90*, 234–243.

Burja, A. M., Banaigs, B., Abou-Mansour, E., Grant, B. J., & Wright, P. C., (2001). Marine cyanobacteria – a prolific source of natural products. *Tetrahedron, 57*, 9347–9377.

Cardozo, K. H. M., Guaratini, T., Barros, M. P., Falcao, V. R., Tonon, A. P., Lopes, N. P., Campos, S., et al., (2007). Metabolites from algae with economical impact. *Comp. Biochem. Physiol. Part C Toxicol. Pharmacol., 146*, 60–78.

Carmichael, W. W., Mahmood, N. A., & Hyde, E. G., (1990). Natural toxins from cyanobacteria (blue-green algae). In: Hall, S., &. Strichartz, G., (eds), *Marine Toxins: Origin, Structure and Molecular Pharmacology* (pp. 87–106). American Chemical Society, Washington DC.

Centella, M. H., Arévalo-Gallegos, A., Parra-Saldivar, R., & Iqbal, H. M. N., (2017). Marine-derived bioactive compounds for value-added applications in bio- and non-bio sectors. *Journal of Cleaner Production., 168*(1), 1559–1565.

Challem, J. J., Passwater, R. A., & Mindell, E. M., (1981). *Spirulina*. New Canaan, CT: Keats Publishing Inc.

Chen, F., (1996). High cell density culture of microalgae in heterotrophic growth. *Trends Biotechnol., 14*, 421–426.

Cho, M., Lee, H., Kang, I., Won, M., & You, S., (2011). Antioxidant properties of extract and fractions from *Enteromorpha prolifera*, a type of green seaweed. *Food Chem., 127*, 999–1006.

Christiansen, G., Yoshida, W. Y., Blom, J. F., et al., (2008). Isolation and structure determination of two microcystins and sequence comparison of the McyABC adenylation domains in *Planktothrix* species. *J. Nat. Prod., 71*, 1881–1886.

Ciferri, O., (1983). *Spirulina*, the edible microorganism. *Microbiol. Rev., 47*, 551–578.

Codd, G. A., Morrison, L. F., & Metcalf, J. S., (2005). Cyanobacterial toxins: Risk management for health protection. *Toxicol. Appl. Pharmacol., 203*, 264–272.

Colleluori, D. M., Tien, D., Kang, F., et al., (2005). Expression, purification, and characterization of recombinant cyanovirin-N for vaginal anti-HIV microbicide development. *Protein Expr. Purif., 39*, 229–236.

Da, B. S., Vaz, J., Moreira, B., De Morais, M. G., & Costa, J. A. V., (2016). Microalgae as a new source of bioactive compounds in food supplements. *Curr. Opin. Food Sci., 7*, 73–77.

Edwards, C., Beattie, K. A., Scrimgeour, C. M., & Codd, G. A., (1992). Identification of anatoxin-a in benthic cyanobacteria (blue-green algae) and in associated dog poisonings at Loch Insh, Scotland. *Toxicon., 30*, 1165–1175.

Ehling-Shulz, M., & Scherer, S., (1999). UV Protection in cyanobacteria. *E. J. Phycol., 34*, 329–338.

Ehlin-Schulz, M., Bilger, W., & Scherer, S., (1997). UV B induced synthesis of photoprotective pigments and extracellular polysaccharides in the terrestrial cyanobacterium *Nostoc commune*. *J. Bacteriol., 179*, 1940–1945.

Eriksen, N. T., (2008). Production of phycocyanin - a pigment with applications in biology, biotechnology, foods and medicine. *Appl. Microbiol. Biotechnol., 80*, 1–14.

Esser, M. T., Mori, T., & Mondor, I., (1999). Cyanovirin-N binds to gp120 to interfere with CD4-dependent human immunodeficiency virus type 1 virion binding, fusion, and infectivity but does not affect the CD4 binding site on gp120 or soluble CD4-induced conformational changes in gp120. *J. Virol., 73*, 4360–4371.

Fleming, E. D., & Castenholz, R. W., (2007). Effects of periodic desiccation on the synthesis of the UV-screening compound, scytonemin, in cyanobacteria. *Environmental Microbiology, 9*, 1448–1455.

Froscio, S. M., Humpage, A. R., Burcham, P. C., & Falconer, I. R., (2003). Cylindrospermopsin-induced protein synthesis inhibition and its dissociation from acute toxicity in mouse hepatocytes. *Environ. Toxicol., 18*, 243–251.

Funari, E., & Testai, E., (2008). Human health risk assessment related to cyanotoxins exposure. *Crit. Rev. Toxicol., 38*, 97–125.

Garcia-Pichel, F., & Belnap, J., (1996). Microenvironments and microscale productivity of cyanobacterial desert crusts. *J. Phycol., 32*, 774–782.

Gehringer, M. M., Adler, L., Roberts, A. A., Moffitt, M. C., Mihali, T. K., Mills, T. J. T., et al., (2012). Nodularin, a cyanobacterial toxin, is synthesized in plant by symbiotic *Nostoc* sp. *ISME J., 6*, 1834–1847.

Gong, M., & Bassi, A., (2016). Carotenoids from microalgae: A review of recent developments. *Biotechnol. Adv., 34*, 1396–1412.

Grossman, A. R., Schaefer, M. R., Chiang, G. G., & Collier, J. L., (1993). The phycobilisomes a light-harvesting complex responsive to environmental conditions. *Microbiol. Rev., 57*, 725–749.

Gupta, D. K., Kaur, P., Leong, S. T., Tan, L. T., Prinsep, M. R., & Hann, C. J. J., (2014). Anti-chikungunya viral activities of aplysiatoxin-related compounds from the marine cyanobacterium *Trichodesmium erythraeum*. *Mar. Drugs, 12*, 115–127.

Hader, D. P., Kumar, H. D., Smith, R. C., & Worrest, R. C., (2007). Effects of solar UV radiation on aquatic ecosystem and interaction with climate change. *Photochem. Photobiol. Sci., 6*, 267–285.

Hayato, M., Masashi, H., Tokutake, S., Nobuyuk, T., Teruo, K., & Kazuo, M., (2006). Fucoxanthin and its metabolite, fucoxanthinol, suppress adipocyte differentiation in 3T3-L1 cells. *Int. J. Mol. Med., 18*, 147–152.

Helle, F., Wychowski, C., & Vu-Dac, N., (2006). Cyanovirin-N inhibits hepatitis C virus entry by binding to envelope protein glycans. *J Biol Chem., 281*, 25177–25183.

Henriksen, P., Carmichael, W. W., An, J., & Moestrup, Ø., (1997). Detection of an anatoxin-a(s)-like anticholinesterase in natural blooms and cultures of cyanobacteria/blue-green algae from Danish lakes and in the stomach contents of poisoned birds. *Toxicon, 35*, 901–913.

Humpage, A. R., Fenech, M., Thomas, P., & Falconer, I. R., (2000). Micronucleus induction and chromosome loss in transformed human white cells indicate clastogenic and aneugenic action of the cyanobacterial toxin, cylindrospermopsin. *Mutat. Res. Genet. Toxicol. Environ. Mutagen., 472*, 155–161.

Jensen, G. S., Ginsberg, D. I., & Huerta, P., (2000). Consumption of *Aphanizomenon flos-aquae* has rapid effects on the circulation and function of immune cells in humans. *JANA, 2*, 50–58.

Jokela, J., et al., (2012). Anabaenolysins, novel cytolytic lipopeptides from benthic *Anabaena* cyanobacteria. *PLoS One, 7*(7), e41222.

Kaebernick, M., & Neilan, B. A., (2001). Ecological and molecular investigations of cyano-toxin production. *FEMS Microbiol. Ecol., 35*, 1–9.

Kannaujiya, V. K., & Sinha, R. P., (2015). Impacts of varying light regimes on phycobiliproteins of *Nostoc* sp. HKAR-2 and *Nostoc* sp. HKAR-11 isolated from diverse habitats. *Protoplasma, 252*, 1551–1561.

Kannaujiya, V. K., Sundaram, S., & Sinha, R. P., (2017). *Phycobiliproteins: Recent Developments and Future Applications*. Springer Nature, Singapore.

Kashihara, N., Toe, S., Nakamura, K., Umezawa, K., Yamamura, S., & Nishiyama, S., (2000). Synthesis and biological activities of hapalosin derivatives with modification at C12 position. *Bioorg. Med. Chem. Lett., 10*, 101–103.

Khalid, M. N., Shameel, M., Ahmad, V. U., Shahzad, S., & Leghari, S. M., (2010). Studies on the bioactivity and phycochemistry of *Microcystis aeruginosa* (*Cyanophycota*) from Sindh. *Pak. J. Bot., 42*, 2635–2646.

Kim, S. M., Jung, Y. H., Kwon, O., Cha, K. H., & Um, B. H., (2012). A potential commercial source of fucoxanthin extracted from the microalga *Phaeodactylum tricornutum. Appl. Biochem. Biotechnol., 166*, 1843–1855.

Kopelman, P. G., (2000). Obesity as a medical problem. *Nature, 404*, 635–643.

Kouzuma, A., & Watanabe, K., (2015). Exploring the potential of algae/bacteria interactions. *Curr. Opin. Biotechnol., 33*, 125–129.

Limón, P., Malheiro, R., Casal, S., Acién-Fernández, F. G., Fernández-Sevilla, J. M., Rodrigues, N., Cruz, R., et al., (2015). Improvement of stability and carotenoids fraction of virgin olive oils by addition of microalgae *Scenedesmus almeriensis* extracts. *Food Chem., 175*, 203–211.

Maeda, H., Hosokawa, M., Sashima, T., Funayama, K., & Miyashita, K., (2007). Effect of medium-chain triacylglycerols on anti-obesity effect of fucoxanthin. *J. Oleo Sci., 56*(12), 615–621.

Manivasagan, P., Venkatesan, J., Sivakumar, K., & Kim, S. K., (2014). Pharmaceutically active secondary metabolites of marine actinobacteria. *Microbiol. Res., 169*, 262–278.

Mankiewicz, J., Malgorzata, T. M., Walter, Z., & Maciej, Z. M., (2003). Natural toxins from cyanobacteria. *Acta Biol. Cracovien. Ser. Bot., 45*, 9–20.

Matsunaga, S., Moore, R. E., Niemczura, W. P., & Carmichael, W. W., (1989). Anatoxin-a(s), a potent anticholinesterase from *Anabaena flos-aquae. J. Am. Chem. Soc., 111*, 8021–8023.

Mazmouz, R., Chapuis-Hugon, F., Mann, S., Pichon, V., Mejean, A., & Ploux, O., (2010). Biosynthesis of cylindrospermopsin and 7-epicylindrospermopsin in *Oscillatoria* sp. strain PCC 6506: Identification of the cyr gene cluster and toxin analysis. *Appl. Environ. Microbiol., 76*, 4943–4949.

Mejean, A., Paci, G., Gautier, V., & Ploux, O., (2014). Biosynthesis of anatoxin-a and analogues (anatoxins) in cyanobacteria. *Toxicon, 91*, 15–22.

Mojaat, M., Foucault, A., Pruvost, J., & Legrand, J., (2008). Optimal selection of organic solvents for biocompatible extraction of β-carotene from *Dunaliella salina. J. Biotechnol., 133*, 433–441.

Mori, T., Gustafson, K. R., Pannell, L. K., Shoemaker, R. H., Wu, L., McMahon, J. B., & Boyd, M. R., (1998). Recombinant production of cyanovirin-N, a potent HIV-inactivating protein derived from cultured cyanobacterium. *Protein Expr. Purif., 12*, 151–158.

Mynderse, J. S., Moore, R. E., Kashiwagi, M., & Norton, T. R., (1977). Antileukemia activity in the *Oscillatoriaceae*: Isolation of debromoaplysiatoxin from *Lyngbya. Science, 196*, 538–540.

Nägeli, C., & Gattungen, E. A., (1849). Physiologisch und systematisch bearbeitet, Neue Denkschrift, Allg. Schweiz. *Natur. Ges., 10*, 1–138.

Neumann, C., Bain, P., & Shaw, G., (2007). Studies of the comparative *in vitro* toxicology of the cyanobacterial metabolite deoxycylindrospermopsin. *J. Toxicol. Environ. Health, 70*, 1679–1686.

O'Keefe, B. R., Smee, D. F., & Turpin, J. A., (2003). Potent anti-influenza activity of cyanovirin-N and interactions with viral hemagglutinin. *Antimicrob Agents Chemother.*, *47*, 2518–2525.

Onodera, H., Oshima, Y., Henriksen, P., & Yasumoto, T., (1997a). Confirmation of anatoxin-a(s) in the cyanobacterium *Anabaena lemmermannii* as the cause of bird kills in Danish lakes. *Toxicon, 35*, 1645–1648.

Oren, A., & Gunde-Cimerman, N., (2007). Mycosporines and mycosporine-like amino acids: UV protectants or multipurpose secondary metabolites? *FEMS. Microbiol. Lett.*, *269*, 1–10.

Panda, D., Deluca, K., Williams, D., Jordan, M. A., & Wilson, L., (1998). Antiproliferative mechanism of action of cryptophycin-52: Kinetic stabilization of microtubule dynamics by high-affinity binding to microtubule ends. *Cell Biol.*, *95*, 9313–9318.

Park, H. D., Watanabe, M. F., Harada, K. I., Nagai, H., Suzuki, M., Watanabe, M., et al., (1993). Hepatotoxin (microcystin) and neurotoxin (anatoxin-a) contained in natural blooms and strains of cyanobacteria from Japanese waters. *Nat. Toxins, 1*, 353–360.

Parveen, S., & Nadumane, V., (2016). Algae as sources of anticancer compounds. *IJBPAS, 5*, 2257–2277.

Perez-Garcia, O., Escalante, F. M. E., de-Bashan, L. E., & Bashan, Y., (2011). Heterotrophic cultures of microalgae: Metabolism and potential products. *Water Res.*, *45*, 11–36.

Plaza, M., Cifuentes, A., & Ibáñez, E., (2008). In the search of new functional food ingredients from algae. *Trends Food Sci. Technol.*, *19*, 31–39.

Qiu, B., Liu, J., Liu, Z., & Liu, S., (2002). Distribution and ecology of the edible cyanobacterium GeXian-Mi (*Nostoc*) in rice fields of Hefeng County in China. *J. Appl Phycol.*, *14*, 423–429.

Raetz, C. R. H., & Whitfield, C., (2002). Lipopolysaccharide endotoxins. *Annu. Rev. Biochem.*, *71*, 635–700.

Rahman, M. A., Kannaujiya, V. K., Rajneesh, Adi, N., Dixit, K., Sinha, R. P., & Sundaram, S., (2016). Impact of ultraviolet-B radiation on photoprotection and pigmentation of *Anabaena cylindrica* and *Synechocystis* PCC 6803: A comparative study. *IJONS, 6*, 36.

Ranga, R. A., Baskaran, V., Sarada, R., & Ravishankar, G. A., (2013). *In vivo* bioavailability and antioxidant activity of carotenoids from microalgal biomass – A repeated dose study. *Food Res. Int.*, *54*, 711–717.

Rastogi, R. P., Sinha, R. P., & Incharoensakdi, A., (2014). The cyanotoxin microcystins: Current overview. *Rev. Environ. Sci. Bio/Technol.*, *13*, 215–249.

Ravindran, B., Gupta, S. K., Cho, W. M., Kim, J. K., Lee, S. R., Jeong, K. H. D., Lee, J., & Choi, H. C., (2016). Microalgae potential and multiple roles – current progress and future prospects – An Overview. *Sustainability, 8*, 1215.

Régnier, P., Bastias, J., Rodriguez-Ruiz, V., Caballero-Casero, N., Caballo, C., Sicilia, D., et al., (2015). Astaxanthin from *Haematococcus pluvialis* prevents oxidative stress on human endothelial cells without toxicity. *Mar. Drugs, 13*, 2857–2874.

Rimbau, V., Camins, A., Pubill, D., et al., (2001). C-phycocyanin protects cerebellar granule cells from low potassium/serum deprivation induced apoptosis. *Arch Pharmacol.*, *364*, 96–104.

Sachindra, N., Sato, E., Maeda, H., Hosokawa, M., Niwano, Y., Kohno, M., et al., (2007). Radical scavenging and singlet oxygen quenching activity of marine carotenoid fucoxanthin and its metabolites. *Agric. Food Chem.*, *55*, 8516–8522.

Safafar, H., Van, W. J., Møller, P., & Jacobsen, C., (2015). Carotenoids, phenolic compounds and tocopherols contribute to the antioxidative properties of some microalgae species grown on industrial wastewater. *Mar. Drugs, 13*, 7339–7356.

Schmid, D., Schurch, C., & Zulli, F., (2006). Mycosporine-like amino acids from red algae potent against premature skin–aging. *Euro Cosmet., 9*, 1–4.

Schwartz, J., & Shklar, G., (1987). Regression of experimental hamster cancer by beta-carotene and algae extracts. *J. Oral. Maxillofac. Surg., 45*, 510–515.

Schwartz, J., Shklar, G., Reid, S., & Trickler, D., (1988). Prevention of experimental oral cancer by extracts of *Spirulina-Dunaliella* algae. *Nutr. Cancer, 11*, 127–134.

Seifert, M., McGregor, G., Eaglesham, G., Wickramasinghe, W., & Shaw, G., (2007). First evidence for the production of cylindrospermopsin and deoxycylindrospermopsin by the freshwater benthic cyanobacterium, *Lyngbya wollei* (Farlow ex Gomont) Speziale and Dyck. *Harmful Algae, 6*, 73–80.

Shao, P., Chen, X., & Sun, P., (2013). *In vitro* antioxidant and antitumor activities of different sulfated polysaccharides isolated from three algae. *Int. J. Biol. Macromol., 62*, 155–161.

Singh, S. P., Häder, D. P., & Sinha, R. P., (2010). Cyanobacteria and ultraviolet radiation (UVR) stress: Mitigation strategies. *Age. Res. Rev., 9*, 79–90.

Singh, S., Kate, B. N., & Banerjee, U. C., (2005). Bioactive compounds from cyanobacteria and microalgae: An overview. *Crit. Rev. Biotechnol., 25*, 73–95.

Sinha, R. P., Lebert, M., Kumar, A., Kumar, H. D., & Hader, D. P., (1995). Spectroscopic and biochemical analyses of UV effects of phycobilisomes of *Anabaena* sp. and *Nostoc carmium. Bot. Acta, 108*, 87–92.

Sivonen, K., & Jones, G., (1999). Cyanobacterial toxins. In: Chorus, I., &. Bartram, J., (eds.), *Toxic cyanobacteria in water: A Guide to their public health consequences, monitoring and management* (pp. 41–111). London: E and FN Spon.

Skulberg, O. M., Carmichael, W. W., Anderson, R. A., Matsunaga, S., Moore, R. E., & Skulberg, R., (1992). Investigations of a neurotoxic oscillatorialean strain (*Cyanophyceae*) and its toxin. Isolation and characterization of homoanatoxin-a. *Env. Toxicol. Chem., 11*, 321–329.

Smith, J. L., Boyer, G. L., & Zimba, P. V., (2008). A review of cyanobacterial odorous and bioactive metabolites: Impacts and management alternatives in aquaculture. *Aquaculture, 280*, 5–20.

Spoof, L., Berg, K. A., Rapala, J., Lahti, K., Lepisto, L., Metcalf, J. S., et al., (2006). First observation of cylindrospermopsin in *Anabaena lapponica* isolated from the boreal environment (Finland). *Environ. Toxicol., 21*, 552–560.

Stamenic, M., Vulic, J., Djilas, S., Misic, D., Tadic, V., Petrovic, S., & Zizovic, I., (2014). Free-radical scavenging activity and antibacterial impact of Greek oregano isolates obtained by SFE. *Food Chem., 165*, 307–315.

Strichartz, G., Rando, T., Hall, S., Gitschier, J., Hall, L., Magnani, B., et al., (1986). On the mechanism by which saxitoxin binds to and blocks sodium channels. *Ann. N. Y. Acad. Sci., 479*, 96–112.

Su, Z., Sheets, M., Ishida, H., Li, F., & Barry, W. H., (2004). Saxitoxin blocks L-type/Ca. *J. Pharmacol. Exp. Therapeut., 308*, 324–329.

Taylor, M. S., Stahl-Timmins, W., Redshaw, C. H., & Osborne, N. J., (2014). Toxic alkaloids in *Lyngbya majuscula* and related tropical marine cyanobacteria. *Harmful Algae, 31*, 1–8.

Thomas, N. V., & Kim, S. K., (2011). Potential pharmacological applications of polyphenolic derivatives from marine brown algae. *Environ. Toxicol. Pharmacol., 32*, 325–335.

Tillett, D., Dittmann, E., Erhard, M., Von, D. H., Borner, T., & Neilan, B. A., (2000). Structural organization of microcystin biosynthesis in *Microcystis aeruginosa* PCC7806: An integrated peptide-polyketide synthetase system. *Chem. Biol., 7*, 753–764.

Um, B. H., & Kim, Y. S., (2009). Review: A chance for Korea to advance algal-biodiesel technology. *Journal of Industrial and Engineering Chemistry, 15*, 1–7.

Vlad, M., Bordas, E., Caseanu, E., Uza, G., Creteanu, E., & Polinicenco, C., (1995). Effect of cuprofilin on experimental atherosclerosis. *Biol. Trace. Elem. Res., 48*, 99–109.

Vo, T. S., Ngo, D. H., & Kim, S. K., (2015). *Handbook of Marine Microalgae: Nutritional and Pharmaceutical Properties of Microalgal Spirulina* (pp. 299–308). Boston: Academic Press.

Volk, R. B., & Furkert, F. H., (2006). Antialgal, antibacterial and antifungal activity of two metabolites produced and excreted by cyanobacteria during growth. *Microbiol. Res., 161*, 180–186.

Vonshak, A., (1997). *Spirulina*: Growth, physiology and biochemistry. In: Vonshak, A., (ed), *Spirulina Platensis (Arthrospira): Physiology, Cell-Biology and Biotechnology* (pp. 43–66). London, UK: Taylor and Francis.

Washida, K., Koyama, T., Yamada, K., Kitab, M., Urmura, D., et al., (2006). Karatungiols A and B two novel antimicrobial polyol compounds, from the symbiotic marine dinoflagellate *Amphidinium* sp. *Tetrahedron Lett., 47*(15), 2521–2525.

Watanabe, F., Takenaka, S., & Kittaka-Katsura, H., (2002). Characterization and bioavailability of vitamin B12-compounds from edible algae. *J. Nutr. Sci. Vitaminol. (Tokyo), 48*, 325–331.

Yabuta, Y., Fujimura, H., Kwak, C. S., Enomoto, T., & Wata-nabe, F., (2010). Antioxidant activity of the phycoerythrobilin compound formed from a dried Korean purple laver (*Porphyra* sp.) during *in vitro* digestion. *Food Sci. Technol. Res., 16*, 347.

CHAPTER 4

Microalgae: A Valuable Source of Natural Colorants for Commercial Applications

CHIDAMBARAM KULANDAISAMY VENIL,[1] MATHESWARAN YAAMINI,[1] and LAURENT DUFOSSÉ[2]

[1]*Department of Biotechnology, Anna University, Regional Campus, Coimbatore – 641046, Tamil Nadu, India*

[2]*Université de la Réunion, CHEMBIOPRO Chimie et Biotechnologie des Produits Naturels, ESIROI Département Agroalimentaire, Sainte-Clotilde F – 97490, Ile de La Réunion, Indian Ocean, France*

ABSTRACT

The harmful effects of synthetic colorants have created an increased awareness among the stakeholders towards natural colorants, especially for applications in the fields of medicine, food, and cosmetics. Such indulgence of the stakeholders has led to the exploration of natural resources like plants, bacteria, algae, etc., for producing natural/organic colorants/products. Of these natural sources, microalgae are one of the valuable and recognized natural sources for producing vitamins, proteins, pigments, fuels, medicines, etc.; and currently, such products are gradually reaching the global markets. Microalgae produce a broad spectrum of attractive coloring pigments like chlorophylls, carotenoids, zeaxanthin, lutein, etc. These algal pigments are eco-friendly and harmless potential compounds to be applied as natural colorants in food, nutraceuticals, cosmetics, and pharma industries. Producing pigments from algae has many advantages,

Microalgal Biotechnology: Bioprospecting Microalgae for Functional Metabolites towards Commercial and Sustainable Applications. Jeyabalan Sangeetha, PhD, Svetlana Codreanu, PhD, & Devarajan Thangadurai, PhD (Eds.)

such as easy extraction, enhanced yields and cheaper production with no seasonal variations. This chapter discusses on the classes of algal pigments, cultivation of algae, factors affecting pigment production and metabolic engineering. In addition, the possible applications of these colorants in food, cosmetics, and health care industries are highlighted too.

4.1 INTRODUCTION

Microalgae, being unicellular or filamentous microbes, are an extremely diverse grouping of simple, plant-like organisms capable of, producing glowing eco-friendly colors (Begum et al., 2016). Globally there is a surging need for natural colorants owing to the positive awareness that has been created among the stakeholders, customers, etc., for safe and biodegradable colorants. By the way, the exploration and exploitation of algae for natural colorants has opened up many exciting possibilities for producing eco-friendly and renewable stable coloring materials. Apart from their coloring properties, they have also many prospective health benefits to humans and animals. Most of the algal metabolites, especially pigments are possessing antiaging, antioxidant, and neuroprotective properties which facilitate them fitting to be utilized in nutraceutical, human make-ups, healthcare, etc. Algal pigments like β-carotene from *Arthrospira* and *Dunaliella* are commercially available as nutritional supplements and they have gained acceptance globally (Koyande et al., 2019). Moreover, algae are being the prospective natural sources to be screened for novel metabolites for various kinds of applications.

Microalgae are extraordinary organisms that can thrive even in extreme environments with the required amount of light and humidity (Kiesenhofer and Fluch, 2018). Owing to their enormous potential for the production of valuable natural products, further researches are emerging in the field of microalgae. Algal pigments are explicitly precious commercial natural dyes, antioxidants, and vitamins in the food and cosmetic industry. Mainly vital pigments like chlorophylls, carotenoids, and phycobiliproteins (PBPs) have extensive broad scope of appliances in diagnostics, biomedical investigation, therapeutics, colorings in make-up products, dairy products, and other foodstuffs; and they are also getting a significant organic identity owing to their non-toxic and non-carcinogenic nature (Sathasivam and Ki, 2018).

In biotechnological point of view, microalgae are incredible sources for pigments because they produce a wide spectrum of colors and various molecules, including chlorophylls (green), carotenoids (red, orange, and yellow) and PBPs (red and blue) (Mulders et al., 2014). Regardless of having all these outstanding advantages, currently microalgae are commercially exploited only for three types of pigments, namely β-carotene (*Dunaliella salina*), astaxanthin (*Haematococcus pluvialis*) and phycocyanin (PC) (*Arthrospira platensis*) (Camacho et al., 2019).

This chapter appraises and emphasizes on the significance of microalgae producing pigments and their possible utilization in food, pharmaceutical, and cosmetics industries for the improvement of health and well-being of humans.

4.2 SWOT ANALYSIS OF ALGAL PIGMENTS

The SWOT (strength, weakness, opportunities, and threats) analysis of algal pigments is represented in Figure 4.1.

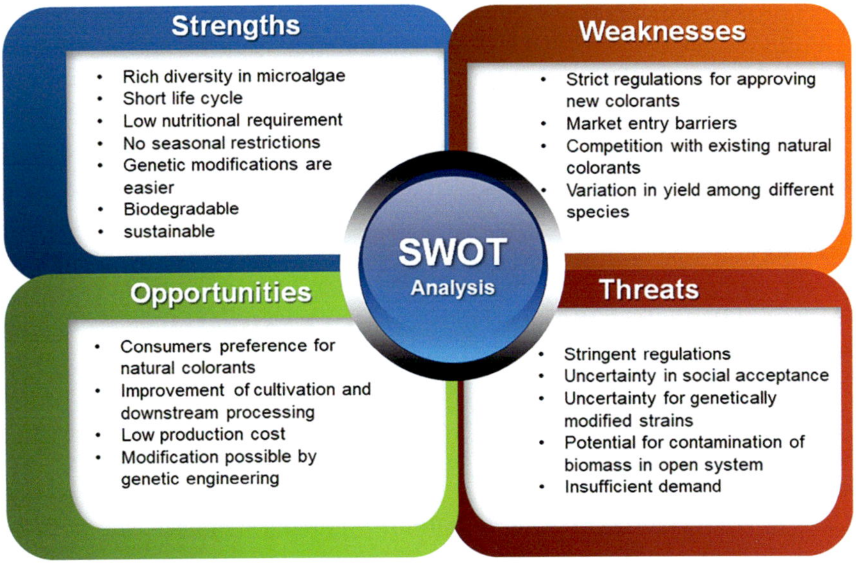

FIGURE 4.1 SWOT (Strength, weakness, opportunities, and threats) analysis of algal pigments.

4.2.1 STRENGTH

Algae produce various shades of coloring pigments without seasonal variations and are continually attracting the interests of multiple markets including food, nutraceuticals, and cosmetics industry. There is a broad opportunity for generating huge markets by the development of optimized microalgal cultivation system and thereby producing valuable algal compounds. Applying of hydrolysates from agricultural residues for microalgal growth is one of the cost-effective methods for algal cultivation. The carbon sources from sweet sorghum, cassava, molasses, rice straw, etc., can be exploited as the low-cost substrates for microalgal cultivation (Gao et al., 2010; Lu et al., 2010; Li et al., 2011; Yan et al., 2011). Current trends for products derived from algae have opened up many new opportunities for the research and development of novel algal products to meet the emerging market needs.

4.2.2 WEAKNESS

The microalgal cultivation is a challenging endeavor as it is a labor intensive one and consists of expensive processes. Further, distinguishing between natural and synthetic pigments is a hard one because of the chirality in isomers. Therefore, it is very easy for the adulterated products to get entry into the market falsely claiming them as organic products. Firm set of laws and slow-moving legislation only may obstruct the route of adulterated products to bazaar, and this factor would slowly open the way to develop new algal products (Novoveska et al., 2019). Moreover, the commercial applicability of algal pigments has also to be checked for their toxicity and quality to get the regulatory endorsement before entering the market.

4.2.3 OPPORTUNITIES

Consumers' preference towards natural pigments increases the need for natural colorants. Exploring potent pigment producing algae may lead to the identification of novel and rare strains. The commercialization of food-grade algal pigments is increasing because of the biotechnological advancement and low capital investment, and this trend is supporting the growth of algal pigment markets globally. The large share of this global pigment market is mainly attributed to the consumer demand for plant-derived alternatives, strict regulation against the use of synthetic colorants and the eco-friendly harmless properties of algal pigments such as high nutritional value, non-toxicity, and their biodegradability.

4.2.4 THREATS

The major threat is the stringent regulations and competition from the already existing commercial natural colorants. Also there is a possibility of contamination when cultured in an open system.

4.3 CLASSES OF ALGAL PIGMENTS

There are three main types of photosynthetic pigments like chlorophylls, carotenoids (carotenes and xanthophylls) and phycobilins in algae. The pigments, chlorophylls, and carotenoids are fat-soluble and phycobilins are water-soluble (Figure 4.2).

4.3.1 CHLOROPHYLLS

Three different types of chlorophylls are there, namely chlorophyll a, b, and c. The chlorophyll particle is the porphyrin containing tetrapyrrole rings. Phorbin is made by the adding of iso-cyclic ring to pyrrole rings. Every pyrrole ring comprises 1 nitrogen and 4 carbon atoms. The central hole contains nitrogen atoms, in which Mg^{2+} metal ions can combine. The formyl group in the second ring of chlorophyll b is interchanged by methyl group in chlorophyll a (Scheer et al., 2004). Because of these structural variations, chlorophyll a has a maximum wavelength at 660–665 nm, with blue-green pigments and chlorophyll b has a maximum wavelength at 642–652 with green-yellow pigments. Owing to the exposure of chlorophyll molecules to heat, light, oxygen, and weak acids, many degraded products are formed (Cubas et al., 2008).

Chlorella contains two key types of chlorophylls (a, b), up to 4.5% of dry weight, and so it may be mainly the smart creation material when developed under most favorable conditions (Cuaresma et al., 2011; Miazek and Ledakowicz, 2013). Alternatively, it has been construed that chlorophyll in microalgal biomass is condensed drastically under stress circumstances (Markou and Nerantzis, 2013).

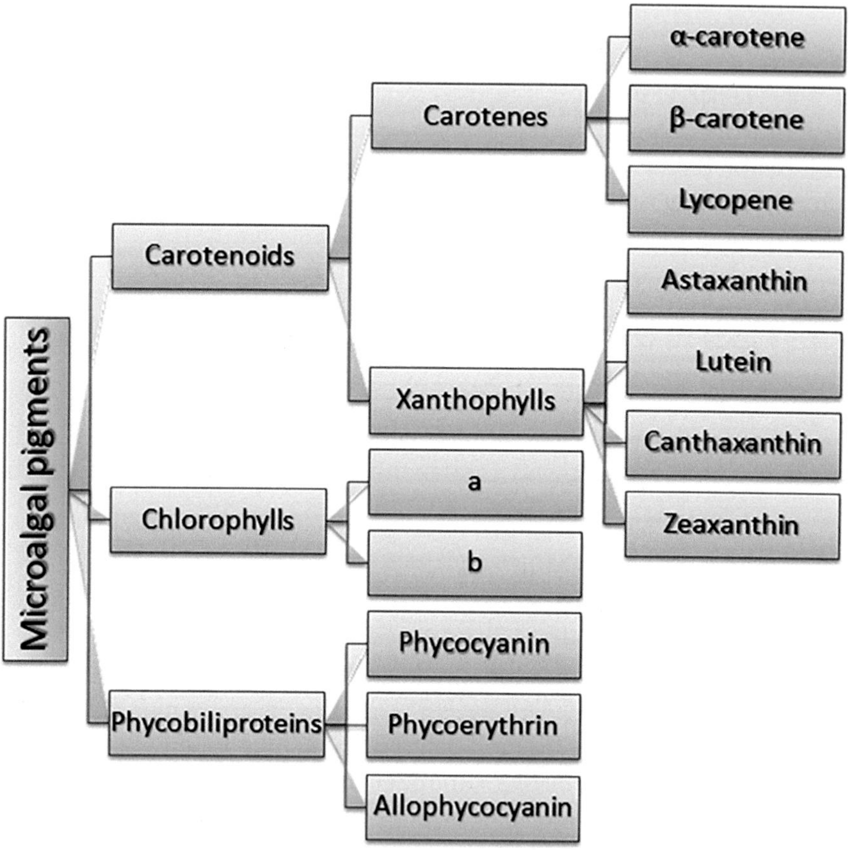

FIGURE 4.2 Natural pigments from microalgae.

4.3.2 *CAROTENOIDS*

Carotenoids are terpenoid pigments derived from 40 carbon polyene chain. It has light absorption characteristics that are crucial for photosynthesis. Carotenoids may have cyclical groups and oxygen possessing performing groups. The hydrocarbon carotenoids are considered as carotenes whereas the oxygenated derivatives are xanthophylls with oxygen being replaced by hydroxyl groups. β-carotene is produced by unicellular algae *Dunaliella salina* under stressed conditions. It is a mixture of 2 stereoisomers, all-trans, and 9-cis in equal amounts with health promoting properties (Del-Campo et al., 2007). The pro-vitamin A is high in β-carotene because every molecule produces 2 molecules of retinol (Christaki et al., 2013). The microalgal pigments possess solid antioxidant activity and intervene in the harmful effects of free radicals by defending the lipophilic part from lipid peroxidation (Lordan et al., 2011; Pangestuti and Kim, 2011). β-carotene acts as a pro-oxidant in the lipid peroxidation process when there is high oxygen pressure and thereby enhances the carotenoid production (Stahl and Sies, 2003).

The freshwater microalgae, *Haematococcus pluvialis* has a high level of astaxanthin, and it is mainly used as coloring compounds in aquaculture; moreover, it has strong antioxidant activity (Daubrawa et al., 2005). Astaxanthin has promising health benefits and aid in curing various diseases like cancer, metabolic syndrome, cardiovascular diseases (CDVs) and it also enhances the immune system by protecting the skin from radiation (Yuan et al., 2011). *H. pluvialis* has four phases and in the aplanospores phase, the astaxanthin is accumulated which is prompted under stress conditions like pH, salinity, temperatures, nutrient depletion, etc. (Li et al., 2017). The mass production of astaxanthin from *H. pluvialis* is carried out by 2-stage batch culture. In the first stage, biomass production occurs, whereas in the second stage, the cultures are stressed to induce astaxanthin production.

Microalgae can synthesize high amount of carotenoids like astaxanthin, β-carotene, canthaxanthin, and echinenone under unfavorable conditions (Singh et al., 2019). *Botryococcus braunii*, *Chlorella* sp., *Chlorococcum* sp., *Coelastrella striolata*, *Haematococcus pluvialis*, *Dunaliella salina*, *Nanochloropsis* sp., *Scenedesmus* sp., and *Arthrospira platensis* are well-known for the production of β-carotene, lutein, canthaxanthin, astaxanthin, and fucoxanthin (Ambati et al., 2018).

The combination of algal β-carotene isomers has a higher accumulation of more than 10 times in comparison with synthetic carotenoids. Carotenoids, like β-carotene, lutein, zeaxanthin, and fucoxanthin involve in photosynthesis process like chlorophylls and absorb blue, violet, and green light of the visible spectrum and thereby reflect yellow, orange, and red colors (Naruka et al., 2019). Gunerken et al. (2015) states that algal carotenoids are given more importance but their commercial use is quite little owing to their high extraction and purification cost.

4.3.3 PHYCOBILINS

PBPs are creative of apoproteins (α and β subunits) and covalently related to prosthetic groups, phycobilins. The two preserved subunits, α and β form a αβ monomer and aggregates to form αβ trimers and αβ hexamers. They are sunlight harvesting pigments found normally in Cyanophyceae and Cryptophyceae. Based on the amino acid sequences and spectroscopic analysis, they are divided into red (phycoerythrin (PE)), blue (phycocyanin) and allophycocyanin (APC). PC, APC, and PE, are consisting of different non-covalently connected subunits with molecular weight of about 16,000–20,000 Daltons, 15,500–17,500 Daltons, and 20,000–22,000 Daltons, respectively. Their constitution varies with species and environmental settings (Chu, 2012). They are used in various fields like markers for electrophoresis, isoelectric focusing, flow cytometry, etc., because of their fluorescent properties and they also have antioxidant properties (Raposo et al., 2013).

It can defuse the reactive oxygen species (ROS) because of their chemical structures and chelating nature (Rodriguez-Sanchez et al., 2012). PBPs are produced commercially from two well-known species: PC from *Arthrospira* sp. and PE from *Porphyridium* sp. (Rodriguez-Sanchez et al., 2012; Borowitzka, 2013). These protein-bound exclusive pigments are found in microalgae and are degraded under stress environments like phosphorus, nitrogen, and sulfur starvation (Eriksen, 2008; Hifney et al., 2013).

PBPs are formed by bonding protein with phycobilins and chromophores. They are colored pigments and found in red algae and cyanobacteria. *Arthrospira platensis* and *Aphanizomenon flos-aquae* are the major sources of phycobiliprotein. They are commercially used as organic dyes in make-ups and in pharmaceutical industries owing to their antioxidant, anti-inflammatory, neuroprotective, and hepatoprotective properties. The volume of global market of PC averages around 5–10 million USD. The commercial producers of PBPs are from the cyanobacteria *Arthrospira* and rhodophyte *Porphyridium*.

4.4 CULTIVATION OF MICROALGAE FOR PIGMENT PRODUCTION

The cultivation methods of microalgae are the main restricting factors for producing microalgae pigments at the industrial level, and developing low-cost cultivation methods for their economical production of pigments is indispensable. There are different modes of cultivation of microalgae: photoautotrophic, heterotrophic, mixotrophic, and photoheterotrophic. In autotrophic cultivation, the microalgae produce raw compounds in daylight to process inert carbon source, CO_2 (Khan et al., 2018). The open pond system is the ancient method for autotrophic cultivation and due to its high susceptibility to contamination and fluctuations in temperature, the open pond system is inappropriate for industrial scale level production.

The heterotrophic cultivation method is depending on the metabolism of organic compounds that afford carbon source and they can grow under dark conditions. This kind of cultivation is a proven method for cell growth and considerably decreasing the cultivation cost. Glucose is used as the main source for this type of cultivation, and it is possible that low-cost alternatives could be explored for use in the heterotrophic method of cultivation. Algae like *Arthrospira*, *Dunaliella*, and *Chlorella* were effectively cultivated for single-cell protein and pigment production (Mata et al., 2010). For the production of pharmaceutical products, closed photobioreactors are preferred for maintenance of pure cultures. The heterotrophic microalgae should be cultivated in conventional bioreactors without light. Bumbak et al. (2011) have testified that the cultivation of heterotrophic microalgae is an encouraging method in the active fermentation set-up for bacteria. The main rewards of cultivating microalgae in heterotrophic conditions are: (i) microalgae are potential to grow even in lack of light; (ii) their cell density increases; (iii) they could be cultivated in industrial-scale fermenters; and (iv) they could be cultivated at reduced cost.

The primary xanthophyll pigment, lutein is produced by microalgal species akin to *Muriellopsis* sp., *Scenedesmus* sp., *Dunaliella* sp. and *Chlorella* sp. at high concentrations. Lutein protects the eye from oxidative damage by blue light and lutein intake decrease the age-linked macular deterioration (Hu et al., 2018). Glucose is the preferred cheap source for heterotrophic cultivation of microalgae *C. protothecoides* and *C. pyrenoidosa* for lutein production. Glycerol has been used for the cultivation of *Scenedesmus* sp. for lutein production (Yen et al., 2011).

Astaxanthin from *H. pluvialis* has antioxidant activity 90 times more than that of synthetic astaxanthin without any toxicity (Régnier et al., 2015). Natural astaxanthin from microalgae are derived mainly from green algae *Haematococcus pluvialis* (Li et al., 2011), and later

Chlorella zofingensis (Liu et al., 2014) and *Chlorococcum* have been testified for their capability to produce astaxanthin. Astaxanthin from *H. pluvialis* has got the sanction of FDA to be used as human dietetic complement, while other sources are mainly utilized as aquaculture feed.

Heterotrophic growth of *H. pluvialis* in dark could assimilate acetate and hence the inclusion of acetate in the medium enhances the astaxanthin production. Ferrous ions enhance the astaxanthin accumulation and when added with acetate, astaxanthin production was greater (Zhang et al., 2016). The astaxanthin production in *H. pluvialis* increases with increasing concentrations of NaCl in the growth medium. Liu et al. (2014) reported that cultivation of *H. pluvialis* is a challenging task due to its slow growth and bacterial contamination in open cultivation system. The addition of glucose, fructose, and sucrose as carbon sources can be considered as an alternative for the heterotrophic cultivation of *Chlorella zofingensis* for astaxanthin production (Liu et al., 2012).

PC has been used as coloring agents in chewing gums, candies, jellies, and dairy products (Sekar and Chandramohan, 2008). The development of PC producing *Arthrospira platensis* is a challenging one in autotrophic cultivation and hence heterotrophic method is preferable. The strain is cultured utilizing glucose as a carbon source for producing PC. The unicellular red alga, *Galdieria sulphuraria*, is a capable strain for heterotrophic PC production. *G. sulphuraria* is a polyextremophilic algae and it can be grown in the dark condition at very low pH. It grows utilizing sugar and sugar alcohols as carbon sources under heterotrophic and mixotrophic conditions (Jain et al., 2014).

Dunaliella salina is cultured in open ponds with high salinity for viable production of β-carotene (Spolaore et al., 2006). This strain *Dunaliella* sp. can grow under heterotrophic conditions utilizing glucose and acetate as the carbon source. There are reports that photoheterotrophic cultivation of *D. salina* for β-carotene with acetate and mixotrophic cultivation with glucose favors carotene production (Mojaat et al., 2008; Morowvat and Ghasemi, 2016). Nevertheless, these were not followed more because of their low growth prospective and low β-carotene production of *D. salina* under sub optimal mixotrophic cultivation. Thus, the major carotenoids lutein, astaxanthin, and PC could be extracted economically and beneficially from heterotrophically cultivated microalgae.

4.5 FACTORS AFFECTING PIGMENT PRODUCTION

The light, temperature, and nutritional requirements are important factors for the growth of microalgae and its pigments.

4.5.1 ENVIRONMENTAL FACTORS

Light and temperature play the vital role in microalgal cultivation. Light is the most important factor for restricting the microalgae growth. Different spectral properties of light influence the comparative pigment composition and act as photomorphogenic signals in microalgae (Kagawa and Suetsugu, 2007). Pisal and Lele (2005) have testified that light intensity plays a major role in the enhanced production of carotenoids from *D. salina*. The duration of light and its' intensity directly influence the growth of algae and photosynthesis. Microalgae require light for photochemical phase to produce adenosine triphosphate (ATP), nicotinamide adenine dinucleotide phosphate-oxidase (NADPH) and require dark to produce required molecules for the growth (Cheirsilp and Torpee, 2012). Al-Qasmi (2012) revealed that the increase in light duration is directly proportional to the increase in cultured microalgae. Light intensity performs a significant part by dominating the pigment build up in algal cells. β-carotene increased sharply with light intensity at higher irradiation (11.2 $\mu mol/m^2/sec$). Astaxanthin production in *H. pluvialis* is higher under light intensity of 546 $\mu mol/m^2/sec$ (Imamoglu et al., 2009). The light intensity greatly influences the culture concentration, cell ripeness and pigment yield. The white light strength for chlorophyll synthesis and PC is higher in *S. fussiformis* (Madhyastha and Vatsala, 2007).

The relationship between temperature and algal growth rate increases exponentially until the optimum temperature is reached. Microalgae can grow in a wide temperature range between 5°C and 40°C. The effect of temperature on the growth of microalgae demonstrated that the growth rate was highest in the temperature range of 27–31°C (Kitaya et al., 2005). The carotenoid production enhances in the blue green microalgae when cultured in higher temperature (Garcia-Gonzalez et al., 2005). The phycobiliprotein from *Anabaena* sp. was higher in the optimum temperature of 30°C whereas the optimum temperature for *S. platensis*, *Anabaena* sp., *Nostoc* sp. was 25, 35 and 36°C, respectively (Hemlata and Fatma, 2009). Bocanegra et al.

(2004) have assessed that the production of astaxanthin from *H. pluvialis* is higher at the optimum temperature of 28°C. Sanchez et al. (2008) found that 55 and 60°C were the extreme temperatures for carotenoid production from *C. vulgaris* and *Nannochloropsis gaditana* correspondingly.

The microalgae have a pH range in which growth is optimal depending on chemical equilibrium and are more accustomed to assimilate. Wang et al. (2010) has reported that the optimal pH range for the growth is between 7 and 9. The phycobiliprotein from the blue-green microalgae, *S. platensis* was enhanced at pH 8 whereas phycobiliprotein from the *Nostoc* sp. was enhanced at pH 9 (Hemlata and Fatma, 2009). Chauhan and Pathak (2010) reported that the chlorophyll production from *S. platensis* was achieved at pH 9. The optimum production of carotenoids from *Scenedesmus almeriensis* was at pH 8 whereas for *Chlorococcum citriforme* and *Neospongiococcus gelatinosum* the optimum production was at pH 7 (Campo et al., 2000).

4.5.2 NUTRIENTS

Nitrogen is essential for the life as it is the central element of protein and an inherent material and it is plentiful in microalgae following carbon, oxygen, and hydrogen (Benavente-Valdes et al., 2016). When microalgal cells grow and divide, they need a supply of nitrogen. Under limited nitrogen supply, the photosynthesis process may continue and the resultant compounds give the minimum amount of pigments. The chlorophyll a and b production from *Chlorella minutissima* increased when nitrogen was added to the medium (Ordog et al., 2012). All through the nitrogen starvation, the microalgae stop splitting as nitrogen is the basic requirements for cell metabolism. Chlorophyll production by *S. platensis* increases with the boost in nitrogen intensity in the culture medium. The phycobiliprotein by *Anabaena* was high in nitrogen free media whereas PBPs by the *Fischerella* sp. was higher in nitrogen supplemented media (Soltani et al., 2007). Nitrogen in the medium increases the pigment production and this strategy can be used for large-scale cultivation.

Phosphorous, the major element in the cell plays an important role in producing ATP for nucleic acid metabolism. Qu et al. (2008) have reported that phosphate assimilation by *Chlorella pyrenoidosa* in biomass production was observed under different culture conditions. The use of phosphorous deprivation as medium constitution reduces process costs in large scale cultivation both in the open and closed systems.

4.6 METABOLIC ENGINEERING

The metabolic pathways can be exploited to develop the pigment substance of microalgae that can be either through upregulation/downregulation of enzymes responsible for pigment synthesis and the formation of metabolic sink (Mulders et al., 2014). The regulation of enzymes can stimulate the synthesis of pigments and thereby increasing the production of the desired pigments. Rosenberg et al. (2008) reports that over expressing enzymes are openly liable for the end production of targeted pigments and this can be used for concurrent boosting of various pigment production. The main challenge linked with the overproduction of pigments is storage space and transportation. For overproduction of microalgal pigments, the application of metabolic engineering procedures like enzyme expression seems to be advantageous. Further research is mandatory to develop the present metabolic engineering strategies for pigment production in microalgae.

Factors like light, temperature, nutrients, etc., influence the algal pigment production. The upregulation of *psy* gene in *Chlorella zogengiensis* under nitrogen limitation resulted in a 4-fold increase in astaxanthin production (Mao et al., 2018). Couso et al. (2012) have reported that exposure of *C. reinhardtii* to high light intensity triggered the expression levels of *psy* and *pds* and increased the chlorophyll production by 2 and 4 folds respectively. The genetic analysis revealed that *Lcyβ* mRNA and the level of carotenoids were increased in *D. salina* under various conditions like light intensity and nutrition (Ramos et al., 2008). The gene *chyB* responsible for zeaxanthin production in *C. reinhardtii* is also upregulated by the light. The production of lutein from *C. reinhardtii* is mediated through 2 cytochrome P450 dependent hydroxylases and under light stress, *cyp97a5* and *cyp97c3* encoding the hydroxylases are up regulated and thereby increasing the lutein production (Couso et al., 2012).

4.7 APPLICATIONS OF ALGAL PIGMENTS

4.7.1 FOOD COLORANTS

Phycobilins are utilized as organic dyes in foodstuff and cosmetic industry. In food products, they are used as colorants in fermented dairy products, ice creams, milkshakes, etc. (Sekar and Chandramohan, 2008). The microalgae like *Arthrospira platensis Chlorella* sp. *Dunaliella terticola, Dunaliella salina*, *Aphanizomenon flos-aquae* are reported to contain nutritional value with high protein content (Soletto et al., 2005). Microalgae, *Arthrospira* sp., *Chlorella* sp., *Nitzchia* sp., *Navicula* sp., *Crypthecodinium* sp., are also sources of food for many animals and are used as feed for marine and earthly animals. Sathasivam et al. (2019) reports that of late microalgae are getting prominence in health-food supermarkets and *Chlorella* and *Arthrospira* are dominating the algal market.

β-carotene is utilized as a pro-vitamin A (retinol) and also it has a main role in the making of healthy foods (Krinsky and Johnson, 2005; Spolaore et al., 2006). β-carotene from *Dunaliella* is utilized to draw the attention of the consumers as foodstuff colorants to increase the look of margarine, cheese, fruit juices, baked goods, dairy products, canned foods, and confectionary. Adding up, β-carotene is also used as a colorant and a pioneer of vitamin A in pet foods (Cantrell et al., 2003). β-carotene from microalgae is most widely used as foodstuff coloring agents in pasta, fruit juices, soft drinks, confectionery, margarine, dairy products, and salad dressings (Christaki et al., 2011, 2013; Guedes et al., 2011). PC, a blue pigment is susceptible to high temperatures and light and it could be used in several foodstuff produce like chewing gums, candies, dairy products, jellies, ice creams, and beverages; Its' color is steady in dried up provisions (Dufossé, 2009; Gouveia, 2014). PE, a red pigment can be utilized for the pigmentation of confectionaries, gelative desserts, and milk produce. Microalgal pigments can be used to enhance the yellowish color of egg yolk and chicken hide thereby improving the appearance of pet foods (Figure 4.3) (Spolaore et al., 2006; Mata et al., 2010; Guedes et al., 2011; Skjanes et al., 2013).

4.7.2 HEALTH BENEFITS OF ALGAL PIGMENTS

4.7.2.1 ANTIOXIDANT ACTIVITY

The microalgal pigments like fucoxanthin, phycoerythrobilin, chlorophyll, and their derivatives comprise potential antioxidant activity. The antioxidant activity of these pigments depends on their structural features (porphyrin ring, phythyl chain and conjugated double bonds). Chlorophyll exhibits antioxidant activity and its' porphyrin ring plays a vital role for its activity. Chlorophyll derivatives which are deficient in Mg^{2+} and phytyl chain possess more antioxidant activity compared to chlorophyll. Chlorophyll derivatives from *Enteromorpha prolifera* have strong antioxidant activity (Pangestuti and Kim, 2011).

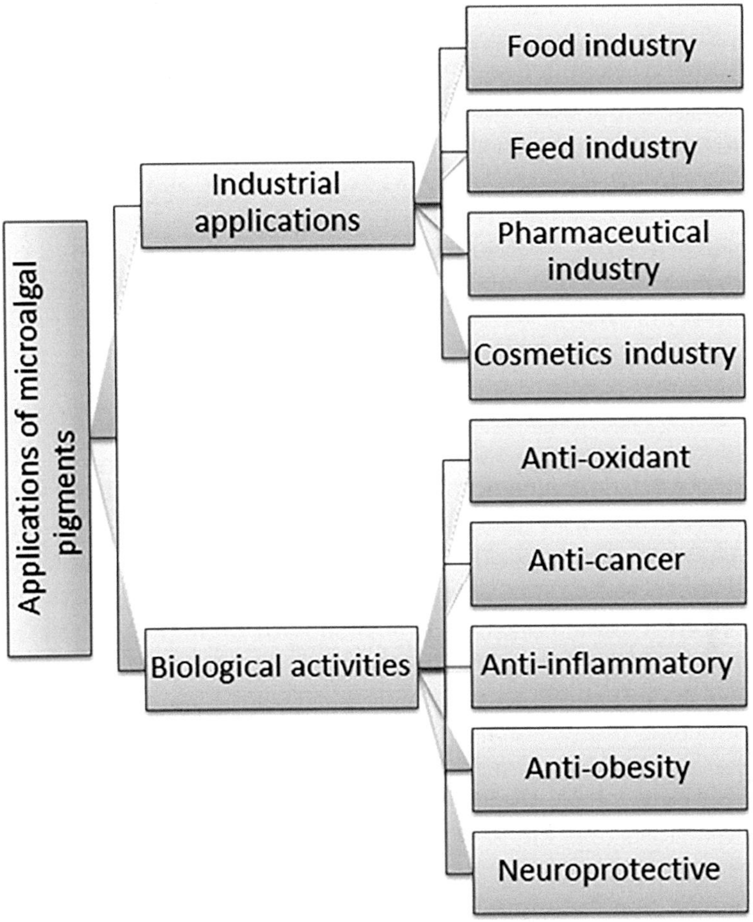

FIGURE 4.3 Applications of natural pigments from microalgae.

Fucoxanthin from *Hijikia fusiformis*, *Undaria pinnatifida* and *Odontella aurita* has major antioxidant activity. Many studies testify that the number of hydroxyl groups on fucoxanthin is linked to the effects of ROS suppression (Yang et al., 2019). β-carotene from *Dunaliella* stimulates antioxidant enzymes like catalase, superoxide dismutase (SOD), peroxidase, and antilipid peroxidase. The algal carotenoids like β-carotene from *Arthrospira platensis*, astaxanthin from *Haematococcus pluvialis*, lutein from *Botryococcus braunii* have significant antioxidant activities by substantially increasing the SOD, peroxidase, and catalase (Cezare-Gomes et al., 2019).

Phycoerythrobilin from *Porphyra* and PBPs from the *Lyngbya* sp. have tough antioxidant activity. The antioxidant properties of PBPs in *Arthrospira* have excellent dietary applications. PC derivatives from *Aphanizomenon flos-aquae* (AFA) is a powerful antioxidant and used against oxidative damage (Benedetti et al., 2004). Because of the stronger antioxidant activity, astaxanthin acts as a super vitamin E and is 10 times higher than β-carotene and 500 times more effective than α-tocoferol (Begum et al., 2016).

4.7.2.2 ANTICANCER ACTIVITY

The cancer cells in humans are stimulated by free radicals and the natural anticancer drugs have got positive impact in the healing of cancer. The algal pigments have radical scavenging activity and they can be applied to indirectly reduce the cancer development. Lutein, β-carotene and chlorophyll a from *Porphyra tenera* showed antimutagenic activity in *Salmonella typhimurium*. Siphonaxanthin from *Codium fragile* is an effective evolution inhibitor against HL-60 cells, better than fucoxanthin. The structure of fucoxanthin have epoxide and allenic bond whereas siphonaxanthin does not have these functional group but have additional hydroxyl group. Fucoxanthin exerts antitumor activity through apoptotic induction (Kotake et al., 2005). Antiproliferative activity of fucoxanthin, fucoxanthinol, β-carotene and astaxanthin evaluated, established that fucoxanthin and fucoxanthinol have notable antiproliferative effects on human T-cell leukemia virus *in vitro* (Ishikawa et al., 2008). Hosokawa et al. (2010) have assessed that fucoxanthin decreases the viability of human colon cancer cell lines and stimulate apoptosis in a dosage and time related manner. Fucoxanthin inhibits the increase of human colon cancer cell lines by induction of cell cycle arrest at G0/G1 phase (Das et al., 2005). Fucoxanthin has significant antiproliferative effects on human urinary bladder cancer cell line by stirring up apoptosis and it was characterized by morphological changes.

PC has significant anticancer properties opposed to human melanoma and human breast adenocarcinoma. It promotes the manifestation of CD59 protein in HeLa cells and has antineoplastic effects on colon carcinogenesis (Jiang et al., 2017). PC contains an anti-cancerous effect by dipping the tumor necrosis factor (TNF-alpha) in the blood serum of mice treated with endotoxin, and it also exhibited neuroprotective effects in the rat cerebella granule cell cultures. Shih et al. (2003) have testified that APC inhibits enterovirus 71-induced cytopathic effects, viral sign development, and viral-induced apoptosis. β-carotene from *Dunaliella* sp. comprises of 40% 9-cis and 50% all-trans stereoisomers that play a decisive role in reducing the incidence of numerous varieties of cancer and degenerative diseases (Begum et al., 2016). *Dunaliella* sp. contains xanthophylls that showed enhanced anti-cancerous activity and upper bioactivity (Roodenburg et al., 2000). β-carotene can induce the immune system and could be involved in curing more than 60 dreadful diseases including cancer, coronary heart diseases, premature aging and arthritis (Mattson, 2004).

4.7.2.3 ANTI-INFLAMMATORY ACTIVITY

The microalgal pigments have promising anti-inflammatory activities that are mainly due to the modulation of macrophage function. Pheophytin α from *E. prolifera* have buried 12-*o*-tetradecanoylphorbol-13-acetate-induced superoxide radicals. Fucoxanthin possesses anti-inflammatory properties both *in vitro* and *in vivo* in response to bacterial lipopolysaccharides (LPSs). The anti-inflammatory properties of fucoxanthin are similar to the commercially available anti-inflammatory steroidal drugs. The inhibition of nuclear factor-kB (NF-kB) and phosphorylation of mitogen activated protein kinases are responsible for the anti-inflammatory properties of fucoxanthin. PC reduces allergic inflammatory response by suppressing antigen-specific IgE antibody (Nemoto-Kawamura et al., 2004). PC is effective in the treatment of acute lung injury by inhibiting inflammatory response and apoptosis in lung tissue (Leung et al., 2013). Phycocyanobilin, are found in blue-green algae and contain potential anti-inflammatory effects.

4.7.2.4 ANTI-OBESITY ACTIVITY

Obesity is one of the greatest health challenges in several modern and developing countries. Wang et al. (2008) have reported that excess growth of adipose tissue from adipocyte hypertrophy and the addition of new adipocytes from precursor cells and regulation of adipogenesis seems to be the best strategy to treat obesity. Fucoxanthin from *U. pinnatifida* inhibits the differentiation of 3T3-L1 preadipocytes into adipocytes. Okada et al. (2008) reported that carotenoids with keto group, expoxy group, hydroxy-carotenoid did not show inhibitory outcome on adipocyte differentiation, whereas the healing with fucoxanthin and neoxanthin has noteworthy anti-obesity effect. The natural pigment, fucoxanthin from marine algae can be used as slimming supplements and medicines in the prevention and management of obesity. Siphonaxanthin, a green algal pigment lowers the lipid accumulation in KK-Ay mice and also lowers the expression of key adipogenesis genes *Cebpa*, *Pparg*, *Fabp4* and *Scd1* (Li et al., 2015). PC stimulates antiobesity effect which might be due to the hypocholesterolemic action. PC inhibits pancreatic lipase resulting in the inhibition of intestinal absorption of dietary fat which can lower serum cholesterol, total cholesterol, triglyceride, and low-density lipids (Pangestuti and Kim, 2011).

4.7.2.5 NEUROPROTECTIVE ACTIVITY

Neurodegenerative diseases surpass cancer and are the next most reason of casualty among elders. Fucoxanthin from *H. fusiformis* inhibited N-myc expression and cell cycle progression of GOTO cells, human neuroblastoma cell line. Khodesevich and Monyer (2010) have found that fucoxanthin reduced cell damage in cortical neurons during hypoxia and oxygen reperfusion. Astaxanthin is a first-rate candidate for testing Alzheimer's diseases and other neurological diseases. Astaxanthin mediated neuroprotection in subarachnoid hemorrhage is credited to the downregulation of augmented nuclear factor kappa B activity and the expression of inflammatory cytokines and intercellular adhesion molecule 1 (Grimmig et al., 2017). PC showed protecting role against hippocampus neuronal cell death and enhanced locomotive behavior in Mongolian Gerbil (Penton-Rol et al., 2011). The neuroprotective activity of PC is accredited towards its antioxidant, anti-inflammatory, and immunomodulatory properties (Wu et al., 2016).

4.7.3 SKIN CARE BENEFITS OF ALGAL PIGMENTS

Microalgal extracts are commonly used in the countenance and skincare products for refreshing, antiaging, anti-irritant, etc. They are also utilized as sun protectants and hair care stuff. Protein rich extort from *Arthrospira* plays an essential role in early skin aging, tightening the skin and *Chlorella vulgaris* stimulates collagen synthesis in skin and supports tissue regeneration and wrinkle reduction. A product has been launched from *Nannochloropsis oculata* with exceptional skin contraction properties and *D. salina* induce cell propagation and optimistically manipulate the energy metabolism of skin. Astaxanthin from *H. pluvialis* exhibited enhanced improvements in skin wrinkles, skin texture, moisture content, etc. (Tominaga et al., 2012). Astaxanthin plays an important role in protecting against photooxidative damage (Camera et al., 2009). Algal pigments showed considerable antioxidant effect and they play a potential role in restricting aging due to the creation of free radicals in the cells. These pigments, particularly are vibrant dyes even at very low levels (parts per million), and possess significant utilities in the pharmaceutical industry (fluorescence-based indicators, biochemical tracers in immune assays) and in the cosmetic industry (skin cream to stimulate collagen synthesis).

4.8 FUTURE PERSPECTIVES AND CONCLUSIONS

For the past about six or seven decades, the world is consuming a lot of non-degradable materials produced from plastics, and discharge them in the environment after their wear and tear. This has caused irreparable damage to the environment and ecology. Confronting with such environmental damages finding alternative eco-friendly, harmless, and degradable materials were imperative to maintain our environment in a sustainable manner. Awareness created in this respect has turned the attention of the stakeholders towards the production of degradable products from natural sources. By the way production of natural colorants from plants, microbes, and algae for application in food, medicine, cosmetic, textile, etc., has emerged and progressed considerably. Among them algae have a major stake and a better prospective natural resource in producing degradable natural products like coloring pigments, etc.

Microalgae have now emerged as a major prospective natural resource for producing pigments such as carotenoids, chlorophylls, and PBPs for commercial applications in the food, pharma, and cosmetics industry. These pigments have proved their incredible capabilities of coloring foods, antioxidant, anticancer, anti-obesity, anti-inflammatory, and neuroprotective activities along with skincare benefits. Yet exploitation of pigment producing algae from unexplored niches would be a fruitful venture to unravel novel high-yielding pigment producers and other metabolites. Metabolic engineering allows manipulating microalgae to have high growth for biomass and pigment production. The exploitation of microalgae for natural colorants can increase the profitability and health benefits overcoming sustainability challenges. Nevertheless, the bottlenecks, like high production costs and low yields, stability of the pigments, etc., need to be addressed before microalgae can be moved from niche markets to large-scale use. The future of algal pigments is bright, colorful, and ecologically sustainable.

CONFLICTS OF INTEREST

The authors declare no conflict of interest.

ACKNOWLEDGMENTS

Dr. C.K. Venil thanks the UGC for awarding the Dr. D.S. Kothari Postdoctoral Fellowship (BL/17-18/0479). Also, the authors thank Anna University, Regional Campus – Coimbatore for providing necessary facilities to carry out the project work. Professor Laurent Dufossé deeply thanks the Conseil Régional de La Réunion, Réunion Island, Indian Ocean, for continuous financial support of research activities dedicated to microbial pigments. This work was supported and funded by the University Grants Commission (UGC), New Delhi under the Dr. D.S. Kothari Post-doctoral Fellowship (BL/17-18/0479) dated 25th September 2018.

KEYWORDS

- **biomass production**
- **cultivation**
- **factors**
- **food colorants**
- **health benefits**
- **microalgae**
- **pigments**
- **skincare**

REFERENCES

Al-Qasmi, M., (2012). A review of effect of light on microalgae growth. *Proceedings of the World Congress on Engineering, 1*, 3–8.

Ambati, R. R., Gogisetty, D., Aswathanarayana, R. G., et al., (2018). Industrial potential of carotenoid pigments from microalgae: Currents trends and future prospects. *Critical Reviews in Food Science and Nutrition, 59*, 1880–1882.

Begum, H., Yusoff, F. M., Banerjee, S., Khatoon, H., & Shariff, M., (2016). Availability and utilization of pigments from microalgae. *Critical Reviews in Food Science and Technology, 56*, 2209–2222.

Benavente-Valdés, J. R., Aguilar, C., Contreras-Esquivel, J. C., Méndez-Zavala, A., & Montañez, J., (2016). Strategies to enhance the production of photosynthetic pigments and lipids in *Chlorophycae* species. *Biotechnology Reports, 10*, 117–125.

Benedetti, S., Benvenuti, F., & Pagliarani, S., (2004). Antioxidant properties of a novel phycocyanin extract from the blue-green alga *Aphanizomenon flos-aquae. Life Science, 75*, 2353–2362.

Bocanegra, D. A. R., Guerrero, L. I., Martinez, J. F., & Tomasini, C. A., (2004). Influence of environmental and nutritional factors in the production of astaxanthin from *Haematococcus pluvialis. Bioresource Technology, 92*(2), 209–214.

Borowitzka, M. A., (2013). High-value products from microalgae – their development and commercialization. *Journal of Applied Phycology, 25*, 743–756.

Bumbak, F., Cook, S., Zachleder, V., Hauser, S., & Kovar, K., (2011). Best practices in heterotrophic high cell density microalgal processes: Achievements, potential and possible limitations. *Applied Microbiology and Biotechnology, 91*, 31–46.

Camacho, F., Macedo, A., & Malcata, F., (2019). Potential industrial applications and commercialization of microalgae in the functional food and feed industries: A short review. *Marine Drugs, 17*(6), 312.

Camera, E., Mastrofrancesco, A., Fabbri, C., et al., (2009). Astaxanthin, canthaxanthin and beta-carotene differently affect UV-A-induced oxidative damage and expression of oxidative stress-responsive enzymes. *Experimental Dermatology, 18*, 222–231.

Campo, D. J. A., Moreno, J., Rodriguez, H., Vargas, M. A., Rivas, J., & Guerrero, M. G., (2000). Carotenoid content of chlorophycean microalgae: Factors determining lutein accumulation in *Muriellopsis* sp. (*Chlorophyta*). *Journal of Biotechnology, 76*, 51–59.

Cantrell, A., McGarvey, D. J., Trustcott, G., Rancan, F., & Bohm, F., (2003). Singlet oxygen quenching by dietary carotenoids in a model membrane environment. *Archives of Biochemistry and Biophysics, 412*, 47–54.

Cezare-Gomes, E. A., Mejia-da-Silva, L. D., Pérez-Mora, L. S., et al., (2019). Potential of microalgae carotenoids for industrial application. *Applied Biochemistry and Biotechnology, 188*, 602–634.

Chauhan, U. K., & Pathak, N., (2010). Effect of different conditions on the production of chlorophyll by *Spirulina platensis*. *Journal of Algal Biomass Utilization, 1*, 89–99.

Cheirsilp, B., & Torpee, S., (2012). Enhanced growth and lipid production of microalgae under mixotrophic culture condition: Effect of light intensity, glucose concentration and fed-batch cultivation. *Bioresource Technology, 110*, 510–516.

Christaki, E., Bonos, E., Giannenas, I., & Florou-Paneri, P., (2013). Functional properties of carotenoids originating from algae. *Journal of the Science of Food and Agriculture, 93*, 5–11.

Christaki, E., Florou-Paneri, P., & Bonos, E., (2011). Microalgae: A novel ingredient in nutrition. *International Journal of Food Science and Nutrition, 62*, 794–799.

Chu, W. L., (2012). Biotechnological applications of microalgae. *International Journal of Medical Science and Education, 6*, 24–37.

Couso, I., Vila, M., Vigara, J., et al., (2012). Synthesis of carotenoids and regulation of the carotenoid biosynthesis pathway in response to high light stress in the unicellular microalga *Chlamydomonas reinhardtii*. *European Journal of Phycology, 47*(3), 223–232.

Cuaresma, M., Janssen, M., Vilchez, C., & Wijffels, R. H., (2011). Horizontal or vertical photobioreactors? How to improve microalgae photosynthetic efficiency. *Bioresource Technology, 102*, 5129–5137.

Cubas, C., Gloria, L. M., & Gonzalez, M., (2008). Optimization of the extraction of chlorophylls in green beans (*Phaseolus vulgaris* L.) by N,N-dimethylformamide using response surface methodology. *Journal of Food Composition and Analysis, 21*, 125–133.

Das, S. K., Hashimoto, T., Shimizu, K., et al., (2005). Fucoxanthin induces cell cycle arrest at G0/G1 phase in human colon carcinoma cells through up-regulation of p21WAF1/Cip1. *Biochim. Biophys. Acta, 1726*(3), 328-335.

Daubrawa, F., Sies, H., & Stahl, W., (2005). Astaxanthin diminishes gap junctional intercellular communication in primary human fibroblasts. *Journal of Nutrition, 135*, 2507–2511.

Del Campo, A. J., Garcia-Gonzalez, M., & Guerrero, M. G., (2007). Outdoor cultivation of microalgae for carotenoid production: Current state and perspectives. *Applied Microbiology and Biotechnology, 74*, 1163–1174.

Dufossé, L., (2009). Microbial and microalgal carotenoids as colorants and supplements. In: Britton, G., Liaaen-Jensen, S., & Pfander, H., (eds.), *Carotenoids, Nutrition and Health* (Vol. 5). Birkhauser Verlag, Basel, Switzerland.

Eriksen, N. T., (2008). Production of phycocyaninda pigment with applications in biology, biotechnology, foods and medicine. *Applied Microbiology and Biotechnology, 80*, 1–14.

Gao, C., Zhai, Y., Ding, Y., & Wu, Q., (2010). Application of sweet sorghum for biodiesel production by heterotrophic microalga *Chlorella protothecoides. Applied Energy, 87*, 756–761.

Garcia-Gonzalez, M., Moreno, J., Manzano, J. C., Florencio, F. J., & Guerrero, M. G., (2005). Production of *Dunaliella salina* biomass rich in 9-cis-β-carotene and lutein in a closed tubular photobioreactor. *Journal of Biotechnology, 115*, 81–90.

Gouveia, L., (2014). From tiny microalgae to huge biorefineries. *Oceanography, 2*, 120.

Grimmig, B., Kim, S. H., Nash, K., Bickford, P. C., & Douglas, S. R., (2017). Neuroprotective mechanisms of astaxanthin: A potential therapeutic role in preserving cognitive function in age and neurodegeneration. *GeroScience, 39*(1), 19–32.

Guedes, A. C., Amaro, H. M., & Malcata, F. X., (2011). Microalgae as sources of carotenoids. *Marine Drugs, 9*, 625–644.

Gunerken, E., D'Hondt, E., Eppink, M. H. M., Garcia-Gonzalez, L., Elst, K., & Wijffels, R. H., (2015). Cell disruption for microalgae biorefineries. *Biotechnology Advances, 33*, 243–260.

Hemlata, & Fatma, T., (2009). Screening of Cyanobacteria for phycobiliproteins and effect of different environmental stress on its yield. *Bull. Environ. Contam. Toxicol., 83*, 509.

Hifney, A. F., Issa, A. A., & Fawzy, M. A., (2013). Abiotic stress-induced production of β-carotene, allophycocyanin and total lipids in *Spirulina* sp. *Journal of Biology and Earth Science, 3*, 54–64.

Hosokawa, M., Miyashita, T., Emi, S., Tsukui, T., Beppu, F., Okada, T., & Miyashita, K., (2010). Fucoxanthin regulates adipocytokine mRNA expression in white adipose tissue of diabetic/obese KK-Ay mice. *Archives of Biochemistry and Biophysics, 504*, 17–25.

Hu, J., Nagarajan, D., Zhang, Q., Chang, J. S., & Lee, D. J., (2018). Heterotrophic cultivation of microalgae for pigment production: A review. *Biotechnology Advances, 36*, 54–67.

Imamoglu, E., Dalay, M. C., & Sukan, F. V., (2009). Influences of different stress media and high light intensities on accumulation of astaxanthin in the green alga *Haematococcus pluvialis. New Biotechnology, 26*, 199–204.

Ishikawa, C., Tafuku, S., Kadekaru, T., Sawada, S., Tomita, M., Okudaira, T., Nakazato, Tet al., (2008). Antiadult T cell leukemia effects of brown algae fucoxanthin and its deacetylated product, fucoxanthinol. *International Journal of Cancer, 123*, 2702–2712.

Jain, K., Krause, K., Grewe, F., Nelson, G. F., Weber, A. P., Christensen, A. C., & Mower, J. P., (2014). Extreme features of the *Galdieria sulphuraria* organellar genomes: A consequence of polyextremophily? *Genome Biology and Evolution, 7*(1), 367–380.

Jiang, L., Wang, Y., Yin, Q., Liu, G., Liu, H., Huang, Y., & Li, B., (2017). Phycocyanin: A potential drug for cancer treatment. *Journal of Cancer, 8*(17), 3416–3429.

Kagawa, T., & Suetsugu, N., (2007). Photometrical analysis with photosensory domains of photoreceptors in green algae. *FEBS Letters, 581*, 368–374.

Khan, M. I., Shin, J. H., & Kim, J. D., (2018). The promising future of microalgae: Current status, challenges and optimization of a sustainable and renewable industry for biofuels, feed and other products. *Microbial Cell Factory, 17*, 36.

Khodosevich, K., & Monyer, H., (2010). Signaling involved in neurite outgrowth of postnatally born subventricular zone neurons *in vitro*. *BMC Neuroscience, 11*, 18.

Kiesenhofer, D. P., & Fluch, S., (2018). The promises of microalgae – still a long way to go. *FEMS Microbiology Letters, 35*, 257.

Kitaya, Y., Azuma, H., & Kiyota, M., (2005). Effects of temperature, CO_2/O_2 concentrations and light intensity on cellular multiplication of microalgae, *Euglena gracilis*. *Advances in Space Research, 35*(9), 1584–1588.

Kotake, N. E., Asai, A., & Nagao, A., (2005). Neoxanthin and fucoxanthin induce apoptosis in PC-3 human prostate cancer cells. *Cancer Letters, 220*, 75–84.

Koyande, A. K., Chew, K. W., Rambabu, K., Tao, Y., Chu, D. T., & Show, P. L., (2019). Microalgae: A potential alternative to health supplementation for humans. *Food Science and Human Wellness, 8*, 1–24.

Krinsky, N. I., & Johnson, E. J., (2005). Carotenoid actions and their relation to health and disease. *Molecular Aspects of Medicine, 26*, 459–416.

Leung, P. O., Lee, H. H., Kung, Y. C., Tsai, M. F., & Chou, T. C., (2013). Therapeutic effect of c-phycocyanin extracted from blue-green algae in a rat model of acute lung injury induced by lipopolysaccharide. *Evidence-Based Complementary and Alternative Medicine*, 916590.

Li, J., Zhu, D. L., Niu, J., Shen, S. D., & Wang, G., (2011). An economic assessment of astaxanthin production by large-scale cultivation of *Haematococcus pluvialis*. *Biotechnology Advances, 29*, 568–574.

Li, K., Cheng, J., Ye, Q., He, Y., Zhou, J., & Cen, K., (2017). *In vivo* kinetics of lipids and astaxanthin evolution in *Haematococcus pluvialis* mutant under 15% CO_2 using Raman microspectroscopy. *Bioresource Technology, 244*(2), 1439–1444.

Li, P., Miao, X., R., & Zhong, J., (2011). *In situ* biodiesel production from fast-growing and high oil content *Chlorella pyrenoidosa* in rice straw hydrolysate. *Journal of Biomedicine and Biotechnology, 2011*, 141201.

Li, Z. S., Noda, K., Fujita, E., Manabe, Y., Hirata, T., & Sugawara, T., (2015). The green algal carotenoid siphonaxanthin inhibits adipogenesis in 3t3-l1 preadipocytes and the accumulation of lipids in white adipose tissue of KK-Ay mice. *The Journal of Nutrition, 145*(3), 490–498.

Liu, J., Huang, J., Jiang, Y., & Chen, F., (2012). Molasses-based growth and production of oil and astaxanthin by *Chlorella zofingiensis*. *Bioresource Technology, 107*, 393–398.

Liu, J., Sun, Z., Gerken, H., Liu, Z., Jiang, Y., & Chen, F., (2014). *Chlorella zofingiensis* as an alternative microalgal producer of astaxanthin: Biology and industrial potential. *Marine Drugs, 12*(6), 3487–3515.

Lordan, S., Paul, R. R., & Stanton, C., (2011). Marine bioactives as functional food ingredients: Potential to reduce the incidence of chronic diseases. *Marine Drugs, 9*, 1056–1100.

Lu, Y., Zhai, Y., Liu, M., &. Wu, Q., (2010). Biodiesel production from algal oil using cassava (*Manihot esculenta* Crantz) as feedstock. *Journal of Applied Phycology, 22*, 573–578.

Madhyastha, H. K., & Vatsala, T. M., (2007). Pigment production in *Spirulina fussiformis* in different photophysical conditions. *Biomolecular Engineering, 24*, 301–305.

Mao, X., Wu, T., Sun, D., Zhang, Z., & Chen, F., (2018). Differential responses of the green microalga *Chlorella zofingiensis* to the starvation of various nutrients for oil and astaxanthin production. *Bioresource Technology, 249*, 791–798.

Markou, G., & Nerantzis, E., (2013). Microalgae for high-value compounds and biofuels production: A review with focus on cultivation under stress conditions. *Biotechnology Advances, 31*, 1532–1542.
Mata, T. M., Martins, A. A., & Caetano, N. S., (2010). Microalgae for biodiesel production and other applications: A review. *Renewable and Sustainable Energy Reviews, 14*, 217–232.
Mattson, M. P., (2004). Pathways towards and away from Alzheimer's disease. *Nature, 430*, 631–639.
Miazek, K., & Ledakowicz, S., (2013). Chlorophyll extraction from leaves, needles and microalgae: A kinetic approach. *International Journal of Agricultural and Biological Engineering, 6*, 107–115.
Mojaat, M., Pruvost, J., Foucault, A., & Legrand, F. J., (2008). Effect of organic carbon sources and Fe^{2+} ions on growth and β-carotene accumulation by *Dunaliella salina. Biochemical Engineering Journal, 39*(1), 177–184.
Morowvat, M. H., & Ghasemi, Y., (2016). Culture medium optimization for enhanced β-carotene and biomass production by *Dunaliella salina* in mixotrophic culture. *Biocatalysis and Agricultural Biotechnology, 7*, 217–223.
Mulders, K. J. M., Lamers, P. P., Martens, D. E., & Wijffels, R. H., (2014). Phototrophic pigment production with microalgae: Biological constraints and opportunities. *Journal of Phycology, 50*(2), 229–242.
Naruka, M., Khadka, M., Upadhayay, S., & Kumar, S., (2019). Potential applications of microalgae in bioproduct production: A review. *Octa Journal of Biosciences, 7*(1), 1–5.
Nemoto-Kawamura, C., Hirahashi, T., Nagai, T., Yamada, H., Katoh, T., & Hayashi, O., (2004). Phycocyanin enhances secretary IgA antibody response and suppresses allergic IgE antibody response in mice immunized with antigen-entrapped biodegradable microparticles. *Journal of Nutritional Science and Vitaminology, 50*(2), 129-136.
Novoveska, L., Ross, M. E., Stanley, M. S., Pradelles, R., Wasiolek, V., & Sassi, J. F., (2019). Microalgal carotenoids: A review of production, current markets, regulations and future directions. *Marine Drugs, 17*, 640.
Okada, T., Nakai, M., Maeda, H., Hosokawa, M., Sashima, T., & Miyashita, K., (2008). Suppressive effect of neoxanthin on the differentiation of 3T3-L1 adipose cells. *Journal of Oleo Science, 57*, 345–351.
Ördög, V., Stirk, W., Bálint, P., Staden, J., & Lovász, C., (2012). Changes in lipid, protein and pigment concentrations in nitrogen-stressed *Chlorella minutissima* cultures. *Journal of Applied Phycology, 24*, 907–914.
Pangestuti, R., & Kim, S. K., (2011). Biological activities and health benefit effects of natural pigments derived from marine algae. *Journal of Functional Foods, 3*, 255–266.
Pentón-Rol, G., Martínez-Sánchez, G., Cervantes-Llanos, M., et al., (2011). C-Phycocyanin ameliorates experimental autoimmune encephalomyelitis and induces regulatory T cells. *International Immunopharmacology, 11*(1), 29–38.
Pisal, D. S., & Lele, S. S., (2005). Carotenoid production from microalgae, *Dunaliella salina. Indian Journal of Biotechnology, 4*, 476–483.
Qu, C., Wu, Z., & Shi, X., (2008). Phosphate assimilation by *Chlorella* and adjustment of phosphate concentration in basal medium for its cultivation. *Biotechnology Letters, 30*, 1735.

Ramos, A., Coesel, S., Marques, A., Rodrigues, M., et al., (2008). Isolation and characterization of a stress-inducible *Dunaliella salina* Lcy- β gene encoding a functional lycopene β-cyclase. *Applied Microbiology and Biotechnology, 79*(5), 819.

Raposo, M. F., De Morais, R. M., & Bernardo De, M. A. M., (2013). Bioactivity and applications of sulphated polysaccharides from marine microalgae. *Marine Drugs, 11*(1), 233–252.

Régnier, P., Bastias, J., Rodriguez-Ruiz, V., et al., (2015). Astaxanthin from *Haematococcus pluvialis* prevents oxidative stress on human endothelial cells without toxicity. *Marine Drugs, 13*(5), 2857-2874.

Rodriguez-Sanchez, R., Ortiz-Butron, R., Blas-Valdivia, V., Hernandez-Garcia, A., & Cano-Europa, E., (2012). Phycobiliproteins or C-phycocyanin of *Arthrospira (Spirulina) maxima* protect against $HgCl_2$-caused oxidative stress and renal damage. *Food Chemistry, 135*, 2359–2365.

Roodenburg, A. J., Leenen, R., Van, H. H. K. H., Weststrate, J. A., & Tijburg, L. B., (2000). Amount of fat in the diet affects the bioavailability of lutein esters but not of alpha-carotene, beta-carotene, and vitamin E in humans. *American Journal of Clinical Nutrition, 71*, 1187–1193.

Rosenberg, J. N., Oyler, G. A., Wilkinson, L., & Betenbaugh, M. J., (2008). A green light for engineered algae: Redirecting metabolism to fuel a biotechnology revolution. *Current Opinion in Biotechnology, 19*(5), 430–436.

Sánchez, J. F., Feranadez, J. M., Acién, F. G., Rueda, A., Pérez-Parra, J., & Molina, E., (2008). Influence of culture conditions on the productivity and lutein content of the new strain *Scenedesmus almeriensis*. *Process Biochemistry, 43*, 398–405.

Sathasivam, R., & Ki, J. S., (2018). A review of the biological activities of microalgal carotenoids and their potential use in healthcare and cosmetic industries. *Marine Drugs, 16*, 26.

Sathasivam, R., Radhakrishnan, R., Hashem, A., & Abd, A. E. F., (2019). Microalgae metabolites: A rich source for food and medicine. *Saudi Journal of Biological Sciences, 26*, 709–722.

Scheer, H., William, J. L., & Lane, M. D., (2004). Chlorophylls and carotenoids. In: Lennarz, W., & Lane, M., (eds.), *Encyclopedia of Biological Chemistry*. Academic Press, New York.

Sekar, S., & Chandramohan, M., (2008). Phycobiliproteins as a commodity: Trends in applied research, patents and commercialization. *Journal of Applied Phycology, 20*, 113–136.

Shih, S. R., Tsai, K. N., & Li, Y. S., (2003). Inhibition of enterovirus 71-induced apoptosis by allophycocyanin isolated from a blue-green alga *Spirulina platensis*. *Journal of Medical Virology, 70*, 119–125.

Singh, D. P., Khattar, J. S., Rajput, A., Chaudhary, R., & Singh, R., (2019). High production of carotenoids by the green microalga *Asterarcys quadricellulare* PUMCC5 under optimized conditions. *PLoS One, 14*(9), e0221930.

Skjanes, K., Rebours, C., & Lindblad, P., (2013). Potential for green microalgae to produce hydrogen, pharmaceuticals and other high value products in a combined process. *Critical Reviews in Biotechnology, 33*, 172–215.

Soletto, D., Binaghi, L., Lodi, A., Carvalho, J. C. M., & Converti, A., (2005). Batch and fed-batch cultivations of *Spirulina platensis* using ammonium sulphate and urea as nitrogen sources. *Aquaculture, 243*, 217–224.
Soltani, N., Khavari-Nejad, R. A., Yazdi, M. T., & Shokravi, S., (2007). Growth and some metabolic features of cyanobacterium *Fischerella* sp. FS18 in different combined nitrogen sources. *Journal of Science Islamic Republic of Iran, 18*, 123–128.
Spolaore, P., Joannis-Cassan, C., Duran, E., & Isambert, A., (2006). Commercial applications of microalgae. *Journal of Bioscience and Bioengineering, 101*, 87–96.
Stahl, W., & Sies, H., (2003). Antioxidant activity of carotenoids. *Molecular Aspects of Medicine, 24*, 345–351.
Tominaga, K., Hongo, N., Karato, M., & Yamashita, E., (2012). Cosmetic benefits of astaxanthin on humans subjects. *Acta Biochimica Polonica, 59*(1), 43.
Wang, C., Li, H., Wang, Q., & Wei, P., (2010). Effect of pH on growth and lipid content of *Chlorella vulgaris* cultured in biogas slurry. *Chinese Journal of Biotechnology, 26*, 1074–1079.
Wang, T., Wang, Y., & Kontani, Y., (2008). Evodiamine improves diet-induced obesity in a uncoupling protein-1-independent manner: Involvement of antiadipogenic mechanism and extracellularly regulated kinase/mitogen-activated protein kinase signaling. *Endocrinology, 149*, 358–366.
Wu, Q., Liu, L., Miron, A., Klimova, B., Wan, D., & Kuca, K., (2016). The antioxidant, immunomodulatory, and anti-inflammatory activities of spirulina: An overview. *Archives of Toxicology, 90*, 1817–1840.
Yan, D., Lu, Y., Chen, Y. F., & Wu, Q., (2011). Waste molasses alone displaces glucose-based medium for microalgal fermentation towards cost-saving biodiesel production. *Bioresource Technology, 102*, 6487–6493.
Yang, G., Jin, L., Zheng, D., et al., (2019). Fucoxanthin alleviates oxidative stress through Akt/Sirt1/$FoxO_3\alpha$ signaling to inhibit HG-induced renal fibrosis in GMCs. *Marine Drugs, 17*, 702.
Yen, H. W., Sun, C. H., & Ma, T. W., (2011). The comparison of lutein production by *Scenesdesmus* sp. in the autotrophic and the mixotrophic cultivation. *Applied Biochemistry and Biotechnology, 164*, 353–361.
Yuan, J. P., Peng, J., Yin, K., & Wang, J. H., (2011). Potential health-promoting effects of astaxanthin: A high-value carotenoid mostly from microalgae. *Molecular Nutrition and Food Research, 55*, 150–165.
Zhang, Z., Wang, B., Hu, Q., Sommerfeld, M., Li, Y., & Han, D., (2016). A new paradigm for producing astaxanthin from the unicellular green algae *Haematococcus pluvialis*. *Biotechnology and Bioengineering, 113*(10), 2088–2099.

CHAPTER 5

Functional Metabolites from Microalgae for Multiple Commercial Streams

AMRITPREET KAUR, SUCHITRA GAUR, and ALOK ADHOLEYA

TERI Deakin Nanobiotechnology Center, Sustainable Agriculture Division, The Energy and Resources Institute, New Delhi, India

ABSTRACT

Food industry is a global and fast-growing industry, with business and research networks still in a phase of transition. Additionally, high-value bioactive compounds (lipids, proteins, and carotenoids) and by-products derived from algae are being investigated for sustainable production of food and nutraceuticals. Moreover, the algal biomass is one of the largest sources of proteins packed with essential amino acids, coenzymes, vitamins, and minerals that makes it an attractive animal feed specifically due to its high protein content-rich (40–60%). However, the utilization of these algal nutrients (proteins and other bioactive extracts) in daily life-style in terms of food products is not yet fully explored. This chapter provides an insight to the commercial products and metabolites derived from algae for commercially viable large production. It outlines the major species used and their key role in food, nutrition, animal feed and medicine. This will help in controlled scale-up production of the target compound for production of targeted molecule via a sustainable source of production.

Microalgal Biotechnology: Bioprospecting Microalgae for Functional Metabolites towards Commercial and Sustainable Applications. Jeyabalan Sangeetha, PhD, Svetlana Codreanu, PhD, & Devarajan Thangadurai, PhD (Eds.)

5.1 INTRODUCTION

During the last 10 years, microalgae have been explored for number of commercial products but due to commercial hurdles in large scale cultivation not many products made it to market and therefore products are very limited and highly unexplored. Microalgae synthesize a wide range of functional metabolites (Figure 5.1) such as polyunsaturated fatty acids (PUFA), pigments, vitamins, and polysaccharides (Ratledge, 2010; Borowitzka, 2013; Sathasivam et al., 2019). The most biotechnologically relevant microalgae are the green algae (Chlorophyceae), *Chlorella vulgaris*, *Haematococcus pluvialis*, *Dunaliella salina,* and *Spirulina* (*Arthrospira*) with high commercial importance as food supplements for humans and animals (Sousa et al., 2008). Moreover, due to high demand for nutritional based product in recent decades, algal products as functional food, feed, health, and personal care sectors has gained huge market (Wells et al., 2017; Barkia et al., 2019). Till date, in Asia-Pacific region 110 algal commercial producer are present with major manufacturer in China, India, and Taiwan (Sathasivam et al., 2019). Currently, most of the commercialized products from microalgae like *Spirulina* (*Arthrospira*) and *Chlorella* are available in markets as health supplements for food and feed in the form of tablets, capsules ad liquids (Pulz and Gross, 2004; Saito and Aono, 2014). The products are capturing the market of carotenoids, PUFAs, sterols, and carbohydrates by their value and volume, i.e., products having low value-high volume whereas high value products turn out to be in low volume. A profitable market for algal carotenoids has been developed due to abundance of nutritive composition and health applications (Paniagua-Michel, 2015). In an estimated report, the global market for algal products for the period 2017–2022 will reach US$ 381.1 million by 2022 at a CAGR of 6.7%. Hence, the exploitation of different metabolites or new molecules is necessary as they contribute significantly in high market demand (Olaizola, 2003). This chapter aims to highlight and provide an insight to the commercial products and metabolites derived from algae in early and late development stages. It outlines the major species used and their key role in food, nutrition, animal feed, and medicine.

5.2 MICROALGAE AS FUNCTIONAL FOOD

A functional food is defined as food that can produce added benefits over its essential nutritious value. Microalgae (freshwater and marine) are gaining importance for food applications; however, an agri-food value chain is not yet fully established (Khan et al., 2018). Furthermore, the assumed nutritional value is species-specific and solely depends upon the prevailing growth conditions such as different light regimes, temperature gradient and other abiotic factors (Camacho et al., 2019). Microalgae for human nutrition are incorporated in snack foods, cereals, and pasta (Spolaore et al., 2006). Many nutritional food products such as soups, noodles, bread, rolls, cookies, etc., are produced from *C. ellipsoidea* (Sathasivam et al., 2019). On the other hand, several reports suggested that *Chlorella* sp. and *Scenedesmus* sp. can be consumed by human as a nutritional supplement however till date the commercial production is limited (Barrow and Shahidi, 2007). In terms of volume for a single product, the dried algal powder of *Spirulina* (*Arthrospira*) and *Chlorella vulgaris* ranks the highest and only a few companies from Asia and US produces these strain's dried powder (Table 5.1). It is estimated that *Spirulina* (*Arthrospira*) and *Chlorella vulgaris* global production in terms of volume is 5,000 and 2,000 tons of dry matter/year respectively (Table 5.1) and their approximate production value is about 40 million $ per year each (Spolaore et al., 2006; Milledge, 2013). The market for *Spirulina* (*Arthrospira*) was estimated to be US$ 220.5 million and which is to be expected to raise high market valuations of over US$ 380 million by the end of 2027, at a CAGR of 5.9% (Future Market Insights). Earthrise Farms located at Calipatria, CA, USA are the largest plant producer of *Spirulina* (*Arthrospira*). *Chlorella* on the other hand used as a coloring agent in food additive (Gouveia et al., 2007). Besides, *Arthrospira*, *Chlorella*, *Dunaliella salina*, *Phaeodactylum,* and *Haematococcus pluvialis* are being worked by the food industry (Table 5.1). *D. tertiolecta* and *E. gracilis* are the most common species of microalgae used in the food industry in Japan. There are reports suggesting that phycobiliproteins (PBPs) such as phycocyanin (PC), a water-soluble pigment produced from commercialized species such as *Porphyridium cruentum*, *Synechococcus* sp., and *Chlorella* sp. (MacColl, 1998) has an emerging application in food such as ice creams, shakes, chewing gum, soft drinks as a colorant due to their nontoxic behaviors (Priyadarshani and Biswajit, 2012; Borowitzka, 2013).

On the other hand, lutein a yellow-colored pigment is considered most important carotenoids in food product (Jin et al., 2003; Sathasivam et al.,

TABLE 5.1 Major industrial microalgal product applicable to food, feed, nutraceutical and pharmaceuticals

Micro-algae source	Product/Bioactive compounds/Active molecules	Application	Function	Alga based company	Annual production cost	References
Arthrospira platensis	Phycocyanin/ allophycocyanin, β-carotene, zeaxanthin, γ-linolenic acid (GLA) 18:3 ω6, 9, 12, PUFAs (n-3) fatty acids, vitamins (C, K, B12, A, and E), protein, lipids, formulated feed ingredient	Food (Milk, Yoghurt, Cheese, Alcohol-free, Beverage, Desserts, Cookies and Biscuits, Miso, Koji, beverages, chips, pasta, liquid extract)	Natural food colorant and human nutrition	Blue Biotech (Germany) Sanda King (Japan) DIC Lifetec (Japan)	3000 tonnes dry weight	Spolaore et al. 2006; Eriksen 2008; Maoka 2011; Stengel et al. 2011; de Jesus Chu et al. 2012; Raposo et al. 2013; Golmakani et al. 2015; Bhattacharjee 2016
		Nutraceutical (powders, extracts, tablets)	Infant formula for full-term infants, improved nutritional properties and immunity, anti-inflammatory and antioxidant activity	Blue Biotech (Germany), Ocean Nutrition (Canada), Hainan Simai Pharmacy Co. (China), Earthrise Nutritionals (California, USA), E.I.D. Parry (India), Necton (Portugal)	-	
		Feed (Coral trout Meat)	Animal nutrition as feed additives, improved fish color and increased growth rates of fish			
		Pharmaceutical	Act as fluorescent markers, reduce risk of certain heart diseases, immunomodulation activity, anticancer activity and hepatoprotective neurosis			
Scenedesmus sp., *Chlorella vulgaris*, *Nannochloropsis salina*	Amino acids(essential)/ peptides	Feed and food	Human and animal nutrition	–	–	Sonnenschein et al. 2014; Templeton et al. 2015

TABLE 5.1 *(Continued)*

Micro-algae source	Product/Bioactive compounds/Active molecules	Application	Function	Alga based company	Annual production cost	References
Chaetoceros muelleri	Chrysolaminarin	Feed	Immuno-stimulatory food additives in aquaculture	–	–	Gugi et al. 2015
Anabaena sp., *Porphyridium* sp., *Phormidium* sp., *Oscillatoria* sp., *Porphyra yezoensis*	Phycoerythrin	Pharmaceutical	Fluorescent agent; tool for biomedical research	–	–	Bagul et al. 2018
A. flos-aquae, *Alexandrium*, *Dinophysis*, *Amphidinium*, *Karenia*, *Gymnodinium*, *Phormidium* spp.	Anatoxin	Pharmaceutical	To investigate neurodegenerative diseases	–	–	Sathasivam et al. 2019
Dunaliella salina, *B. braunii*, *Chlamydomonas nivalis*, *Dunaliella bardawil*, *Chlorococcum* spp., *Chlorella sorokiniana*, *Dunaliella tertiolecta*, *Paeonia obovata*, *Tetraselmis* spp.	β-carotene	Nutraceutical	Antioxidant, pro-vitamin A, food additive E160a, coloration of egg yolk	–		Borowitzka et al. 1988; Walker et al. 2005; Abe et al. 2007; Coesel et al. 2008; Garbayo et al. 2008; Chu et al. 2012; Sathasivam et al. 2012; Sathasivam et al. 2014; Wu et al. 2016
		Pharmaceutical	Cancer preventive properties, prevent light blindness, prevent liver fibrosis			
Dunaliella spp., *Dunaliella salina*, *Botryococcus braunii*, *Nannochloropsis oculata*, *Nannochloropsis gaditana*, *Chlorella ellipsoidea*	Zeaxanthin	Food	Food additive	–		Cardozo et al. 2007; Hu et al. 2008; Leya et al. 2009; Gierhat and Fox 2013

TABLE 5.1 *(Continued)*

Micro-algae source	Product/Bioactive compounds/Active molecules	Application	Function	Alga based company	Annual production cost	References
Scenedesmus spp., *Chlorella* spp., *Dunaliella* spp., *Botryococcus braunii*, *Chlamydomonas nivalis*, *Chlorococcum* spp., *Chlorella sorokiniana*, *Chlorella fusa*	Lutein	Food	Natural colorant	–		Ben Amotz et al. 1986; Egeland et al. 1995; Rao et al. 2006; Wu et al. 2007; Leya et al. 2009; Gierhat and Fox 2013
		Feed	Animal nutrition			
Chlamydomonas nivalis, *Chlorococcum* spp., *Chamydocapsa* spp., *Chlorella emersonii*, *Chlorella fusa*, *Chlorella vulgaris*, *Chlorella zofingiensis*, *Chlorella striolata*, *Haematococcus lacustris*, *Neospongiococcum* spp., *Nannochloropsis oculata*, *Nannochloropsis salina*, *Nannochloropsis graditana*, *Haematococcus pluvialis*	Canthaxanthin	Nutraceutical and food	Anti-inflammatory, food additive, anti-oxidant property	–		Gouveia et al. 1996; Huertas et al. 2000; Yuan et al. 2002; Mendes et al. 2003; Pelah et al. 2004; Bernstein 2005; Abe et al. 2007; Chattopadhyay et al. 2008; Leya et al. 2009; Plaza et al. 2009; Bhosale and Pasquet et al. 2011
Brown algae, *Chlorella gracilis*, *Chlorella calcitrans*, *O. aurita*, *Cylindrotheca closterium*, *Nitzchia* spp., *P tricornutum*, *Ochromonas danica*, *Ochromonas* spp., *Chromulina ochromonoides*, *A. marina*, *Isochrysis* spp., *Botryococcus braunii*, *Phaedactylum tricornutum*	Fucoxanthin	Pharmaceutical	Anti-obesity, antioxidant property, anti-adiposity	–		Rebolloso-Fuentes et al. 2001; Raja et al. 2007; Kim et al. 2012; Crupi et al. 2013; Xia et al. 2013; Eilers et al. 2016
		Feed	Nutrition in aquaculture	–		

TABLE 5.1 *(Continued)*

Micro-algae source	Product/Bioactive compounds/Active molecules	Application	Function	Alga based company	Annual production cost	References
Pavlova spp., *Nannochloropsis* spp., *Monodus* spp., *Chlorella minutissima*, *Chlorella vulgaris*, *Tetraselmis* spp., *Tetraselmis suecia*, *Isochrysis galbana*, *Phaedodactylum tricornutum*	Eicosapentaenoic acid (EPA)	Pharmaceutical	Heart diseases, blood clotting, blood pressure, lowering plasma level, anti-thrombosis	–		Seto et al. 1984; Belarbi et al. 2000; Jude et al. 2006; Patil et al. 2007; Tzovenis et al. 2009; Chuu et al. 2012;
		Nutraceutical	Nutritional supplements			
Isochrysis galbana, *Crypthecodiuimu* spp., *Schizochytrium* spp., *Pyramimonas* spp.	Docosahexaenoic acid (DHA)	Nutraceutical	Infant formula, nutritional supplements, antioxidant activity	–		Gouveia et al. 1996; Tzovenis et al. 2009; Chu 2012
		Pharmaceutical	Reduce risk of anaemia, improves blood pressure			
		Feed	Aquaculture feed			
Nannochloris atomus, *Pediastrum oryanum*, *Parietochoris incisia*	Arachidonic acid	Nutraceutical	Nutritional supplement, infant formula	–		Reitan et al. 1994; Zhang et al. 2002; De Jesus Raposo et al. 2015
Scenedesmus spp.	Amino acids (Ile, Leu)	Nutraceutical	Dietary supplement, anti-oxidant	–		Vizcaíno et al. 2019
Chlorella vulgaris, *Chlorella minutissima*, *Virginica grossii*	Linolenic acid, glycoprotein, lutein, Cu, chlorophyll, sulfated polysaccharides, sterol, formulated feed ingredient	Nutraceutical (powder, paste, capsules or tablets)	Nutritional properties, anticancer, anti-inflammatory activity, anti-photoaging	Necton (Portugal) Buggypower (Portugal) E.I.D. Parry (India) Phycom (Netherlands) Chlorella Co. (Taiwan) Blue BioTech Int. (Germany) Nikken Sohonsha Taiwan Chlorella Manufacturing and Co. Taiwan Earthrise (US)	2000 tonnes dry weight	Tanaka et al. 1998; Hasegawa et al. 2002; Skjånes et al. 2013; Enzing et al. 2014; Nikolaevna 2015; Maliwat et al. 2017; Beyre et al. 2017
		Feed	Aquaculture, increased specific growth rate, enhanced immune response (total haemocyte count and prophenol oxidase activity)		-	

TABLE 5.1 *(Continued)*

Micro-algae source	Product/Bioactive compounds/Active molecules	Application	Function	Alga based company	Annual production cost	References
		Food	Biomass for health food	Phycom (Netherlands)	-	
		Pharmaceutical	Immunomodulation and cancer prevention, antiviral, reduce total and LDL cholesterol	-	-	
Dunaliella salina, *Dunaliella* sp., *Dunaliella tertiolecta*	β-carotene, lutein, zeaxanthin, chlorophylls a, b, formulated feed ingredient	Food (Miso, Pasta)	Biomass for health food, as food colorant and antioxidant in food	Cognis Nutrition and Health (Hutt Lagoon and Whyalla, Australia) Cyanotech (Kona, Hawaii, USA) Inner Mongolia Biological Eng. (Inner Mongolia, China) Nature Beta Technologies (Eilat, Israel) Tianjin Lantai Biotechnology (Tianjin, China) Parry Nutraceuticals India Proalgen Biotech India Nature Beta Technologies Ltd. Israel Seambiotic Israel Solazyme, Inc. (San Francisco)	1200 tonnes dry weight	Marques et al. 2006; Madhumathi and Rengasamy 2011; Skjånes et al. 2013; Enzing et al. 2014; Shenbaga et al. 2018

TABLE 5.1 *(Continued)*

Micro-algae source	Product/Bioactive compounds/Active molecules	Application	Function	Alga based company	Annual production cost	References
		Feed	Strongly enhanced the immunological and antioxidant factors	–	–	-
		Nutraceutical	Antioxidant activity, improved immunity, improved immune system	–	–	-
Haematocococcus pluvialis	Astaxanthin, canthaxanthin, lutein, chlorophylls a, b and PUFAs (n-3) fatty acid, formulated feed ingredient	Food	Biomass for health food such as desserts, cookies and biscuits	AlgaTech (Israel) Blue Biotech (Germany Fuji Chemicals (Japan) Mera Pharma (USA/ Hawaii) BioReal (Sweden) Cyanotech corp. (USA)	300 tonnes dry weight	Sheikhzadeh et al. 2012; Enzing et al. 2014
		Feed (Almon and trout meat)	Feed supplement (pigment enhancer for fish), no adverse effects in shrimp performance and improved shrimp pigmentation	–		-
		Pharmaceutical	Cancer prevention and reduce risk of certain heart diseases	-		-
		Nutraceutical	Astaxanthin, antioxidant activity	Cyanotech (US, Hawaii) EID Parry (India) Mera Pharma (US/ Hawaii) BioReal (Sweden) US Nutra (US) Parry Nutraceuticals (India)	-	-

TABLE 5.1 *(Continued)*

Micro-algae source	Product/Bioactive compounds/Active molecules	Application	Function	Alga based company	Annual production cost	References
Crypthecodinium cohnii	DHA oil, phycocyanin, vitamin C, Se, Zn, Fe, Mg, docosahexaenoic acid (DHA) 22: 6 ω3, 6, 9, 12, 15, 18	Pharmaceutical	Anti-inflammatory activities, improve cholesterol levels)	Martek/DSM (USA/NL)	240 tonnes DHA oil	Spolaore et al. 2006
		Nutraceutical	Infant formulas for full-term/ preterm infants, nutritional supplements	–		
		Feed	Pharmaceutical			
Shizochytrium sp.	DHA oil, PUFA-ω-3, EPA, alpha linoleic acid, rotifer and Artemia live prey (dry product form)	Pharmaceutical	Improved cardiovascular, brain and eye systems	Flora Health (USA) Xiamen Huison Biotech Co. (China)	10 tonnes DHA oil	Mordenti et al. 2010; Dalle Zotte et al. 2014; Kousoulaki et al. 2015; Sarker et al. 2016a; Smith et al. 2017
		Feed (meat)	No signs of toxicity, negative effect on fillet quality diet and improved protein efficiency	–		
		Nutraceutical	Anti-inflammatory activity, increased nutritional properties	Xiamen Huison Biotech Co. (China)	-	
Porphyridium cruentum	Polysaccharides, EPA, DHA and phycoerythrin and PUFA	Nutraceutical	Health nutrition	Innova IG (France)	-	Plaza et al. 2009; Rodriguez-Sanchez et al. 2012; Borowitzka 2013; Saeid et al. 2016
		Pharmaceutical	Antiviral, antibacterial activity/ antitumor activities	–		

TABLE 5.1 *(Continued)*

Micro-algae source	Product/Bioactive compounds/Active molecules	Application	Function	Alga based company	Annual production cost	References
Isochrysis galbana	DHA/ 1-[hydroxyl-diethyl malonate]-isopropyl dodecenoic acid, protein, PUFA	Pharmaceutical	Prevention of cardiovascular disease	–		Urrutia et al. 2016
		Feed (meat)	Animal nutrition as living feed (aquaculture)			
Phaedodactylum tricornutum	EPA, fucoxanthin, polysaccharides, fatty acids (EPA, DHA), phycoerythrin, sulfated polysaccharides, vitamin E	Food (cookies and biscuits)	Nutrition	A4F-Algae 4 Future (Portugal) AlgaTech (Israel)	–	Ginzberg et al. 2000
		Nutraceutical	Improved nutritional properties, antioxidant activity	–		-
		Pharmaceutical	Antiviral, antibacterial activity, immunomodulation, antitumor activities, Drag-reducing effect			-
Muriellopsis sp.	Lutein/ food and feed	Nutraceutical, food and feed and pharmaceutical	Health food, food supplement, feed, antioxidant, immunomodulation and cancer prevention	–		Del Campo et al. 2007; Fernandez-Sevilla et al. 2010; Ibanez and Cifuentes 2013; Enzing et al. 2014; Saleh et al. 2017

TABLE 5.1 *(Continued)*

Micro-algae source	Product/Bioactive compounds/Active molecules	Application	Function	Alga based company	Annual production cost	References
Nannochloropsis sp., *Nannochloropsis oculata*	DHA (ω-3), Eicosapentaenoic acid (EPA) 20: 5 ω3, 6, 9, 12, 15, green water for finfish larvae, rotifer live prey	Nutraceutical	Nutritional and food supplements	AstraReal Co. (Japan) AlgaTech (Israel) Cyanotech (US, Hawaii) E.I.D. Parry (India) Feed Additive Blue Biotech (Germany) Innovative Aqua (Canada)	–	
		Feed	Innovative aqua product algal paste, algal biomass food for larval and juvenile marine fish, improved growth of fish	–		
Pontederia cordata, *Tetraselmis suecica*	Sterols	Pharmaceutical	Antidiabetic, anticancer, anti-inflammatory, lowers the blood cholesterol	–		Ahmed and Schenk 2017

2019). There are many algae enriched cereal products with enhanced nutrition value (Wells et al., 2017). They are as follows and listed in Table 5.1.

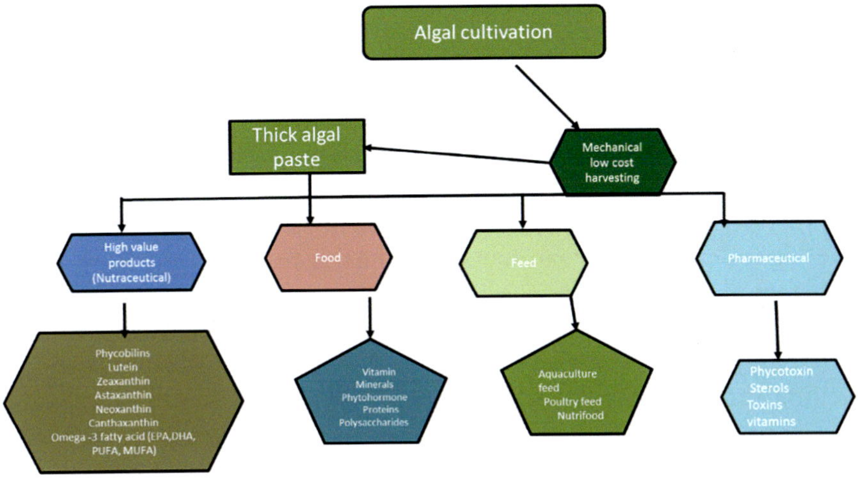

FIGURE 5.1 High-value products from microalgae.

1. **Algae Rich Pasta:** This product contains a good number of amino acids, carbohydrates, vitamins, and iron with an improved variety of fatty acids (FAs) and fucoxanthin (Domínguez, 2013).
2. **Noodles:** These are considered as main food source in Asia and denotes about 40% of total wheat products (Nethravathy et al., 2019).
3. **Bread:** Algae as bread are consumed worldwide and considered as high nutritive product components in breads (Keservani et al., 2010).
4. **Cookies:** The algae-based cookies from *C. vulgaris* and *A. platensis* have shown high antioxidant property with shows significantly high protein content (Batista et al., 2017).

The global demand for microalgal and macroalgal seaweed-derived functional food is growing and buyers are turning towards the algae-derived foods as seaweeds include 10 to 100 times more vitamins (A, D,

E, K, C, B_1, B_2, B_9, B_{12}) and minerals (calcium, iron, iodine, magnesium, phosphorus, potassium, zinc, copper, manganese, selenium, and fluoride) than the conventional crops (Rupérez, 2002; Misurcova, 2012; Qin, 2018). Algal functional food also provides a replacement for cobalamin (vitamin B_{12}) in the food intake than the higher plants as they cannot synthesize cobalamin (Croft et al., 2005). The reports suggest that the microalgae in the nutrition of the global population plays an important role in coming decades being rich in high protein contents, lipids, and many high value compounds such as β-carotene, PC, astaxanthin, EPA (eicosapentaenoic acid), and DHA (docosahexaenoic acid) compared to various plant and floral species. There is significant evidence on the algal-derived food products with health benefits, but there persist substantial challenges in measuring these gains, there is less knowledge on algal nutritional composition throughout algal species. Nevertheless, microalgal functional foods have not flooded the market, but the high selling price of microalgae-based products with active molecules than the conventional food additions may overall improve the economic feasibility (Richmond, 2008). Therefore, a recognized market value for algal-based products is very difficult to prove basically for the food products such as β-carotene, astaxanthin, and PC. Many reports suggested that adding microalgae into pasta and doughnuts increased its nutritional quality in terms of mineral, fiber, and protein (Rabelo, 2013). Corbion (with nestle) has demonstrated the value of algae in high-value food and feed that can deliver highly nourishing products (Zitter, 2019). Thus, it is important to focus on acquiring the products from algae with functional food properties for industrial purpose and human nutrition.

5.3 MICROALGAE AS HUMAN NUTRITION/NUTRACEUTICAL

Nutraceuticals are food products sold as a food supplement and functional foods with health and medical benefits (Khapre and Samantaray, 2019) for their antioxidant properties (Pulz and Gross, 2004). To date, 30,000 microalgal strains have been explored (Richmond, 2008) but not much are exploited for the production of carotenoids and new bioactive compounds (Sharma et al., 2017). There are many edible microalgae such as *Scenedesmus*, *Dunaliella*, *Haematococcus*,

Chlorella, *Chlamydomonas*, and *Muriellopsis* (Table 5.1) which are well established and commercialized as potential nutraceutical (Herrero et al., 2006; Spolaore et al., 2006; Hu et al., 2008; Mata et al., 2010). *Dunaliella* sp. is one of the well-known producers of β-carotene. β-carotene is considered an active carotenoid due to the presence of active form of Pro-Vitamin A which is used as an additive in the health and food industry (Sathasivam et al., 2019). β-carotene market is projected to increase from $438 million (2014) to $532 million (2020) with global production of 1,200 tons dry weight per annum (Gouveia et al., 2007). There are well-established commercial plants for cultivation of *Dunaliella* in countries (Table 5.1) such as China, the USA, and Israel (Garcia-Gonzalez et al., 2005), Australia, Cyanotech Corp., and many others (Benedetti et al., 2004). Besides, β-carotene, *Dunaliella* cells produce phytoene and phytofluene with several health applications such as preventing oxidative damage and protection against UV which leads to anti-aging (Fujii et al., 1993). *Arthrospira platensis* has highest recorded protein content of any whole food, which is used as a food and dietary supplement (Table 5.1) and display antioxidant and anti-inflammatory properties (Zaid et al., 2015). On the other hand, DHA is considered an important FA and provided as supplements in food to nursing and pregnant women (Kroes et al., 2003; Ward and Singh, 2005). *S. platensis* is known to be a rich source of DHA with 9.1% of total FA (Yukino et al., 2005). Moreover, astaxanthin from *Haematococcus* has strong antioxidant property which are 100-fold higher than α-tocopherol (Higuera-Ciapara et al., 2006) and 10 times stronger than β-carotene (Lorenz, 2000). The use of synthetic by-products has been influential in increasing the market mandates for natural form of astaxanthin (Camacho et al., 2019). For consumptions of human, the astaxanthin market price is $2,500/g (dry wt.) (Ambati et al., 2014). Cyanotech Corp., Parry Nutraceutical, BioReal Inc. and others are the major player for commercialized astaxanthin production (Table 5.1). However, till date, astaxanthin from *Haematococcus* has been used as a dietary supplement in the market for human consumption worldwide (Lorenz and Cysewski, 2000). It is best stated that *Chlorella vulgaris*, *Spirulina* (*Arthrospira*) contain high amount of protein content up to 62% (natural source of vitamins A, B_1, B_2, B_{12}, carotenoids, and xanthophyll) compared to other microalgal species with dominating market demand as food supplement (Spolaore et al., 2006). The market statistics for microalgae-based protein was the highest and is further expected to witness tremendous gains due to increased demand in the production of dietary and functional food markets. Moreover, the use of microalgae as a food source is still not developed in Europe due to high production cost, acceptance at the consumer level and food safety (Henchion et al., 2017).

5.3.1 CHEMICAL STRESS AND CULTIVATION CONDITIONS TO INCREASE BY-PRODUCTS PRODUCTION

Microalgal biochemicals have huge potential in human health, aquaculture feed, nutraceuticals, and many more (Nethravathy et al., 2019). To produce various biomolecules from algae in a very cost-effective way, one of the challenges that need to be addressed is the low productivity of specific bioproducts molecules (Liang et al., 2019).

To solve the issues, various attempts are undertaken by applying elicitors to enhance the productivity of the biomolecules in microalgae. There are many algal species when cultivated under environmental stress factors accumulates high value products such as pigments, lipids, and others that can be useful in food, feed, and nutraceuticals sectors (Skjanes et al., 2013). One well-considered example is pigment astaxanthin whose production was enhanced under various stress conditions, such as high light, salinity, and nutrient stress (Tripathi et al., 2002). During the production, certain obstacles are also faced which decrease the value of biomass and production of important products like carotenoids, lipids, and FAs. Microalgae are adaptable to various abiotic stresses such as high light regimes, temperature, and salinity to produce value-added by-products (Paliwal et al., 2017). Besides abiotic stresses, several chemicals additive and phytohormones such as auxin, gibberellin, and others enhanced the algal-based productivity of several products as mentioned in Table 5.2. This chapter mainly focuses on the commercial algal strains recent developments in using chemical additives and cultivation conditions for the production of the algal-based bioproducts.

TABLE 5.2 Chemicals and Different Cultivation Conditions on the Production of High-Value Bio Products from Microalgae

Enhancement Approaches	Product	Microalgal Strain	References
High irradiance (400 μmol/m²/s)	Total carotenoids content	*Haematococcus pluvialis*	Gwak et al. (2014)
Temperature, salinity, nutrient, light stress	Lipids (EPA, DHA, ARA, and GLA), biomass, metabolites	*Schizochytrium* sp., *Nannochloropsis gadianta*, *Botryococcus braunii*, *Phaeodactylum tricornutum*, *Arthrospira* sp., *Pavlova lutheri* and *Isochrysis* sp.	Chisti (2007); Ahmed and Schenk (2017); Dixon and Wilken (2018); Kim et al. (2018); Sun et al. (2018); Kumar et al. (2019); Sun et al. (2019)
Improved culture conditions	Proteins	*Arthospira maxima* and *Chlorella* sp.	Lam et al. (2018)
Light stress and nutrient depletion	Pigments, phycocyanin, phycoerythrin, vitamins (C, K, B_{12}, A, and E)	*Dunaliella* sp., *Haematococcus* sp., *Chlorella* sp., *Arthrospira* sp., *P. cruentum*	Dixon and Wilken (2018); Khan et al. (2018)
Jasmonic acid, salicylic acid, gibberellic acid, Fe, sodium acetate, NaClO, citrate, pyruvate, with acetate, fulvic acid	Astaxanthin	*Haematococcus pluvialis*, *Chlorella zofingiensis*	Wang et al. (2004); Ip and Chen (2005); Chen et al. (2009); Lu et al. (2010); Gao et al. (2012); Zhao et al. (2015)
Fe, citrate, diphenylamine	β-carotene	*Dunaliella salina*, *Haematococcus pluvialis*	Mojaat et al. (2008); Vidhyavathi et al. (2009)
Brassinosteriods, salicylic acid, polyamines, zeatin	Biomass	*Chlorella vulgaris*	Czerpak et al. (2002); Pitrowska and Czerpak (2009)
H_2O_2	Total carotenoids	*Arthrospira platensis*	Abd El-Baky et al. (2009)
Sodium acetate, malic acid	DHA	*Schizochytrium* sp.	Ren et al. (2009)
Acetate, ethanol, sesamol	DHA	*C. cohnii*	Sijtsma et al. (2010); Liu et al. (2015)

TABLE 5.2 *(Continued)*

Enhancement Approaches	Product	Microalgal Strain	References
Nicotine, 2-methylimidazole	Lycopene, lutein	*Dunaliella salina*	Fazeli et al. (2009); Yildirim et al. (2017)
Citrate, malic acid	Increased astaxanthin	*Chlorella zofingiensis*	Chen et al. (2009)

5.4 MICROALGAE AS AN ALTERNATIVE FEED FOR THE AQUACULTURE INDUSTRY

Worldwide demand for animal-derived protein will likely to be doubled by 2050 (Henchion et al., 2017) thus raising concerns regarding (sustained) security and safety. Microalgae denote the biggest market for animal feed. Globally, approximately 40 algal strains are being used in aquaculture (Pulz and Gross, 2004). The most commonly used microalgal genera for animal feed are *Chlorella*, *Tetraselmis*, *Isochrysis*, *Pavlova*, *Phaeodactylum*, *Chaetoceros*, *Nannochloropsis*, *Skeletonema*, and *Thalassiosira* (Sathasivam et al., 2019). Those commonly used are listed in Table 5.1.

The proteinaceous algae are considered a good source for animal feed which improves the growth development of animal feed (Vigani et al., 2015) and can replace the current conventional sources of proteins (Spolaore et al., 2006; Becker, 2007). Ginzberg et al. (2000) reported that feeding the animal with *Porphyridium* algae lowered the cholesterol content in egg yolks by 10% with change in color due to accumulation of carotenoids. Microalgae have also been with time verified either as value-added biocommodity by the feed industry for improving the immunity of the animals (Spolaore et al., 2006). Wijffels (2008) reported that the cost of algal biomass is 100€/kg for human consumption, 5–20€/kg for animal and fish feed. For aquaculture feed, there are many challenges faced while using algae, firstly the toxicity is a major concern and secondly is the high nutrition quality (Renaud et al., 2002). Aquaculture has been expanding rapidly around the world in the last few decades, at an average rate of 8–10% per year; however, there is a constraint in the growth of the aquaculture industry worldwide – the production of fishmeal has not risen. Aquaculture feeds currently use over 80% of the world's fishmeal, which are extracted from small ocean-caught fish. A further problem in aquaculture is that fish food can cost up to 30–60% of the farm's operating costs and this is predicted to keep rising. Fishmeal is a protein-rich food and extensively used in the aquaculture industry to feed a variety of cultured fishes. Identification of suitable alternate protein sources for inclusion in fish feeds becomes imperative to counter the scarcity of fishmeal. Responsible expansion of aqua feed requires finding sustainable alternatives to fishmeal and fish oil ingredients that are economically viable. There has been a considerable shift towards modernizing and intensifying fish farming.

Many reports recommend that it is better to use the consortium of algal strains to provide a better nutrition quality to animal feed (Becker, 2004). Deoiled algae is a protein-rich (40–60%) by-product that could find valuable application as algal meal/animal feed supplement similar to soy meal. In fact, the algal meal is rated to be more nutritious given the superior nutrient composition of algae. The size of animal feed market is smaller and worth about $20 billion. On the other hand fish meal has gained three times more market compared with animal feed in terms of per t basis ($1,200/T vs $300/T for animal feed) (http://www.thefishsite.com/articles/552/fish-and-fishery-products-a-global-market-analysis). The microalgal animal feed market is estimated to be around 300 million US$ (https://making-biodiesel-books.com/algae-bioproducts/algae-products). Globally the algal biomass is sold for animal feed applications (Becker, 2007) by Ingrepro BV – as feed and omega-3 FAs. *Spirulina* (*Arthrospira*) is mostly used as feed supplement globally. The use of microalgae as a substitute for fishmeal will increase the rate of growth of the cultured fish species due to its better digestibility than the land vegetative proteins. Several algal biomasses of *Dunaliella* is consumed as animal and fish feed (Dufossé et al., 2005). β-carotene and lutein from *Dunaliella* and *Muriellopsis* sp. has a great role in the growth of fish larvae (Del Campo, 2000). Moreover, astaxanthin from *Haematococcus* (Table 5.1) is being largely used as pigments in salmon fish feed (Lorenz and Cysewski, 2000). The Astaxanthin aquaculture market per year is projected at $200 million, with an average price of $2,500/kg (Spolaore et al., 2006). Furthermore, astaxanthin from *H. pluvialis* and *Chlorella zofingiensis* (Liu et al., 2014) accounts for major natural forms with strong and growing market demands as aquaculture feed. Omega-3 PUFA, on the other hand, are synthesized by phytoplankton and algae, transferred through the food web, and incorporated into the lipids of fish and marine mammals (Shahidi, 2009). Most of the omega-3

FAs (polyunsaturated long-chain fatty acids, especially EPA and DHA) used in functional foods, supplements, and drugs are derived from fish oil (Table 5.1). The products based on algal source for food/feed and nutrition are owned by many private companies (Figure 5.2) very little information is available at each product level.

5.5 MICROALGAE IN MEDICINES

Marine algae are considered an important source for the production of omega-3 FAs with huge application in the prevention of cancer, asthma, and many others. There are numerous microalgae such as *Chlorella* sp. and *Dunaliella* sp. which can accumulate Vitamin C (ascorbic acid) having antioxidant property. Vitamin C plays an important role as a neurotransmitter and in the prevention of atherosclerosis (Running et al., 2002; Barbosa et al., 2005). There are several algal toxins in algae which are potent neurotoxic compounds such as okadaic acid (Table 5.1) produced by *Dinophysis* sp. and are used for studies on neurodegenerative diseases and in the treatment of schizophrenia (He et al., 2005). The astaxanthin proven to have antioxidant properties which protect the cells against the oxidative damages (Dufossé et al., 2005). The yellow-colored carotenoids lutein plays an important role in our vision to function normally (Skjanes et al., 2013). Astaxanthin from *Haematococcus* has wide application as a food additive and colorant. Due to its antioxidant property, astaxanthin in encapsulated form has been sold for human consumptions (Higuera-Ciapara et al., 2006) and proven to be effective against diabetes, neurogenerative disorders and others (Yuan et al., 2011). In India, Israel, and the United States, production of *Haematococcus* (dry wt.) is around 300 tons in a year (Irianto and Austin, 2002).

The omega-3 FA market was worth \$690 million in 2004, and the Asian Omega-3 PUFA market was expected to be worth \$596.6 million in 2012 (Seambiotic, 2013). The global market for EPA and DHA is estimated at \$300 million and \$1.5 billion, respectively. In the near future, the market for PUFA particularly omega 3) is expected to grow. PUFA [particularly, γ-linolenic acid (GLA, 18:3 ω-6), (EPA, 20:5 ω-3), arachidonic acid (ARA, 20:6 ω-6), docosapentaenoic acid (22:5 ω-3), and (DHA, 22:6 ω-3)] is gaining interest because of their application in health industry (Fraeye

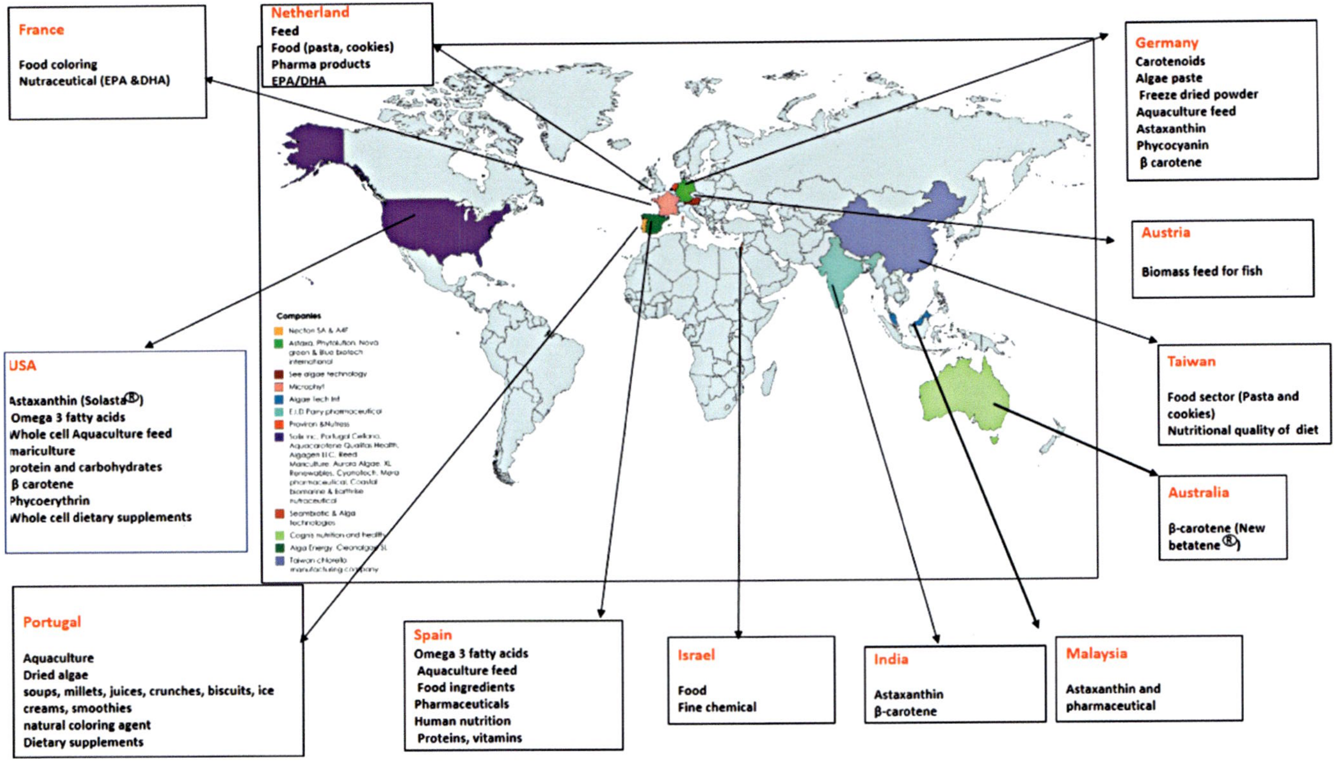

FIGURE 5.2 Global map showing the food, nutraceutical, feed products distribution derived from micro-algae with key player.

et al., 2012). Besides β-carotene, *Dunaliella* cells produce phytoene and phytofluene with several health applications such as preventing oxidative damage and protection against UV, which leads to anti-aging (Ben-Amotz, 1993). Furthermore, obtaining bioactive molecules/color from microorganisms will not improve the market value but also acquire anticancer, antioxidant, anti-inflammatory activities (Venil and Lakshmanaperumalsamy, 2009).

5.6 ROLE OF SYMBIONTS FOR PRODUCTION OF COMMERCIALLY VALUABLE METABOLITES

In nature, algae live in symbiotic relation with algae and secrete the dissolved organic carbon that is readily available to bacteria (Yao et al., 2019). Because of the high demand for microalgae in a circular based economy, better understanding of the risk and benefit involved with algal–microbial interaction is need of hour in the industry sector (Lian et al., 2018). Biofilm and single microorganisms associated with microalgae could play a significant role for each other in algal cultivation systems, e.g., through growth-enhancing interactions or the alleviation or induction of external stresses. A new realization from a recent review by Ines et al. (2017) is that associated organism with microalgae may have distinct and significant function for expressed metabolic/functional behavior by microalgae. There are many surfaces associated microorganism with microalgae known to switch off the gene expression involved in motility and then work as elicitors for microalgae involved in adhesion and biofilm development. Some of the known functions are absorption of heavy metals and denaturing other toxic organic compounds, nutrient regeneration (Decho, 2000), element cycling and biodegradation of xenobiotic compounds (Davey and Toole, 2000). It is well known that heterotrophic bacteria promote plant growth by nutrient exchange mechanism besides decomposing the organic matter (Philippot et al., 2013). There are reports confirming "commensalism" however its duration and stages of function are unknown. This is of significant interest and probably can help enhancing the system efficiency if understood well. There occurs increased excretion by fixed organic carbon due to increased exposure to light, especially when inorganic carbon is non-limiting and serves as a sink to prevent photoinhibition (Cherrier et al., 2014). There are many bacteria associated with algae by degrading complex polysaccharides and offer beneficial traits to algae.

Both algae and associated bacteria cope with the extreme environmental conditions by secreting increased levels of exopolysaccharides (EPS) (Ramanan et al., 2016). A recent study proved that indole acetic acid (IAA) was transferred to algae in exchange for organosulfur compounds by Sulfitobacter, another member of Roseobacter clade. Number of many growth-promoting molecules and algicidal molecules produced by bacteria, such as quinolones, pyrroles, alkaloids, and enzymes result in cell wall disruption, pore formation increasing carbohydrate content and extraction of lipids without using organic solvents (Fuentes et al., 2016). The relationship between algae and bacteria can also be mutualistic. It has been described by Yoch (2002) that algae produce dimethylsulfoniopropionate (DMSP), which is a rich source of sulfur and carbon for the bacteria, and in turn produces growth hormones for microalgae.

It would be interesting to find the function and role of microalgae producing metabolites of importance for application in agriculture and as food additives. The study on unknown but associated organism will be useful to better understand to necessity of natural existing association and functional role for exploitable benefit.

5.7 CONCLUSION

Recently, the significance of production of microalgae-based high valued products has been proven by their important functions in human nutrition, animal health, cosmetics, and pharmaceuticals. The potential of microalgae for the production of proteins, essential amino acids, carotenoids, FAs, and lipids makes them factory of valued biomass. The highly commercial products have been reported to come out from a microalga which contributes to the market trend in billions. Investigation for potential microalgae species are needed to be studied which may have high economic value and produces commercial products which are in demand in the health care sector and animal nutrition field. The production and optimization of algal biomass are the solution for the accomplishment of commercial scale-up of algae. Emerging applications of microbial symbiosis in aquaculture and for biotechnological application is gaining more attention. In summary, as described in this chapter, different metabolites produced from diverse groups of microalgae can be further exploited to be used in developing commercial products having health benefits for humans and animals. The interaction between algae and microbes can range from valuable to harmful for algal cultivation; therefore, regulating some of these interactions may function as a very powerful tool. Considering the above facts, microalgae-based food products still face commercial difficulties in the market. Nevertheless, this situation is expected to be changed due to the increasing body of scientific information made by start-ups and many multinational companies.

KEYWORDS

- **carotenoids**
- **feed**
- **food**
- **lipids**
- **microalgae**
- **nutrition**
- **omega-3 fatty acids**

REFERENCES

Abe, K., Hattori, H., & Hirano, M., (2007). Accumulation and antioxidant activity of secondary carotenoids in the aerial microalga *Coelastrella striolata* var. *multistriata*. *Food Chemistry, 100*(2), 656–661.

Abu, Z. A. A., Hammad, D. M., & Sharaf, E. M., (2015). Antioxidant and anticancer activity of *Spirulina platensis* water extracts. *International Journal of Pharmacology, 11*(7), 846–851.

Ahmed, F., & Schenk, P. M., (2017). UV–C radiation increases sterol production in the microalga *Pavlova lutheri*. *Phytochemistry, 139*, 25–32.

Alavi, N., & Golmakani, M. T., (2017). Improving oxidative stability of virgin olive oil by addition of microalga *Chlorella vulgaris* biomass. *Journal of Food Science and Technology, 54*(8), 2464–2473.

Ambati, R. R., Phang, S. M., Ravi, S., & Aswathanarayana, R. G., (2014). Astaxanthin: Sources, extraction, stability, biological activities and its commercial applications: A review. *Marine Drugs, 12*(1), 128–152.

Bagul, S., Tripathi, S., Chakdar, H., et al., (2018). Exploration and characterization of cyanobacteria from different ecological niches of India for phycobilins production. *International Journal of Current Microbiology and Applied Sciences, 7*(12), 2822–2834.

Barbosa, M. J., Zijffers, J. W., Nisworo, A., Vaes, W., Van, S. J., & Wijffels, R. H., (2005). Optimization of biomass, vitamins, and carotenoid yield on light energy in a flat-panel reactor using the A-stat technique. *Biotechnology and Bioengineering, 89*(2), 233–242.

Barkia, I., Saari, N., & Manning, S. R., (2019). Microalgae for high-value products towards human health and nutrition. *Marine Drugs, 17*(5), 304.

Barrow, C., & Shahidi, F., (2007). *Marine Nutraceuticals and Functional Foods*. CRC Press, Boco Raton, Florida.

Batista, A. P., Niccolai, A., Fradinho, P., et al., (2017). Microalgae biomass as an alternative ingredient in cookies: Sensory, physical and chemical properties, antioxidant activity and *in vitro* digestibility. *Algal Research, 26*, 161–171.

Becker, E. W., (2007). Micro-algae as a source of protein. *Biotechnology Advances, 25*(2), 207–210.

Becker, W., (2004). Microalgae for aquaculture. *Handbook of Microalgal Culture: Biotechnology and Applied Phycology, 380*.

Ben-Amotz, A., (1993). Production of β-carotene and vitamins by the halotolerant alga *Dunaliella*. In: *Pharmaceutical and Bioactive Natural Products* (pp. 411–417). Springer, Boston, MA.

Ben-Amotz, A., Mokady, S., & Avron, M., (1988). The β-carotene-rich alga *Dunaliella bardawil* as a source of retinol in a rat diet. *British Journal of Nutrition, 59*(3), 443–449.

Benedetti, S., Benvenuti, F., Pagliarani, S., Francogli, S., Scoglio, S., & Canestrari, F., (2004). Antioxidant properties of a novel phycocyanin extract from the blue-green alga *Aphanizomenon flos-aquae*. *Life Sciences, 75*(19), 2353–2362.

Bhattacharjee, M. E., (2016). Pharmaceutically valuable bioactive compounds of algae. *Asian J. Pharm. Clin. Res., 9*, 43–47.

Bhosale, P., & Bernstein, P. S., (2005). Microbial xanthophylls. *Applied Microbiology and Biotechnology, 68*(4), 445–455.

Borowitzka, M. A., & Borowitzka, L. J., (1988). *Micro-algal Biotechnology*. Cambridge University Press.

Borowitzka, M. A., (2013). High-value products from microalgae – their development and commercialization. *Journal of Applied Phycology, 25*(3), 743–756.

Camacho, F., Macedo, A., & Malcata, F., (2019). Potential industrial applications and commercialization of microalgae in the functional food and feed industries: A short review. *Marine Drugs, 17*(6), 312.

Cardozo, K. H., Guaratini, T., Barros, M. P., et al., (2007). Metabolites from algae with economical impact. *Comparative Biochemistry and Physiology Part C: Toxicology and Pharmacology, 146*(1,2), 60–78.

Chattopadhyay, P., Chatterjee, S., & Sen, S. K., (2008). Biotechnological potential of natural food-grade biocolorants. *African Journal of Biotechnology, 7*(17).

Chen, W., Zhang, C., Song, L., Sommerfeld, M., & Hu, Q., (2009). A high throughput Nile red method for quantitative measurement of neutral lipids in microalgae. *Journal of Microbiological Methods, 77*(1), 41–47.

Cherrier, J., Valentine, S., Hamill, B., Jeffrey, W. H., & Marra, J. F., (2015). Light-mediated release of dissolved organic carbon by phytoplankton. *Journal of Marine Systems, 147*, 45–51.

Chisti, Y., (2007). Biodiesel from microalgae. *Biotechnology advances, 25*(3), 294–306.

Chu, W. L., (2012). Biotechnological applications of microalgae. *IeJSME, 6*(1), S24–S37.

Coesel, S. N., Baumgartner, A. C., Teles, L. M., et al., (2008). Nutrient limitation is the main regulatory factor for carotenoid accumulation and for Psy and Pds steady-state transcript levels in *Dunaliella salina* (*Chlorophyta*) exposed to high light and salt stress. *Marine Biotechnology, 10*(5), 602–611.

Croft, M. T., Lawrence, A. D., Raux-Deery, E., Warren, M. J., & Smith, A. G., (2005). Algae acquire vitamin B 12 through a symbiotic relationship with bacteria. *Nature, 438*(7064), 90–93.

Crupi, P., Toci, A. T., Mangini, S., et al., (2013). Determination of fucoxanthin isomers in microalgae (*Isochrysis* sp.) by high-performance liquid chromatography coupled with diode-array detector multistage mass spectrometry coupled with positive electrospray ionization. *Rapid Communications in Mass Spectrometry, 27*(9), 1027–1035.

Czerpak, R., Bajguz, A., Gromek, M., Kozłowska, G., & Nowak, I., (2002). Activity of salicylic acid on the growth and biochemism of *Chlorella vulgaris* Beijerinck. *Acta Physiologiae Plantarum, 24*(1), 45–52.

Dalle, Z. A., Cullere, M., Sartori, A., et al., (2014). Dietary *Spirulina* (*Arthrospira platensis*) and Thyme (*Thymus vulgaris*) supplementation to growing rabbits: Effects on raw and cooked meat quality, nutrient true retention and oxidative stability. *Meat Science, 98*(2), 94–103.

Davey, M. E., & O'toole, G. A., (2000). Microbial biofilms: From ecology to molecular genetics. *Microbiology and Molecular Biology Reviews, 64*(4), 847–867.

De Jesus, R. M. F., De Morais, R. M. S. C., & De Morais, A. M. M. B., (2013). Health applications of bioactive compounds from marine microalgae. *Life Sciences, 93*(15), 479–486.

Decho, A. W., (2000). Exopolymer microdomains as a structuring agent for heterogeneity within microbial biofilms. In: *Microbial Sediments* (pp. 9–15). Springer, Berlin, Heidelberg.

Del Campo, J. A., García-González, M., & Guerrero, M. G., (2007). Outdoor cultivation of microalgae for carotenoid production: Current state and perspectives. *Applied Microbiology and Biotechnology, 74*(6), 1163–1174.

Del Campo, J. A., Moreno, J., Rodríguez, H., Vargas, M. A., Rivas, J., & Guerrero, M. G., (2000). Carotenoid content of chlorophycean microalgae: Factors determining lutein accumulation in *Muriellopsis* sp. (*Chlorophyta*). *Journal of Biotechnology, 76*(1), 51–59.

Dixon, C., & Wilken, L. R., (2018). Green microalgae biomolecule separations and recovery. *Bioresources and Bioprocessing, 5*(1), 1–24.

Domínguez, H., (2013). Algae as a source of biologically active ingredients for the formulation of functional foods and nutraceuticals. In: *Functional Ingredients from Algae for Foods and Nutraceuticals* (pp. 1–19). Woodhead Publishing.

Dufossé, L., Galaup, P., Yaron, A., et al., (2005). Microorganisms and microalgae as sources of pigments for food use: A scientific oddity or an industrial reality? *Trends in Food Science and Technology, 16*(9), 389–406.

Dyerberg, J. H., Bang, H. O., Stoffersen, E., Moncada, S., & Vane, J. R., (1978). Eicosapentaenoic acid and prevention of thrombosis and atherosclerosis?. *The Lancet, 312*(8081), 117–119.

Egeland, E. S., Johnsen, G., Eikrem, W., Throndsen, J., & Liaaen-Jensen, S., (1995). Pigments of *Bathycoccus prasinos* (*Prasinophyceae*): Methodological and chemosystematic implications. *Journal of Phycology, 31*(4), 554–561.

Eilers, U., Bikoulis, A., Breitenbach, J., Büchel, C., & Sandmann, G., (2016). Limitations in the biosynthesis of fucoxanthin as targets for genetic engineering in *Phaeodactylum tricornutum*. *Journal of Applied Phycology, 28*(1), 123–129.

Enzing, C., Ploeg, M., Barbosa, M., & Sijtsma, L., (2014). Microalgae-based products for the food and feed sector: An outlook for Europe. *JRC Scientific and Policy Reports*, 19–37.

Ercan, E., Yildirim, P., Hacisa, M., Metin, C., & Bahrioğlu, E., (2018). Algae uses for organic life. *New Knowledge Journal of Science, 7*(2), 141–148.

Eriksen, N. T., (2008). Production of phycocyanin – a pigment with applications in biology, biotechnology, foods and medicine. *Applied Microbiology and Biotechnology, 80*(1), 1–14.

Fazeli, M. R., Tofighi, H., Madadkar-Sobhani, A., et al., (2009). Nicotine inhibition of lycopene cyclase enhances accumulation of carotenoid intermediates by *Dunaliella salina* CCAP 19/18. *European Journal of Phycology, 44*(2), 215–220.

Fraeye, I., Bruneel, C., Lemahieu, C., Buyse, J., Muylaert, K., & Foubert, I., (2012). Dietary enrichment of eggs with omega-3 fatty acids: A review. *Food Research International, 48*(2), 961–969.

Fuentes, J. L., Garbayo, I., Cuaresma, M., Montero, Z., González-del-Valle, M., & Vílchez, C., (2016). Impact of microalgae-bacteria interactions on the production of algal biomass and associated compounds. *Marine Drugs, 14*(5), 100.

Fujii, Y., Sakamoto, S., Ben-Amotz, A., & Nagasawa, H., (1993). Effects of beta-carotene-rich algae *Dunaliella bardawil* on the dynamic changes of normal and neoplastic mammary cells and general metabolism in mice. *Anticancer Research, 13*(2), 389–393.

Furbeyre, H., Van, M. J., Mener, T., Gloaguen, M., & Labussière, E., (2017). Effects of dietary supplementation with freshwater microalgae on growth performance, nutrient digestibility and gut health in weaned piglets. *Animal, 11*(2), 183–192.

Gao, Z., Meng, C., Zhang, X., et al., (2012). Induction of salicylic acid (SA) on transcriptional expression of eight carotenoid genes and astaxanthin accumulation in *Haematococcus pluvialis*. *Enzyme and Microbial Technology, 51*(4), 225–230.

Garbayo, I., Cuaresma, M., Vílchez, C., & Vega, J. M., (2008). Effect of abiotic stress on the production of lutein and β-carotene by *Chlamydomonas acidophila*. *Process Biochemistry, 43*(10), 1158–1161.

García-González, M., Moreno, J., Manzano, J. C., Florencio, F. J., & Guerrero, M. G., (2005). Production of *Dunaliella salina* biomass rich in 9-cis-β-carotene and lutein in a closed tubular photobioreactor. *Journal of Biotechnology, 115*(1), 81–90.

Gierhart, D. L., & Fox, J. A., (2012). *U.S. Patent No. 8088363*. Washington, DC: U.S. Patent and Trademark Office.

Ginzberg, A., Cohen, M., Sod-Moriah, U. A., Shany, S., Rosenshtrauch, A., & Arad, S. M., (2000). Chickens fed with biomass of the red microalga *Porphyridium* sp. have reduced blood cholesterol level and modified fatty acid composition in egg yolk. *Journal of Applied Phycology, 12*(3), 325–330.

Gouveia, L., Batista, A. P., Miranda, A., Empis, J., & Raymundo, A., (2007). *Chlorella vulgaris* biomass used as coloring source in traditional butter cookies. *Innovative Food Science and Emerging Technologies, 8*(3), 433–436.

Gügi, B., Le Costaouec, T., Burel, C., Lerouge, P., Helbert, W., & Bardor, M., (2015). Diatom-specific oligosaccharide and polysaccharide structures help to unravel biosynthetic capabilities in diatoms. *Marine Drugs, 13*(9), 5993–6018.

Gwak, Y., Hwang, Y. S., Wang, B., et al., (2014). Comparative analyses of lipidomes and transcriptomes reveal a concerted action of multiple defensive systems against photooxidative stress in *Haematococcus pluvialis*. *Journal of Experimental Botany, 65*(15), 4317–4334.

Henchion, M., Hayes, M., Mullen, A. M., Fenelon, M., & Tiwari, B., (2017). Future protein supply and demand: Strategies and factors influencing a sustainable equilibrium. *Foods, 6*(7), 53.

Herrero, M., Cifuentes, A., & Ibañez, E., (2006). Sub- and supercritical fluid extraction of functional ingredients from different natural sources: Plants, food-by-products, algae and microalgae: A review. *Food Chemistry, 98*(1), 136–148.

Higuera-Ciapara, I., Felix-Valenzuela, L., & Goycoolea, F. M., (2006). Astaxanthin: A review of its chemistry and applications. *Critical Reviews in Food Science and Nutrition, 46*(2), 185–196.

Hu, Y. R., Guo, C., Xu, L., et al., (2014). A magnetic separator for efficient microalgae harvesting. *Bioresource Technology, 158*, 388–391.

Hu, Z., Li, Y., Sommerfeld, M., Chen, F., & Hu, Q., (2008). Enhanced protection against oxidative stress in an astaxanthin-overproduction *Haematococcus* mutant (*Chlorophyceae*). *European Journal of Phycology, 43*(4), 365–376.

Huertas, E., Montero, O., & Lubián, L. M., (2000). Effects of dissolved inorganic carbon availability on growth, nutrient uptake and chlorophyll fluorescence of two species of marine microalgae. *Aquacultural Engineering, 22*(3), 181–197.

Ibañez, E., & Cifuentes, A., (2013). Benefits of using algae as natural sources of functional ingredients. *Journal of the Science of Food and Agriculture, 93*(4), 703–709.

Ip, P. F., & Chen, F., (2005). Production of astaxanthin by the green microalga *Chlorella zofingiensis* in the dark. *Process Biochemistry, 40*(2), 733–738.

Irianto, A., & Austin, B., (2002). Probiotics in aquaculture. *Journal of Fish Diseases, 25*(11), 633–642.

Jeyanthi, S., Santhanam, P., & Devi, A. S., (2018). Halophilic benthic diatom *Amphora coffeaeformis*: a potent biomarker for lipid and biomedical application. *Indian Journal of Experimental Biology, 56*, 698–701.

Jin, E., Feth, B., & Melis, A., (2003). A mutant of the green alga *Dunaliella salina* constitutively accumulates zeaxanthin under all growth conditions. *Biotechnology and Bioengineering, 81*(1), 115–124.

Keservani, R. K., Kesharwani, R. K., Vyas, N., Jain, S., Raghuvanshi, R., & Sharma, A. K., (2010). Nutraceutical and functional food as future food: A review. *Der Pharmacia Lettre, 2*(1), 106–116.

Khan, M. I., Shin, J. H., & Kim, J. D., (2018). The promising future of microalgae: Current status, challenges, and optimization of a sustainable and renewable industry for biofuels, feed, and other products. *Microbial Cell Factories, 17*(1), 1–21.
Khapre, I., & Samantaray, S. M., (2019). Microalgal nutraceutics: Opportunity for nutritional market. *International Journal of Science and Research, 8*(10), 1201–1206.
Kousoulaki, K., Østbye, T. K. K., Krasnov, A., Torgersen, J. S., Mørkøre, T., & Sweetman, J., (2015). Metabolism, health and fillet nutritional quality in Atlantic salmon (*Salmo salar*) fed diets containing n-3-rich microalgae. *Journal of Nutritional Science, 4*.
Kroes, R., Schaefer, E. J., Squire, R. A., & Williams, G. M., (2003). A review of the safety of DHA45-oil. *Food and Chemical Toxicology, 41*(11), 1433–1446.
Krohn-Molt, I., Alawi, M., Förstner, K. U., et al., (2017). Insights into microalga and bacteria interactions of selected phycosphere biofilms using metagenomic, transcriptomic, and proteomic approaches. *Frontiers in Microbiology, 8*, 1941.
Kumar, B. R., Deviram, G., Mathimani, T., Duc, P. A., & Pugazhendhi, A., (2019). Microalgae as rich source of polyunsaturated fatty acids. *Biocatalysis and Agricultural Biotechnology, 17*, 583–588.
Leah, Z., (2019). Nestlé, corbion eye microalgae for plant-based food. *Aquaculture.*
Leya, T., Rahn, A., Lütz, C., & Remias, D., (2009). Response of arctic snow and permafrost algae to high light and nitrogen stress by changes in pigment composition and applied aspects for biotechnology. *FEMS Microbiology Ecology, 67*(3), 432–443.
Li, X., Xu, H., & Wu, Q., (2007). Large-scale biodiesel production from microalga *Chlorella protothecoides* through heterotrophic cultivation in bioreactors. *Biotechnology and Bioengineering, 98*(4), 764–771.
Lian, J., Wijffels, R. H., Smidt, H., & Sipkema, D., (2018). The effect of the algal microbiome on industrial production of microalgae. *Microbial Biotechnology, 11*(5), 806–818.
Liang, M. H., Wang, L., Wang, Q., Zhu, J., & Jiang, J. G., (2019). High-value bioproducts from microalgae: Strategies and progress. *Critical Reviews in Food Science and Nutrition, 59*(15), 2423–2441.
Liu, J., Sun, Z., Gerken, H., Huang, J., Jiang, Y., & Chen, F., (2014). Genetic engineering of the green alga *Chlorella zofingiensis*: A modified norflurazon-resistant phytoene desaturase gene as a dominant selectable marker. *Applied Microbiology and Biotechnology, 98*(11), 5069–5079.
Liu, J., Sun, Z., Zhong, Y., Huang, J., Hu, Q., & Chen, F., (2012). Stearoyl-acyl carrier protein desaturase gene from the oleaginous microalga *Chlorella zofingiensis*: Cloning, characterization and transcriptional analysis. *Planta, 236*(6), 1665–1676.
Lorenz, R. T., & Cysewski, G. R., (2000). Commercial potential for *Haematococcus* microalgae as a natural source of astaxanthin. *Trends in Biotechnology, 18*(4), 160–167.
MacColl, R., (1998). Cyanobacterial phycobilisomes. *Journal of Structural Biology, 124*(2, 3), 311–334.
Maliwat, G. C., Velasquez, S., Robil, J. L., et al., (2017). Growth and immune response of giant freshwater prawn *Macrobrachium rosenbergii* (De Man) postlarvae fed diets containing *Chlorella vulgaris* (Beijerinck). *Aquaculture Research, 48*(4), 1666–1676.
Maoka, T., (2011). Carotenoids in marine animals. *Marine Drugs, 9*(2), 278–293.

Marques, A., Thanh, T. H., Sorgeloos, P., & Bossier, P., (2006). Use of microalgae and bacteria to enhance protection of gnotobiotic *Artemia* against different pathogens. *Aquaculture, 258*(1–4), 116–126.

Mata, T. M., Martins, A. A., & Caetano, N. S., (2010). Microalgae for biodiesel production and other applications: A review. *Renewable and Sustainable Energy Reviews, 14*(1), 217–232.

Mendes, R. L., Nobre, B. P., Cardoso, M. T., Pereira, A. P., & Palavra, A. F., (2003). Supercritical carbon dioxide extraction of compounds with pharmaceutical importance from microalgae. *Inorganica Chimica Acta, 356*, 328–334.

Milledge, J. J., & Heaven, S., (2013). A review of the harvesting of micro-algae for biofuel production. *Reviews in Environmental Science and Bio/Technology, 12*(2), 165–178.

Mišurcová, L., Ambrožová, J., & Samek, D., (2011). Seaweed lipids as nutraceuticals. *Advances in Food and Nutrition Research, 64*, 339–355.

Mišurcová, L., Škrovánková, S., Samek, D., Ambrožová, J., & Machů, L., (2012). Health benefits of algal polysaccharides in human nutrition. *Advances in Food and Nutrition Research, 66*, 75–145.

Mobin, S., & Alam, F., (2017). Some promising microalgal species for commercial applications: A review. *Energy Procedia, 110*, 510–517.

Mojaat, M., Foucault, A., Pruvost, J., & Legrand, J., (2008). Optimal selection of organic solvents for biocompatible extraction of β-carotene from *Dunaliella salina*. *Journal of Biotechnology, 133*(4), 433–441.

Mordenti, A. L., Sardi, L., Bonaldo, A., et al., (2010). Influence of marine algae (*Schizochytrium* spp.) dietary supplementation on doe performance and progeny meat quality. *Livestock Science, 128*(1–3), 179–184.

Mu, N., Mehar, J. G., Mudliar, S. N., & Shekh, A. Y., (2019). Recent advances in microalgal bioactives for food, feed, and healthcare products: Commercial potential, market space, and sustainability. *Comprehensive Reviews in Food Science and Food Safety, 18*(6), 1882–1897.

Nikolaevna, O. M., (2015). *Cheese Product Production Method*. RU Patent, 2542479(1).

Olaizola, M., (2003). Commercial development of microalgal biotechnology: From the test tube to the marketplace. *Biomolecular Engineering, 20*(4–6), 459–466.

Paliwal, C., Mitra, M., Bhayani, K., et al., (2017). Abiotic stresses as tools for metabolites in microalgae. *Bioresource Technology, 244*, 1216–1226.

Paniagua-Michel, J., (2015). Microalgal nutraceuticals. In: *Handbook of Marine Microalgae* (pp. 255–267). Academic Press.

Pasquet, V., Chérouvrier, J. R., Farhat, F., et al., (2011). Study on the microalgal pigments extraction process: Performance of microwave assisted extraction. *Process Biochemistry, 46*(1), 59–67.

Patil, V., Källqvist, T., Olsen, E., Vogt, G., & Gislerød, H. R., (2007). Fatty acid composition of 12 microalgae for possible use in aquaculture feed. *Aquaculture International, 15*(1), 1–9.

Pelah, D., Sintov, A., & Cohen, E., (2004). The effect of salt stress on the production of canthaxanthin and astaxanthin by *Chlorella zofingiensis* grown under limited light intensity. *World Journal of Microbiology and Biotechnology, 20*(5), 483–486.

Philippot, L., Raaijmakers, J. M., Lemanceau, P., & Van, D. P. W. H., (2013). Going back to the roots: The microbial ecology of the rhizosphere. *Nature Reviews Microbiology, 11*(11), 789–799.
Plaza, M., Herrero, M., Cifuentes, A., & Ibanez, E., (2009). Innovative natural functional ingredients from microalgae. *Journal of Agricultural and Food Chemistry, 57*(16), 7159–7170.
Priyadarshani, I., & Rath, B., (2012). Commercial and industrial applications of microalgae – A review. *Journal of Algal Biomass Utilization, 3*(4), 89–100.
Pulz, O., & Gross, W., (2004). Valuable products from biotechnology of microalgae. *Applied Microbiology and Biotechnology, 65*(6), 635–648.
Qin, Y., (2018). Applications of bioactive seaweed substances in functional food products. In: *Bioactive Seaweeds for Food Applications* (pp. 111–134). Academic Press.
Rabelo, S. F., Lemes, A. C., Takeuchi, K. P., Frata, M. T., Carvalho, J. C. M. D., & Danesi, E. D. G., (2013). Development of cassava doughnuts enriched with *Spirulina platensis* biomass. *Brazilian Journal of Food Technology, 16*(1), 42–51.
Raja, R., Hemaiswarya, S., & Rengasamy, R., (2007). Exploitation of *Dunaliella* for β-carotene production. *Applied Microbiology and Biotechnology, 74*(3), 517–523.
Ramanan, R., Kim, B. H., Cho, D. H., Oh, H. M., & Kim, H. S., (2016). Algae–bacteria interactions: Evolution, ecology and emerging applications. *Biotechnology Advances, 34*(1), 14–29.
Ratledge, C., Anderson, A. J., & Kanagachandran, K., (2010). *U.S. Patent No. 7674609*. Washington, DC: U.S. Patent and Trademark Office.
Rebolloso-Fuentes, M. M., Navarro-Pérez, A., Ramos-Miras, J. J., & Guil-Guerrero, J. L., (2001). Biomass nutrient profiles of the microalga *Phaeodactylum tricornutum*. *Journal of Food Biochemistry, 25*(1), 57–76.
Reitan, K. I., Rainuzzo, J. R., & Olsen, Y., (1994). Effect of nutrient limitation on fatty acid and lipid content of marine microalgae. *Journal of Phycology, 30*(6), 972–979.
Ren, X., Akdag, A., Kocer, H. B., Worley, S. D., Broughton, R. M., & Huang, T. S., (2009). N-Halamine-coated cotton for antimicrobial and detoxification applications. *Carbohydrate Polymers, 78*(2), 220–226.
Renaud, S. M., Thinh, L. V., Lambrinidis, G., & Parry, D. L., (2002). Effect of temperature on growth, chemical composition and fatty acid composition of tropical Australian microalgae grown in batch cultures. *Aquaculture, 211*(1–4), 195–214.
Richmond, A., (2008). *Handbook of Microalgal Culture: Biotechnology and Applied Phycology*. John Wiley and Sons.
Running, J. A., Severson, D. K., & Schneider, K. J., (2002). Extracellular production of L-ascorbic acid by *Chlorella protothecoides*, *Prototheca* species, and mutants of *P. moriformis* during aerobic culturing at low pH. *Journal of Industrial Microbiology and Biotechnology, 29*(2), 93–98.
Rupérez, P., (2002). Mineral content of edible marine seaweeds. *Food Chemistry, 79*(1), 23–26.
Saeid, A., Chojnacka, K., Korczyński, M., Korniewicz, D., & Dobrzański, Z., (2013). Biomass of *Spirulina maxima* enriched by biosorption process as a new feed supplement for swine. *Journal of Applied Phycology, 25*(2), 667–675.

Saito, H., & Aono, H., (2014). Characteristics of lipid and fatty acid of marine gastropod *Turbo cornutus*: High levels of arachidonic and n-3 docosapentaenoic acid. *Food Chemistry, 145*, 135–144.

Sarker, P. K., Kapuscinski, A. R., Lanois, A. J., Livesey, E. D., Bernhard, K. P., & Coley, M. L., (2016). Towards sustainable aquafeeds: Complete substitution of fish oil with marine microalga *Schizochytrium* sp. improves growth and fatty acid deposition in juvenile Nile tilapia (*Oreochromis niloticus*). *PLoS One, 11*(6), e0156684.

Sathasivam, R., Kermanee, P., Roytrakul, S., & Juntawong, N., (2012). Isolation and molecular identification of β-carotene producing strains of *Dunaliella salina* and *Dunaliella bardawil* from salt soil samples by using species-specific primers and internal transcribed spacer (ITS) primers. *African Journal of Biotechnology, 11*(102), 16677–16687.

Sathasivam, R., Praiboon, J., Chirapart, A., et al., (2014). Screening, phenotypic and genotypic identification of β-carotene producing strains of *Dunaliella salina* from Thailand. *Indian Journal of Geo-Marine Sciences, 43*(12), 2198–2216.

Sathasivam, R., Radhakrishnan, R., Hashem, A., & Abd_Allah, E. F., (2019). Microalgae metabolites: A rich source for food and medicine. *Saudi Journal of Biological Sciences, 26*(4), 709–722.

Seto, A., Wang, H. L., & Hesseltine, C. W., (1984). Culture conditions affect eicosapentaenoic acid content of *Chlorella minutissima*. *Journal of the American Oil Chemists' Society, 61*(5), 892–894.

Shahidi, F., (2009). Nutraceuticals and functional foods: Whole versus processed foods. *Trends in Food Science and Technology, 20*(9), 376–387.

Sharma, P., & Sharma, N., (2017). Industrial and biotechnological applications of algae: A review. *Journal of Advances in Plant Biology, 1*(1), 1.

Sheikhzadeh, N., Tayefi-Nasrabadi, H., Oushani, A. K., & Enferadi, M. H. N., (2012). Effects of *Haematococcus pluvialis* supplementation on antioxidant system and metabolism in rainbow trout (*Oncorhynchus mykiss*). *Fish Physiology and Biochemistry, 38*(2), 413–419.

Skjånes, K., Rebours, C., & Lindblad, P., (2013). Potential for green microalgae to produce hydrogen, pharmaceuticals and other high value products in a combined process. *Critical Reviews in Biotechnology, 33*(2), 172–215.

Smith, D. M., & Smith, D. M., (2016). *Feeding Algae to Cattle at Low Doses to Produce High Omega 3 Levels in Beef.* International Patent No. WO2016065024A1.

Sonnenschein, M. F., (2021). *Polyurethanes: Science, Technology, Markets, and Trends*. John Wiley and Sons.

Sousa, I., Gouveia, L., Batista, A. P., Raymundo, A., & Bandarra, N. M., (2008). Microalgae in novel food products. *Food Chemistry Research Developments*, 75–112.

Spolaore, P., Joannis-Cassan, C., Duran, E., & Isambert, A., (2006). Commercial applications of microalgae. *Journal of Bioscience and Bioengineering, 101*(2), 87–96.

Stengel, D. B., Solène, C., & Popper, Z. A., (2011). Algal chemodiversity and bioactivity: Sources of natural variability and implications for commercial application. *Biotechnology Advances, 29*, 483–501.

Sun, X. M., Ren, L. J., Zhao, Q. Y., Ji, X. J., & Huang, H., (2018). Microalgae for the production of lipid and carotenoids: A review with focus on stress regulation and adaptation. *Biotechnology for Biofuels, 11*(1), 1–16.

Sun, X. M., Ren, L. J., Zhao, Q. Y., Ji, X. J., & Huang, H., (2019). Enhancement of lipid accumulation in microalgae by metabolic engineering. *Biochimica et Biophysica Acta (BBA) – Molecular and Cell Biology of Lipids, 1864*(4), 552–566.

Templeton, D. W., & Laurens, L. M., (2015). Nitrogen-to-protein conversion factors revisited for applications of microalgal biomass conversion to food, feed and fuel. *Algal Research, 11*, 359–367.

Tripathi, U., Sarada, R., & Ravishankar, G. A., (2002). Effect of culture conditions on growth of green alga – *Haematococcus pluvialis* and astaxanthin production. *Acta Physiologiae Plantarum, 24*(3), 323–329.

Tzovenis, I., Fountoulaki, E., Dolapsakis, N., et al., (2009). Screening for marine nanoplanktic microalgae from Greek coastal lagoons (Ionian Sea) for use in mariculture. *Journal of Applied Phycology, 21*(4), 457–469.

Urrutia, O., Mendizabal, J. A., Insausti, K., Soret, B., Purroy, A., & Arana, A., (2016). Effects of addition of linseed and marine algae to the diet on adipose tissue development, fatty acid profile, lipogenic gene expression, and meat quality in lambs. *PLoS One, 11*(6), e0156765.

Venil, C. K., & Lakshmanaperumalsamy, P., (2009). An insightful overview on microbial pigment, prodigiosin. *Electronic Journal of Biology, 5*(3), 49–61.

Vidhyavathi, R., Sarada, R., & Ravishankar, G. A., (2009). Expression of carotenogenic genes and carotenoid production in *Haematococcus pluvialis* under the influence of carotenoid and fatty acid synthesis inhibitors. *Enzyme and Microbial Technology, 45*(2), 88–93.

Vigani, M., Parisi, C., Rodríguez-Cerezo, E., et al., (2015). Food and feed products from micro-algae: Market opportunities and challenges for the EU. *Trends in Food Science and Technology, 42*(1), 81–92.

Vizcaíno, A. J., Sáez, M. I., Martínez, T. F., Acién, F. G., & Alarcón, F. J., (2019). Differential hydrolysis of proteins of four microalgae by the digestive enzymes of gilthead sea bream and Senegalese sole. *Algal Research, 37*, 145–153.

Walker, T. L., Purton, S., Becker, D. K., & Collet, C., (2005). Microalgae as bioreactors. *Plant Cell Reports, 24*(11), 629–641.

Ward, O. P., & Singh, A., (2005). Omega-3/6 fatty acids: Alternative sources of production. *Process Biochemistry, 40*(12), 3627–3652.

Wells, M. L., Potin, P., Craigie, J. S., et al., (2017). Algae as nutritional and functional food sources: revisiting our understanding. *Journal of Applied Phycology, 29*(2), 949–982.

Xia, S., Wang, K., Wan, L., Li, A., Hu, Q., & Zhang, C., (2013). Production, characterization, and antioxidant activity of fucoxanthin from the marine diatom *Odontella aurita*. *Marine Drugs, 11*(7), 2667–2681.

Yao, S., Lyu, S., An, Y., Lu, J., Gjermansen, C., & Schramm, A., (2019). Microalgae–bacteria symbiosis in microalgal growth and biofuel production: A review. *Journal of Applied Microbiology, 126*(2), 359–368.

Yoch, D. C., (2002). Dimethylsulfoniopropionate: Its sources, role in the marine food web, and biological degradation to dimethylsulfide. *Applied and Environmental Microbiology, 68*(12), 5804–5815.

Yuan, J. P., Chen, F., Liu, X., & Li, X. Z., (2002). Carotenoid composition in the green microalga *Chlorococcum*. *Food Chemistry, 76*(3), 319–325.

Yuan, J. P., Peng, J., Yin, K., & Wang, J. H., (2011). Potential health-promoting effects of astaxanthin: A high-value carotenoid mostly from microalgae. *Molecular Nutrition and Food Research, 55*(1), 150–165.

Yukino, T., Hayashi, M., Inoue, Y., Imamura, J., Nagano, N., & Murata, H., (2005). Preparation of docosahexaenoic acid fortified *Spirulina platensis* and its lipid and fatty acid compositions. *Nippon Suisan Gakkaishi, 71*(1), 74–79.

Zhang, K., Kurano, N., & Miyachi, S., (2002). Optimized aeration by carbon dioxide gas for microalgal production and mass transfer characterization in a vertical flat-plate photobioreactor. *Bioprocess and Biosystems Engineering, 25*(2), 97–101.

CHAPTER 6

Role of Microalgae in Agriculture

RAYANEE CHAUDHURI and PARAMASIVAN BALASUBRAMANIAN

Agriculture and Environmental Biotechnology Laboratory, Department of Biotechnology and Medical Engineering, National Institute of Technology, Rourkela, Odisha – 769008, India

ABSTRACT

Due to the increase in the worldwide population, the demand for freshly produced vegetables has increased significantly. In order to meet this additional food demand, chemical fertilizers are added to the soil to enhance soil fertility and plant yield, which has several adverse effects in the long run. Moreover, an uncontrolled application of such synthetic fertilizers can lead to the scarcity of limited nutrient resources like phosphorus. Microalgae can be used as a potential alternative for the commercially available synthetic fertilizer, as they can enhance the soil fertility by fixing the nitrogen biologically and thus increase the nutrient availability for the plant through enhanced mobility and solubilization capacity. Furthermore, several bioactive compounds (such as phytohormones or extracellular polysaccharides) produced by microalgal strains have shown the potential to induce plant growth along with protecting them against different biotic and abiotic stresses. This chapter highlights the potential of microalgae as biofertilizers, biostimulants, and biopesticides, in sustainable agricultural practice to increase the overall yield.

Microalgal Biotechnology: Bioprospecting Microalgae for Functional Metabolites towards Commercial and Sustainable Applications. Jeyabalan Sangeetha, PhD, Svetlana Codreanu, PhD, & Devarajan Thangadurai, PhD (Eds.)

6.1 INTRODUCTION

Microalgae denote a diverse group of organisms, which include both eukaryotes and prokaryotes. They can do photosynthesis and produce bio-fuel and several bioactive compounds. The strategic cultivation of microalgae for the treatment of nutrient-rich wastewater discharged from different industries is not new. Microalgae can utilize these nutrients to grow their biomass and make use of their unique structural characteristic of a high surface-to-volume ratio to absorb the nutrients on their surface. This nutrient-rich microalgal biomass may be recovered later and used subsequently as biofertilizer. These biofertilizers are proven to be a good source of nitrogen and phosphorous that may be used to promote plant growth. Compared to the chemical fertilizers, the biomass exhibits greater bio-availability and a slower rate of nutrient release providing an environment-friendly way of supplying nutrients (Coppens et al., 2016). They help in promoting plant growth without increasing the risk of runoff pollution. This makes the microalgal biofertilizer superior to conventional chemical fertilizers, which are well known for their implication in environmental pollution.

Microalgae can also produce bioactive compounds that perform some specialized functions. Bio-stimulants derived from microalgae have the potential to improve the tolerance level of plants against different biotic and abiotic stresses. They may also induce nutrient uptake and stimulate plant growth. Finally, there are microalgae-derived bioactive compounds that have the potential to be used as biopesticides. Microalgae produce them to counteract the reactive oxygen species (ROS), generated in their body under different biotic and abiotic stresses. Bio-compounds are very different from their chemical counterparts both in nature and the way they act. The bioactive compounds derived from microalgae can protect the crops against harmful pests by inducing the plants' defense mechanism. This approach to plant protection is radically different from the conventional approach, which encourages the use of chemical pesticides. Therefore, biopesticides not only save the crops but also take care of the ecosystem by eliminating the possible hazards associated with conventional chemical pesticides. The potential of biopesticides and biofertilizers in promoting sustainable agriculture has been evidenced in recent years (Balasubramanian and Karthickumar, 2017).

This chapter highlights the role of microalgae in sustainable agricultural practice by covering all those areas stated above: the production of bio-fertilizer and bio-stimulators in enhancing nutrient availability, inducing plant's self-defense to withstand biotic and abiotic stress, soil reclamation, and plant growth promotion. It will give a brief overview of the current state of knowledge in this area, indicating the trends and challenges, and will help all those following the progress being made in this field with interest.

6.2 MICROALGAE AS BIOFERTILIZERS AND BIOSTIMULANTS

To fulfill the increasing food demand by the continuously growing population, in the post-industrial revolutionary period, the green revolution took place, which increased the crop yield per unit area. Currently, the land area used for agriculture has slightly decreased because of urbanization (Martellozzo et al., 2015), and with the growing population, demand for plant-derived nutrients has increased. As a combined effect of these two issues, the pressure on arable land is increasing day by day. This calls for the application of modern agriculture practice, with the adoption of new technologies, which will be more effective in respect of yield enhancement but less impactful on the environment. One possible approach in this regard is the enrichment of soil with nutrients which depletes over time when the same land is used again and again for the cultivation of crops. The nutrient restoration occurs either through natural decomposition by the microbes present in the soil (which is quite slow) or through the external addition of nutrients in the form of fertilizer.

The nutrient availability of the plants can be categorized as macronutrients and micronutrients based on the need and availability. Nitrogen, phosphorous, and potassium are called primary macronutrients and whereas calcium, magnesium, and sulfur come under secondary macronutrients. Other components like copper, iron, manganese, zinc come under micronutrients.

Fertilizer is a natural or synthetic compound, which contains at least two of these three primary nutrients – nitrogen (N), phosphorous (P), and potassium (K) at a minimum level of 5% (w/w). Fertilizer provides all these macro- and micro-nutrients to plants and increases the yield.

To achieve the highest possible yield, in the early 19th Century, the production of synthetic fertilizer and pesticides started to boost the yield and protect the crop from pests. Due to readily available nitrogen, potassium, and phosphorous, synthetic fertilizers became popular. Though it is well established that these synthetic fertilizers can increase the yield by restoring the soil fertility so that the growth of the plant is no longer limited by the nutrient deficiency of the soil (Chandini et al., 2019), it has been observed through studies that continuous application of these synthetic fertilizers can reduce the soil quality by soil acidification, hampers the ecosystem and also adversely effects on human health (Geisseler and Scow, 2014). For example, when synthetic fertilizers are applied to the soil, only 50% of the nitrogen fertilizer gets used by the plants and the rest remains in the environment, which interacts with the soil, surface-, and groundwater (Shaviv and Mikkelsen, 1993). This excess nitrogen and phosphorus leaches to the surface water and causes eutrophication (Liu et al., 2010), which is excess growth of aquatic flora in the water bodies. This ultimately disturbs the aquatic ecosystem, leaving an impact on biodiversity. Chemical fertilizers also contribute to the emission of greenhouse gas through the emission of CO_2 and NO_2 from the soil. Moreover, prolonged use of chemical fertilizers can also leave an impact on the soil. Excessive use of chemical fertilizer can cause soil acidification, which disturbs the natural organic matter, humic-like substances. A decrease in soil pH reduces the phosphate intake process by the plants and increases the concentration of the toxic ions in the soil, which adversely affects crop growth (Chandini et al., 2019). For all these reasons nowadays, people are adopting the new agricultural technique, called organic farming, due to which, agrochemical industries and growers are inclining towards the industrial production of microalgal biofertilizers and biostimulants, aiming to the sustainable production of food (Elarroussi et al., 2016).

Biofertilizer refers to natural compounds (can be living organisms too), which can improve soil fertility, by changing its chemical and biological properties, ultimately inducing the plant growth. Whereas biostimulants are organic compounds that are complex in nature, and when present in small quantities, they stimulate plant growth and play an important role in tackling different biotic and abiotic stresses. Microalgae secrets humic substances, N-containing substances and complex organic material, and these can act as biostimulants. Protein hydrolysates (PHs) are a mixture of polypeptides, oligopeptides, and amino acids, produced by partial hydrolysis of protein, which can act as plant biostimulants (Colla et al., 2016).

As microalgae show both biofertilizer and biostimulant properties, they can be considered as an attractive alternative for synthetic fertilizer. Biofertilizer and biostimulant derived from microalgae can be used alone or along with the synthetic fertilizer, in order to get higher yield under normal as well as stressed conditions (i.e., drought or high salinity). These compounds can increase the production rate with a very low environmental impact (Bulgari et al., 2015; Ertani et al., 2015). The different microalgal strains used in agricultural purpose and their working mechanism is discussed in further sections.

6.3 USE OF MICROALGAL BIOFERTILIZERS FOR INCREASING SOIL FERTILITY

Microalgae has shown the potential to be used as an attractive alternative of the synthetic biofertilizers as they can fix several nutrients like, nitrogen, and phosphate, and able to increase the carbon content of the soil which helps in increasing the soil fertility. They can also increase the nutrient solubility by increasing their mobility in the soil. Here in this chapter, all those above-mentioned points were discussed in an elaborated manner.

6.3.1 *FIXATION OF NUTRIENTS*

Among all the different kinds of biofertilizers, two microorganisms: microalgae and blue-green cyanobacteria, which are capable of photosynthesis, has gained much attention due to their contribution in increasing soil fertility, resulting in better crop yield. They increase soil fertility by fixing nutrients like N, P in the soil, increasing their bioavailability, changing soil structure, and reclaiming saline or heavy metal contaminated soil. All these microalgal activities are briefly discussed here.

6.3.1.1 CARBON SEQUESTERING BY MICROALGAE

Being a primary producer in the food chain, cyanobacteria, and other microalgae, contribute to photosynthesis, which is almost 50% of the total photosynthesis process on earth. So, they are directly involved in carbon capturing. During photosynthesis, they fix atmospheric carbon dioxide (CO_2) into their biomass and increase soil's organic carbon (SOC) content. One of the main components of SOC is soil organic matter (SOM). It acts as an indicator of soil health, as it provides nutrients for plants and also increases the water-holding capacity of the soil. SOC increases the porosity, which facilitates aeration and water diffusion into the soil. Moreover, it also influences the cation exchange capacity, maintains nutrient concentration in the rhizosphere, increases buffering capacity, forms aggregate, and sometimes from chelate of metal ions (Degens et al., 2000). This enriches soil structure and enhances soil fertility, increases the yield, and ultimately helps to achieve our sustainable development goal. The extracellular polymeric substance excreted by microalgae consists of different polysaccharides, which promote the growth of the crop. Algal proteoglycans help in cell adhesion to the solid surface, and act as an add-on in the soil aggregation process. It has been observed by Lewis et al. (2017) that the application of residual algal biomass, after lipid extraction can enhance the soil aggregation and carbon storage in soil. However, the authors also concluded from the results that over-application can increase the Na^+ ion concentration in the soil which can affect adversely on plant growth. It has been found cyanobacteria and microalgae can increase the microbe's population and enhance the microbial activity resulting in an increment in the organic carbon content in the soil.

6.3.1.2 NITROGEN FIXATION BY MICROALGAE

After carbon, nitrogen is considered to be the next most important nutrient as its concentration in different biomasses can be as high as 10%. Nitrogen is important for the plant also. Nitrogen deficiency is reflected in a reduced level of chlorophyll and lipid accumulation in biomass. Plants cannot utilize nitrogen from the atmosphere directly as it is present in elemental form. So, the fixation of nitrogen to the soil for its enrichment is needed to make it available to the plants. Cyanobacteria can do this by using an oxygen-sensitive catalyst known as 'nitrogenase.' Some cyanobacteria possess a specialized group of cells known as 'heterocyst.' Cyanobacteria having these specialized cells are called heterocystous cyanobacteria. They are a group of bacteria that have been studied most widely. Cyanobacteria can fix nitrogen through these cells in an anaerobic environment where nitrogen is in short supply. The structure of heterocyst cells is quite different from the normal vegetative cell. These cells exhibit a high rate of respiration, though they restrict the diffusion of oxygen because of the presence of a thick cell envelope. Another distinguishing characteristic of heterocyst cells is that they lack PS(II) activity and hence they do not produce oxygen. It is this feature, which helps these cells create an anaerobic microenvironment necessary for nitrogenase to function properly in an otherwise aerobic microorganism. Heterocysts and vegetative cells work in a symbiotic manner. In the presence of light vegetative cells produce sucrose by photosynthesis. This sucrose gets transferred to heterocysts (in a cyanobacterial cell), which in turn converts it into NADP. The NADP then produces ATP through PS(I) cycle. The ATP captures atmospheric nitrogen (N_2) and transforms it into ammonia through a reaction catalyzed by the enzyme nitrogenase (Masukawa et al., 2012). This ammonia generates ammonium ions making it available for uptake by the plants. Prasanna and Kaushik (1995) experimented with five heterocystous cyanobacterial strains, namely *Tolypothrix ceylonica*, *Scytonematopsis* sp., *Mastigocladus* sp., *Scytonema* sp. and *Fischerella* sp. Among these *Tolypothrix ceylonica* has been found to exhibit the highest activity. *Anabaena variabilis* (a filamentous cyanobacterium) is one of the well-characterized strains of heterocystous cyanobacteria which can function under different environmental conditions due to the presence of different nitrogenase gene clusters. These gene clusters encode three different nitrogenases: one V-nitrogenase encoded by vnf genes, which can work in the molybdenum (Mo) deficient environment, and two different Mo-nitrogenases encoded by Nif1 and Nif2 genes, where the filaments grow under aerobic and anaerobic conditions respectively. Non-heterocystous cyanobacteria are also able to fix atmospheric nitrogen. Cyanobacterium *Oscillatoria* sp. strain is an example of this group of bacteria, which is capable of fixing nitrogen in aerobic conditions.

Overall cyanobacteria contribute very significantly to the global nitrogen cycle. Globally, the annual addition of nitrogen to the soil is 180 million tones and two-third of this is accounted for by microbial activities (Kaushik, 2014). Cyanobacteria have the potential to reduce the requirement of synthetic nitrogenous fertilizer by as much as 25–40%. The application of cyanobacteria-based biofertilizers in rice farming is quite popular throughout Asia. Thus, cyanobacteria improve the soil fertility if inoculated at an appropriate level depending on the characteristics of the agro-economical regions. Such an application of biofertilizer based on cyanobacteria will open up an opportunity to improve soil health while also mitigating GHG emissions and degradation of the aquatic ecosystem. Pereira et al. (2009) conducted a study to assess the impact of cyanobacteria-based biofertilizer on rice production. Results from the study indicate that cyanobacteria can cut down the quantum of chemical fertilizer by around 50% without compromising the quality and quantity of the crop yield. The study reveals that *Nostoc commune* strain has the highest nitrogen-fixing ability among the various strains tested.

6.3.1.3 INCREASE PHOSPHORUS UPTAKE

Phosphorous (P) is another major nutrient which shares 0.2% of the plant's dry weight. Phosphorous remains present in ATP, nucleic acids, phospholipids as a key constituent. In soil, P can occur in both inorganic and organically bound forms. While the organically bound category accounts for 20–80% of the total phosphorous present in the soil (Richardson, 1994), inorganic phosphorous (P_i) makes up the remainder. The latter variety P_i, is bio-available and is therefore taken up by the plants for their growth. Compared to the requirement of the plant, the natural concentration of bio-available P in the soil is quite low. And because of this, plants have specialized transporter for P_i uptake inside the cell, where the concentration of P_i is much higher than outside. So, inorganic phosphorous, present in soil remains dissociated in two different bio-available forms: $H_2PO_4^-$ and HPO_4^{2-}, depending on the pH of the solution (Schachtman et al., 1998). When P_i is added to the soil, it brings about depolarization of the membrane and also acidification of cytoplasm pointing to the occurrence of P_i uptake via co-transportation with a cation H^+ (Ullrich and Novacky, 1990).

The mobility of P_i is poor. Because of its reactive nature, phosphorous ion binds readily with different minerals present in the soil-forming water-insoluble compounds. This makes the P_i unavailable for uptake by the plants. However, acidification of soil and the presence of chelating metal ions in the soil around the rhizosphere area can reverse this process. This makes the P_i ions free, increases their mobility and increases its availability to the plants.

Microalgae secrete extracellular compounds like phosphatases that are needed for phosphorous absorption. When added to the soil, microalgae release acidic metabolites and increase the soil acidity enhancing the bioavailability of phosphorous. Thus, the inoculation of agricultural land with microalgae facilitates the uptake of phosphorous by plants and impacts the fertility of the soil positively. It has also been observed that the uptake rate of phosphorous derived from microalgae is high compared to that of inorganic phosphate. This has been attributed to the greater bioavailability of phosphorous and microalgae's ability to temporarily convert the soil phosphorous into biomass, making possible the sustained release of phosphorous as a nutrient for a prolonged period (Fuller and Rogers, 1952). Microalgae can improve soil fertility by releasing nitrogen also. *Chlorella vulgaris* has been found to be capable of releasing both N and P_i from microalgal biomass (Schreiber et al., 2018). Two more nitrogen-fixing cyanobacterial strains: *Westiellopsis prolifica* and *Anabaena variabilis* have been found to take part in phosphorous solubilization (Yandigeri et al., 2010). These strains transform phosphorous from its insoluble to soluble form and thereby release it into the medium while using merely a fraction of it for their development and metabolic activities.

6.3.2 NUTRIENT SOLUBILIZATION AND INCREASE THEIR MOBILITY

Different organic acids, humic acid-like substances, exo-polysaccharides, derived from microalgae, have great importance in agriculture. Humic acid constitutes around 60% of the total organic matter in the soil. It is considered as a very important component of the ecosystem as it plays a vital role in nutrient solubilization and mobilization in soil. It stays for a very long time in the soil as it cannot be degraded easily by the microorganism due to its complexity. While interacting with the organic compounds as well as different minerals, metal ions, and even toxic pollutants, it forms complexes that can be dissolved in soil or water. This can be helpful in the removal of bulk heavy metals without causing any further soil contamination. Reports show that humic acid can increase nutrient availability in the soil when they are present in a very limited amount (Li et al., 2019). They also can influence nutrient uptake, different metabolic and signaling pathway and also act as a plant growth stimulator. For example, due to the presence of auxin like substance in humic acid, lateral root formation can be observed in *Arabidopsis thaliana* (Trevisan et al., 2010). It has been found that humic acid induces cell division and differentiation in the first stage of growth, which is typically the function of auxin. However, the specific target molecules and the signaling pathways through which this humic acid-like substance and the plant cells interact with each other are needed to be explored further, for its commercial application.

6.3.3 SOIL RECLAMATION

The presence of salts in the upper layer of soil lowers the water permeability and makes them less productive. Soil affected by the salt can show two properties: highly alkaline or presence of soluble salt in high concentration. In the case of alkaline soil, the soil aeration rate is influenced negatively, due to the high pH and presence of a significant amount of sodium and carbonate ions. This increases clay dispersal, clogging the soil pores, which ultimately disturbs the natural microbial symbiotic system in the soil. Whereas the presence of soluble salts in high concentrations increases the osmotic pressure, which ultimately disturbs the nutrient uptake process.

Moisander et al. (2002) have reported that one cyanobacterial strain *Anabaena aphanizomenoides* has a salinity tolerance limit up to 15 g/L whereas another strain *Anabaenopsis* sp. can tolerate up to 20 g/L. Apart from treating saline soil, microalgae have shown the potential for treating heavy metal contaminated soil. Wang and Mulligan (2009), in their studies, have indicated that pre-treatment with humic acid, in heavy metal contaminated soil can be beneficial as humic acid increases the mobility of arsenic and other heavy metals.

6.4 BIOSTIMULANT ACTIVITY OF MICROALGAE

Microalgae produce bioactive compounds, mostly secondary metabolites, which work as plant growth inducers, by assimilating the nutrients. When applied in a small amount, they also induce some structural changes in plants to mitigate the effect of different biotic and abiotic stress. The application of different microalgal extracts, for pre-treatment of seed or as a foliar spray, at different growth stages of the plant, has shown the potential to be an attractive eco-friendly alternative to the chemical fertilizers used for promoting plant growth. However, compared to macroalgae, microalgae, due to their wide variety, are not much explored as a source of bioactive compounds and the research indicates that microalgae could be a more sustainable option in this regard (Behera et al., 2021).

6.4.1 *GROWTH STIMULATING ACTIVITIES BY PHYTOHORMONES*

Phytohormones are important in modern agriculture practice as they promote plant growth by interfering in the plant's metabolism process and improve the defense system against different abiotic stress. They increase the crop yield by not only inducing the growth of the plants but also by controlling weeds and protecting plants from various soil-borne pathogens (Grover et al., 2013; Cho et al., 2015). Phytohormones are small bioactive molecules, which mainly include auxin, cytokinin, gibberellins, abscisic acid (ABA) and ethylene. All these can be found in microalgal strains, influencing the plant growth by working as bio-chemical messengers for regulating cellular activities like improving metabolic influx, nutrient uptake, synthesis of nucleic acid, etc. Phytohormones can be classified into two major classes depending on their action: growth promoters and growth inhibitors. Auxins, cytokinin, and gibberellic acid (GA) come under the first category, whereas ABA and ethylene fall into the second category. Microalgae, including cyanobacteria, usually produce these hormones inside the microalgal cell and in some cases, microalgal strains excrete these hormones in the culture media.

Being growth-promoting phytohormones, auxin, and cytokinin both plays a vital role in the overall growth and development of various parts of the crop. It has been suggested by Finet and Jaillais (2012), that auxin is important for the evolution of plants as they play an important role in elongation of the stem, regulates the branching and induce the formation of lateral organs. It induces the root growth by increasing cell growth and division rate helps in meristem maintenance (Woodward and Bartel, 2005; Perrot-Rechenmann, 2010). Indole-3-acetic acid (IAA) is the most commonly found biologically active auxin, which promotes cell division and growth. One of the main functions of IAA is to activate the H^+-ATPase present in plasmalemma to regulate the growth by elongation. Whereas cytokinin mostly induces shoot growth by increasing cell division rate and plays an important role in the regulation of different natural plant activities like leaf senescence, seed germination, nutrient metabolism, etc. (Haberer and Kieber, 2002). Studies show that the production of these biostimulants can vary depending upon the cultivation condition of microalgae.

For example, IAA is found in the extracellular product of two green microalgal strains *Chlorella pyrenoidosa* and *Scenedesmus armatus* (Mazur et al., 2001). *Scenedesmus armatus* culture content higher concentration of IAA compare to the other one. The study also showed that *S. armatus* culture showed fast growth and increase in the biomass when 2% CO_2/air mixture was supplied constantly but the IAA concentration recorded was lower than the slowly grown culture. Another study conducted by Stirk et al. (2014) showed the effect of light on the endogenous production of phytohormones in a green microalgal strain, *Chlorella minutissima*. The study showed an increase in IAA concentration when the culture is kept in a continuous dark environment with 5 g/L glucose concentration while culture grew under 14:10 h, light: dark condition has given highest and cytokinin concentration. GA, also known as Gibberellin, is a tetracyclic diterpenoid growth stimulator produced in the plant. In the extracellular metabolites secreted by a cyanobacterial strain *Scytonema hofmanni*, GA-like biostimulator was found which increases the tolerance against salinity in rice seedlings (Rodríguez et al., 2006). ABA plays an important role in response to different abiotic stress like drought or high salinity. Some green microalgae and cyanobacteria produce ABA as an extracellular product, under different abiotic stress. For example, microalgal strains like *C. vulgaris*, *Stichococcus bacillaris* produce extracellular ABA, which is induced by different abiotic stress like high saline conditions or drought by almost 5 to 10 times (Maršálek et al., 1992). Stress due to saline water can induce extracellular ABA production, after 6 days of cultivation of cyanobacterial strains like *Nostoc muscorum*, *Synechococcus leopolensis*, and *Trichormus versicolor* (Liu et al., 2016).

Brassinosteroid is a newly recognized compound that acts as a growth-promoting hormone and also induces the stress response. They can act synergistically with auxin, supplement the effect of GA and induce ethylene production. Reports indicate that there exists an opposing effect between the level of gibberellin and the level of ABA and cytokinin (Tsavkelova et al., 2006; Stirk et al., 2009), whereas auxin acts in collaboration with gibberellin by positively regulating its level in biomass. Jasmonic acid, another phytohormone, comes under the oxylipins group, which are natural oxygenated components, formed during the formation of fatty acids (FAs), commonly found in microalgal strains. The presence of jasmonic acid has been detected in several microalgal strains: *Chlorella* sp., *Dunaliella salina*, *Dunaliella tertiolecta*, including cyanobacterial strain *Spirulina* sp. (Tarakhovskaya et al., 2007).

A recent study was conducted with two algal strains: *Arthrospira* sp. and *Scenedesmus* sp. showed, in *Scenedesmus* sp., the concentration of phytohormones including auxin, cytokinins, GA, ABA, and salicylic acid, is higher compared to the other strain (Plaza et al., 2018). Whereas *Arthrospira* sp. increases the dry weight of root, moisture content when applied as a foliar spray on *Petunia* × *hybrida*.

6.4.2 STRUCTURAL CHANGES AND INDUCED DEFENSE MECHANISM BY MICROALGAL POLYSACCHARIDES AND OTHER BIOACTIVE COMPOUNDS

Exopolysaccharides (EPS) are a group of high molecular weight biopolymers, produced by microalgae (mostly red algae and cyanobacteria) during the growth phase. They are either loosely bound with the cell wall or secreted into the environment. They play an important role in interactions between cells, adhesion, and protecting the organism under different abiotic stress by creating a biofilm. EPS released in the culture medium are easy to separate, and due to their properties like antibacterial and antioxidant, they have gained much priorities in the pharmaceutical industries as well.

This microalgal polysaccharide can increase the overall plant growth by inducing different physiological changes like enhancing root and shoot growth. When polysaccharide extracted from a cyanobacterial strain like *Spirulina platensis* is applied to tomato and pepper for 30 days at different stages of their growth, increased root weight and size of the plants have been observed (Elarroussi et al., 2016). In most of the studies, the biostimulation effect of cyanobacteria on rice cultivation showed the induced effect on seed germination and a significant increase in root and shoot (Shariatmadari et al., 2013). Studies also showed that the co-existence of cyanobacteria increases the chlorophyll content in leaves (which increase the photosynthesis rate) and the root dry weight, in wheat by secreting extracellular substances (Gantar et al., 1995a, b). It was observed by Supraja et al. (2020a) that tomato seeds pre-treated with 40% microalgal extract (which consists of different proteins and carbohydrates, around 26% and 41%, respectively) induces plant growth by acting as precursors for the different bioactive compounds. It has also been observed that the application of 60% microalgal extract as a foliar spray can induce the plant's biomass as well as chlorophyll content significantly. Apart from this, it was observed that microalgae secrete several extracellular bioactive compounds, which enhances plant growth when added to a hydroponic cultivation setup (Supraja et al., 2020b).

However, most of these studies conducted are on a lab-scale or pilot scale. Field trials need to be carried out to check the effectiveness of the process. Moreover, for commercial application, the dose and concentration of such bioactive compounds depend on the plant species, microalgal strain and its cultivation condition (Behera et al., 2021). More detailed studies are required to shed light on this aspect.

6.5 ROLE OF MICROALGAE AS BIOPESTICIDE

Previously the capability of different microalgal strains as a biofertilizer and biostimulant have been discussed. But in organic sustainable agricultural practice, to obtain greater yield, the crops need to be prevented from different pests. Synthetic pesticides that are available in the market have several negative impacts on consumers' (both human and animal) health and can cause environmental pollution. So, in order to maintain a sustainable agricultural practice, it is important to find an alternative to these synthetic pesticides. Microalgae is found to be an excellent choice in this scenario as it can be grown in wastewater, without any arable land, and can produce quite high biomass per unit area.

Biopesticides are natural compounds or bio-reactive molecules produced by an organism and can be used as an alternative to synthetic pesticides, used for protecting plants. They protect the plants from pests by secreting enzymes like endotoxins, for cell-wall degradation and sometimes by inducing viral infection (Thakore, 2006). Interestingly it is predicted that the resistance against this biopesticide will develop in a very slower manner due to their complexity in nature. Biopesticides can be segregated into three classes depending on their origin: biochemical-, botanical-, and microbial biopesticides.

Biochemical pesticides protect the plants by nontoxic process. Plant growth regulators (PGRs) and other pheromones come under this category. For example, auxin like PGR is found to be the most useful herbicide as it shows selective detoxification speed, which works especially against dicot weeds, in different higher cereal producing plants (Grossmann, 2010). PGRs acts as a stimulator when present at a small concentration. They stimulate cell division, growth, and elongation in plants. But at higher concentrations, the same PGR acts as a biopesticide by causing abnormalities in weeds, which leads to membrane degradation, vascular system failure and ultimately to the death of the weed (Grossmann, 1998).

Botanical pesticides are plant extracted compounds, mostly secondary metabolites, which are used to prevent or destroy different pests. Bioactive compounds like phenolics, terpenoids, some essential oils come under this category. The use of these oils as insecticides for food storage has been a long practice. Studies have shown that these essential oils attack the nervous system of the insect but do not have any effect on mammals and fish, and thus are called pesticides with reduced health risk (Koul et al., 2008).

Bioactive components derived from bacteria, fungi, and microalgae having the ability to protect the plant from pests are called microbial biopesticides. Among all the microalgae, it has been found that cyanobacteria are very effective against different fungal strains and soil-borne pathogens, as they can produce different secondary metabolites (generally low molecular weight in nature), which shows anti-fungal, antibacterial characteristics. These compounds induce different modifications (both, functional, and structural), which ultimately leads to membrane disruption, inactivation of different enzymes and hampers the protein synthesis process (Swain et al., 2017; Renuka et al., 2018). Different bioactive compounds like phenolic compounds, tocopherols, polysaccharides, allelochemicals, polyunsaturated fatty acids (PUFA), are responsible for these cyanobacterial activities. The first isolated bioactive compound, which showed bactericidal properties, was chlorellin, isolated from a very common green microalgal strain *Chlorella vulgaris*. Though it is not accepted for large-scale commercial production, it showed an inhibitory effect against both gram-positive (*Staphylococcus aureus*, *Streptococcus pyogenes*, and *Bacillus subtilis*) and gram-negative bacteria (*Pseudomonas aeruginosa*) (Shannon and Abu-Ghannam, 2016). Studies show that cyanobacteria can enhance plant defense systems by stimulating the production of different antioxidant compounds like chitinase, peroxidase, catalase, endoglucanase, etc. (Renuka et al., 2018). Cyanobacteria present in the rhizosphere area, colonize, and by promoting the release of different hydrolytic enzymes and bioactive metabolites having antimicrobial capacity, induce the plant defense activity (Babu et al., 2014). When a cyanobacterial strain, *Oscillatoria chlorina*, was tested against *Meloidogyne arenaria*, a nematode, causing root-knot, in tomato plants, exponentially beneficial effect was observed with increasing *O. chlorina* concentration up to 0.8% (w/w) (Khan et al., 2007). It was also observed from the same experiment that, at the highest dose of 1% (w/w) of cyanobacterial powder application, the *Meloidogyne arenaria* population decreased by 97.6%. Biondi et al. (2004) showed that the extraction of *Nostoc* strain ATCC 53789, by methanol can act against a wide variety of fungi, nematodes, insects, and herbs. Victor and Reuben (2000) showed when chemical fertilizer, urea is applied to the field it causes an increase in the mosquito population in a rice field whereas cyanobacterial biofertilizer can fix nitrogen without increasing the mosquito population.

6.6 CONCLUSION AND FUTURE ASPECTS

The use of microalgae in the food and agricultural sector has a very high potential. Microalgal biofertilizers can act as an attractive alternative to synthetic fertilizers for sustainable development. When applied to the soil, they improve the soil structure, increase the nutrient uptake, help in reclaiming saline or heavy metal contaminated soil, induce plant growth and improve plant's defense mechanisms against different biotic and abiotic stresses. Some microalgae-derived compounds exhibit antioxidant properties that can mitigate oxidative stress in plants, protecting them against several diseases. All these properties make them very suitable for use in agriculture. Moreover, the cultivation of microalgae is less resource-intensive as it can be grown without soil and freshwater.

But despite having so many advantages, there are still a few shortcomings that restrict the large-scale production and commercialization of microalgal biofertilizers. Compared to chemical fertilizer, microalgae-based biofertilizers examined so far are deficient in nutrients, due to which, for the same amount of nutrients, a significantly huge amount of microalgal biomass is needed, compared to synthetic fertilizer. Thus, depending solely on the microalgal biofertilizer increases the overall production cost. Therefore, it is necessary to explore and identify more microalgal strains capable of producing low-volume, nutrient-rich bio-fertilizer to meet the demand. Strain enrichment through genetic engineering could be of great help in this endeavor. Moreover, different microalgae-derived complex bio-active compounds, which can work as biostimulator or biopesticides, are not widely explored. Therefore, extensive research is required in this aspect to identify those compounds, along with their reaction mechanisms, so that they can be used more efficiently. Additionally, the impact of the microalgae culture techniques and extraction methods of such metabolites, on the yield requires a thorough assessment to improve the economic viability.

KEYWORDS

- **bioactive compounds**
- **biofertilizers**
- **biopesticides**
- **biostimulants**
- **cyanobacteria**
- **extracellular polysaccharides**
- **microalgae**
- **nutrient availability**
- **phytohormones**
- **soil fertility**

REFERENCES

Babu, S., Prasanna, R., Bidyarani, N., & Singh, R., (2015). Analyzing the colonization of inoculated cyanobacteria in wheat plants using biochemical and molecular tools. *J. Appl. Phycol., 27*(1), 327–338. doi: 10.1007/s10811-014-0322-6.

Balasubramanian, P., & Karthickumar, P., (2017). Biofertilizers and biopesticides: A holistic approach for sustainable agriculture. In: *Sustainable Utilization of Natural Resources* (pp. 255–284). CRC Press, Florida, USA.

Behera, B., Supraja, K. V., & Paramasivan, B., (2021). Integrated microalgal biorefinery for the production and application of biostimulants in circular bioeconomy. *Bioresource Technology, 339*, 125588. doi: 10.1016/j.biortech.2021.125588.

Biondi, N., Piccardi, R., Margheri, M. C., Rodolfi, L., Smith, G. D., & Tredici, M. R., (2004). Evaluation of *Nostoc* strain ATCC 53789 as a potential source of natural pesticides. *Appl. Environ. Microbiol., 70*(6), 3313–3320. doi: 10.1128/AEM.70.6.3313-3320.2004.

Bulgari, R., Cocetta, G., Trivellini, A., Vernieri, P., & Ferrante, A., (2015). Biostimulants and crop responses: A review. *Biological Agriculture and Horticulture, 31*(1), 1–17. doi: 10.1080/01448765.2014.964649.

Chandini, K. R., Kumar, R., & Prakash, O., (2019). The impact of chemical fertilizers on our environment and ecosystem. In: Poonam, S., (ed.), *Research Trends in Environmental Sciences* (pp. 69–86). Delhi, India: AkiNik Publications.

Cho, S. T., Chang, H. H., Egamberdieva, D., Kamilova, F., Lugtenberg, B., & Kuo, C. H., (2015). Genome analysis of *Pseudomonas fluorescens* pcl1751: A rhizobacterium that controls root diseases and alleviates salt stress for its plant host. *PLoS One, 10*(10). doi: 10.1371/journal.pone.0140231.

Colla, G., Rouphael, Y., Lucini, L., et al., (2016). Protein hydrolysate-based biostimulants: Origin, biological activity and application methods. *Acta Horticulturae, 1148*, 27–34. doi: 10.17660/actahortic.2016.1148.3.

Coppens, J., Grunert, O., Van Den, H. S., et al., (2016). The use of microalgae as a high-value organic slow-release fertilizer results in tomatoes with increased carotenoid and sugar levels. *Journal of Applied Phycology, 28*(4), 2367–2377. doi: 10.1007/s10811-015-0775-2.

Degens, B. P., Schipper, L. A., Sparling, G. P., & Vojvodic-Vukovic, M., (2000). Decreases in organic C reserves in soils can reduce the catabolic diversity of soil microbial communities. *Soil Biol. Biochem., 32*(2), 189–196. doi: 10.1016/S0038-0717(99)00141-8.

Elarroussi, H., Elmernissi, N., Benhima, R., et al., (2016). Microalgae polysaccharides a promising plant growth biostimulant. *Journal of Algal Biomass Utilization, 7*(4), 55–63.

Ertani, A., Sambo, P., Nicoletto, C., Santagata, S., Schiavon, M., & Nardi, S., (2015). The use of organic biostimulants in hot pepper plants to help low input sustainable agriculture. *Chem. Biol. Technol. Agric., 2*(11). doi: 10.1186/s40538-015-0039-z.

Finet, C., & Jaillais, Y., (2012). Auxology: When auxin meets plant evo-devo. *Developmental Biology, 369*(1), 19–31. doi: 10.1016/j.ydbio.2012.05.039.

Fuller, W. H., & Roger, R. N., (1952). Utilisation of the phosphorus of algal cells as measured by the Neubauer technique. *Soil Sci., 74*, 417–429. doi: 10.1097/00010694-195212000-00002.

Gantar, M., Kerby, N. W., Rowell, P., Obreht, Z., & Scrimgeour, R., (1995a). Colonization of wheat (*Triticum vulgare* L.) by N_2-fixing cyanobacteria. IV. Dark nitrogenase activity and effects of cyanobacteria on natural ^{15}N abundance on plants. *New Phytol., 129*, 337–343. doi: 10.1111/j.1469-8137.1995.tb04304.x.

Gantar, M., Rowell, P., Kerby, N. W., & Sutherland, I. W., (1995b). Role of extracellular polysaccharide in the colonization of wheat (*Triticum vulgare* L.) roots by N_2-fixing cyanobacteria. *Biology and Fertility of Soils, 19*(1), 41–48. doi: 10.1007/bf00336345.

Geisseler, D., & Scow, K. M., (2014). Long-term effects of mineral fertilizers on soil microorganisms: A review. *Soil Biol. Biochem., 75*, 54–63.

Grossmann, K., (1998). Quinclorac belongs to a new class of highly selective auxin herbicides. *Weed Sci., 46*, 707–716. doi: 10.1017/S004317450008975X.

Grossmann, K., (2009). Auxin herbicides: Current status of mechanism and mode of action. *Pest Management Science, 66*, 113–120. doi: 10.1002/ps.1860.

Grover, A., Mittal, D., Negi, M., & Lavania, D., (2013). Generating high temperature tolerant transgenic plants: Achievements and challenges. *Plant Sci., 20*, 38–47. doi: 10.1016/j.plantsci.2013.01.005.

Haberer, G., & Kieber, J. J., (2002). Cytokinins. New insights into a classic phytohormone. *Plant Physiology, 128*(2), 354–362. doi: 10.1104/pp.010773.

Kaushik, B. D., (2014). Developments in cyanobacterial biofertilizer. *Proceedings of the Indian National Science Academy, 80*(2), 379–388. doi: 10.16943/ptinsa/2014/v80i2/55115.

Khan, Z., Kim, Y., Kim, S., & Kim, H., (2007). Observations on the suppression of root-knot nematode (*Meloidogyne arenaria*) on tomato by incorporation of cyanobacterial powder (*Oscillatoria chlorina*) into potting field soil. *Bioresource Technology, 98*(1), 69–73. doi: 10.1016/j.biortech.2005.11.029.

Koul, O., Walia, S., & Dhaliwal, G. S., (2008). Essential oils as green pesticides: Potential and constraints. *Biopestic. Int., 4*(1), 63–84.

Lewis, K., Foster, J., Hons, F., & Boutton, T., (2017). Initial aggregate formation and soil carbon storage from lipid-extracted algae amendment. *AIMS Environmental Science, 4*(6), 743–762. doi: 10.3934/environsci.2017.6.743.

Liu, L., Pohnert, G., & Wei, D., (2016). Extracellular metabolites from industrial microalgae and their biotechnological potential. *Mar. Drugs, 14*(10), 191. doi: 10.3390/md1410019.

Liu, W., Zhang, Q., & Liu, G., (2010). Lake eutrophication associated with geographic location, lake morphology and climate in China. *Hydrobiologia, 644*, 289–299. doi: 10.1007/s10750-010-0151-9.

Maršálek, B., Zahradníčková, H., & Hronková, M., (1992). Extracellular production of abscisic acid by soil algae under salt, acid or drought stress. *Z. Naturforschung C., 47*, 701–704.

Martellozzo, F., Ramankutty, N., Hall, R. J., Price, D. T., Purdy, B., & Friedl, M. A., (2014). Urbanization and the loss of prime farmland: A case study in the Calgary-Edmonton corridor of Alberta. *Regional Environmental Change, 15*(5), 881–893. doi: 10.1007/s10113-014-0658-0.

Masukawa, H., Kitashima, M., Inoue, K., Sakurai, H., & Hausinger, R. P., (2012). Genetic engineering of cyanobacteria to enhance biohydrogen production from sunlight and water. *AMBIO, 41*(S2), 169–173. doi: 10.1007/s13280-012-0275-4.

Mazur, H., Konop, A., & Synak, R., (2001). Indole-3-acetic acid in the culture medium of two axenic green microalgae. *Journal of Applied Phycology, 13*, 35–42. doi: 10.1023/A:1008199409953.

Moisander, P. H., McClinton, E., & Paerl, H. W., (2002). Salinity effects on growth, photosynthetic parameters and nitrogenase activity in estuarine planktonic cyanobacteria. *Microbial. Ecol., 43*(4), 432–442. doi: 10.1007/s00248-001-1044-2.

Pandey, K. D., Shukla, P. N., Giri, D. D., & Kashyap, A. K., (2005). Cyanobacteria in alkaline soil and the effect of cyanobacteria inoculation with pyrite amendments on their reclamation. *Biol. Fertil. Soils, 41*, 451–457. doi: 10.1007/s00374-005-0846-7.

Pereira, I., Ortega, R., Barrientos, L., Moya, M., Reyes, G., & Kramm, V., (2009). Development of a biofertilizer based on filamentous nitrogen-fixing cyanobacteria for rice crops in Chile. *Journal of Applied Phycology, 21*, 135–144. doi: 10.1007/s10811-008-9342-4.

Perrot-Rechenmann, C., (2010). Cellular responses to auxin: Division versus expansion. *Cold Spring Harbor Perspectives in Biology, 2*(5). doi: 10.1101/cshperspect.a001446.

Plaza, B. M., Gómez-Serrano, C., Acién-Fernández, F. G., & Jimenez-Becker, S., (2018). Effect of microalgae hydrolysate foliar application (*Arthrospira platensis* and *Scenedesmus* sp.) on *Petunia* × *hybrida* growth. *Journal of Applied Phycology, 30*(4), 2359–2365. doi: 10.1007/s10811-018-1427-0.

Prasanna, R., & Kaushik, B. D., (1995). Nitrogen fixation and *nif* gene organization in branched heterocystous cyanobacteria: Variation in the presence of *xisA*. *Folia Microbiologica, 40*(2), 176–180. doi: 10.1007/bf02815418.

Renuka, N., Guldhe, A., Prasanna, R., Singh, P., & Bux, F., (2018). Microalgae as multi-functional options in modern agriculture: Current trends, prospects and challenges. *Biotechnol. Adv., 36*, 1255–1273. doi: 10.1016/j.biotechadv.2018.04.004.

Richardson, A. E., (1994). Soil microorganisms and phosphorus availability. In: Pankhurst, C. E., Doube, B. M., Gupta, V. V. S. R., & Grace, P. R., (eds.), *Soil Biota: Management in Sustainable Farming Systems* (pp. 50–62). Melbourne: CSIRO, Australia.

Rodríguez, A., Stella, A., Storni, M., Zulpa, G., & Zaccaro, M., (2006). Effects of cyanobacterial extracellular products and gibberellic acid on salinity tolerance in *Oryza sativa* L. *Saline Systems, 2*(1), 7. doi: 10.1186/1746-1448-2-7.

Schachtman, D. P., Reid, R. J., & Ayling, S. M., (1998). Phosphorus uptake by plants: From soil to cell. *Plant Physiology, 116*(2), 447–453. doi: 10.1104/pp.116.2.447.

Schreiber, C., Schiedung, H., Harrison, L., et al., (2018). Evaluating potential of green alga *Chlorella vulgaris* to accumulate phosphorus and to fertilize nutrient-poor soil substrates for crop plants. *Journal of Applied Phycology, 30*, 2827–2836. doi: 10.1007/s10811-018-1390-9.

Shannon, E., & Abu-Ghannam, N., (2016). Antibacterial derivatives of marine algae: An overview of pharmacological mechanisms and applications. *Marine Drug, 14*(4), 81. doi: 10.3390/md14040081.

Shariatmadari, Z., Riahi, H., Seyed, H. M., Ghassempour, A. R., & Aghashariatmadary, Z., (2013). Plant growth-promoting cyanobacteria and their distribution in terrestrial habitats of Iran. *Soil Science and Plant Nutrition, 59*(4), 535–547. doi: 10.1080/00380768.2013.782253.

Shaviv, A., & Mikkelsen, R. L., (1993). Controlled-release fertilizers to increase efficiency of nutrient use and minimize environmental degradation: A review. *Fertilizer Research, 35*, 1–12. doi: 10.1007/bf00750215.

Singh, R. N., (1961). *Role of Blue-Green Algae in Nitrogen Economy of Indian Agriculture*. New Delhi: Indian Council of Agricultural Research.

Stirk, W. A., Balint, P., Tarkowska, D., et al., (2014). Effect of light on growth and endogenous hormones in *Chlorella minutissima* (Trebouxiophyceae). *Plant Physiol. Biochem., 79*, 66–76. doi: 10.1016/j.plaphy.2014.03.005.

Stirk, W. A., Novák, O., Hradecká, V., et al., (2009). Endogenous cytokinins, auxins and abscisic acid in *Ulva fasciata* (*Chlorophyta*) and *Dictyota humifusa* (Phaeophyta):

Towards understanding their biosynthesis and homeostasis. *European Journal of Phycology, 44*(2), 231–240. doi: 10.1080/09670260802573717.

Supraja, K. V., Behera, B., & Balasubramanian, P., (2020a). Efficacy of microalgal extracts as biostimulants through seed treatment and foliar spray for tomato cultivation. *Industrial Crops and Products, 151*, 112453. doi: 10.1016/j.indcrop.2020.112453.

Supraja, K. V., Behera, B., & Balasubramanian, P., (2020b). Performance evaluation of hydroponic system for co-cultivation of microalgae and tomato plant. *Journal of Cleaner Production, 272*, 122823. doi: 10.1016/j.jclepro.2020.122823.

Swain, S. S., Paidesetty, S. K., & Padhy, R. N., (2017). Antibacterial, antifungal and antimycobacterial compounds from cyanobacteria. *Biomedicine and Pharmacotherapy, 90*, 760–776. doi: 10.1016/j.biopha.2017.04.030.

Tarakhovskaya, E. R., Maslov, Y. I., & Shishova, M. F., (2007). Phytohormones in algae. *Russian Journal of Plant Physiology, 54*(2), 163–170. doi: 10.1134/S1021443707020021.

Thakore, Y., (2006). The biopesticide market for global agricultural use. *Ind. Biotechnol., 2*(3), 194–208. doi: 10.1089/ind.2006.2.194.

Trevisan, S., Pizzeghello, D., Ruperti, B., et al., (2009). Humic substances induce lateral root formation and expression of the early auxin-responsive IAA19 gene and DR5 synthetic element in *Arabidopsis*. *Plant Biol., 12*, 604–614. doi: 10.1111/j.1438-8677.2009.00248.x.

Tsavkelova, E. A., Klimova, S. Y., Cherdyntseva, T. A., & Netrusov, A. I., (2006). Hormones and hormone-like substances of microorganisms: A review. *Applied Biochemistry and Microbiology, 42*(3), 229–235. doi: 10.1134/s000368380603001x.

Ullrich, C. I., & Novacky, A. J., (1990). Extra- and intracellular pH and membrane potential changes induced by K, Cl, H_2PO_4, and NO_3 uptake and fusicoccin in root hairs of *Limnobium stoloniferum*. *Plant Physiol., 94*(4), 1561–1567. doi: 10.1104/pp.94.4.1561.

Victor, T. J., & Reuben, R., (2000). Effects of organic and inorganic fertilizers on mosquito populations in rice fields of southern India. *Medical and Veterinary Entomology, 14*(4), 361–368. doi: 10.1046/j.1365-2915.2000.00255.x.

Wang, S., & Mulligan, C. N., (2009). Enhanced mobilization of arsenic and heavy metals from mine tailings by humic acid. *Chemosphere, 74*(2), 274–279. doi: 10.1016/j.chemosphere.2008.09.040.

Woodward, A. W., (2005). Auxin: Regulation, action, and interaction. *Annals of Botany, 95*(5), 707–735. doi: 10.1093/aob/mci083.

Yandigeri, M. S., Yadav, A. K., Meena, K. K., & Pabbi, S., (2010). Effect of mineral phosphates on growth and nitrogen fixation of diazotrophic cyanobacteria *Anabaena variabilis* and *Westiellopsis prolifica*. *Antonie van Leeuwenhoek, 97*(3), 297–306. doi: 10.1007/s10482-009-9411-y.

CHAPTER 7

Potential Application of Microalgae in Aquaculture

PANKAJ KUMAR SINGH and ARCHANA TIWARI

Diatom Research Laboratory, Amity Institute of Biotechnology, Amity University, Noida, Uttar Pradesh, India

ABSTRACT

The increased aquaculture demand calls for innovative intervention to boost the productivity on a sustainable scale round the globe. To make aquaculture more productive, nutritious, and value-added, the use of microalgae is quite beneficial. Algae are used for the purpose of fishmeal in aquaculture and utilized as live feed also, besides this a large quantity of wastewater is produced in aquaculture and algae have efficiency to remediate the wastewater biologically because they possess the metabolic efficiency to remediate the wastewater. The diverse algae play a key role in aquaculture to combat microbe-borne diseases, utilization of feed, to enhance the major nutrients like protein, lipid, and carbohydrate content, enhanced growth rate, response against stress, and so on. In this chapter, the prolific role of algae in aquaculture has been elaborated, which can prove to be a way forward towards sustainability.

7.1 INTRODUCTION

In the present scenario, the aquaculture is a fast-growing and rapidly developing sector and constantly increasing its valuable production.

Microalgal Biotechnology: Bioprospecting Microalgae for Functional Metabolites towards Commercial and Sustainable Applications. Jeyabalan Sangeetha, PhD, Svetlana Codreanu, PhD, & Devarajan Thangadurai, PhD (Eds.)

Microalgae find a myriad of applications in aquaculture, owing to their unique ability to grow fast and a diverse metabolic potential (Bhattacharjya et al., 2021). Many microalgae like *Chlorella*, *Tetraselmis*, *Skeletonema*, *Chaetoceros*, *Scenedesmus*, *Nannochloropsis*, *Phaeodactylum*, *Isochrysis*, *Thalassiosira*, and *Pavlova* have been reported in aquaculture application (Saxena et al., 2021). These species have the ability to grow fast and a rapid growth rate. In hatchery system, microalgae are stable in culture to viable disparity in nutrients, temperature, and light as may take place. Because of microalgae should have quality nutrient composing counting non-appearance of toxicant that may be passing on up to the food chain (Brown, 2002). Due to the ever-increasing human population, fishmeal is a flavored ingredient as they are protein-rich, digestible easily and eatable and supply a completely stabilize source of different amino acids which are essential, omega-3 fatty acids like Phospholipids, DHA, and EPA (Bhattacharjya et al., 2020). The contributions of aquaculture to meeting upcoming food require will need extra feasible implementation to help duo the terrestrial and aquatic ecosystem (Sirakov et al., 2015). Aquacultures have been developing rapidly in the recent years due to more and more demand for aquatic products or aquatic animals. The freshwater aquaculture products in China were reached up to 28.02 million tons, of which the pond aquaculture output reached at 19.89 million tons. As a food delivering area, aquaculture offers abundant chances to mitigate neediness, craving, and hunger, creates financial development and guarantees better utilization of common assets (FAO, 2017).

As per FAO (2010), an aquaculture creation is anticipated to ascend from 40 million tons by 2008 to 82 million tons in 2050. Because of expanding request per capita in corresponding to the expansion of worldwide populace to build the aquaculture creation has been activated. In any case, the advancement of an economical aquaculture industry is especially tested by the restricted accessibility of regular assets just as the effect of the business on the earth (Verdegem, 2013). Aquaculture, in global food security, plays an important role, which is an analytical and provocation for the 21st Century. The demand of food increases from 60–100% above 2005 levels, due to forecast increase of the global population from 7.6 to 9.8 billion by the year 2050 (Tilman et al., 2011). The demand of 110% for high quality protein, rising affluence is predicted (Tilman et al., 2011). Now a day, there are approx. 57% of the total global protein supply is occurring with the help of source like plants which includes terrestrial also, and the rest of the 43% of total global protein supply is covering from the other animal's sources like seafood's, dairy, eggs, poultry, red meat and other products (FAO, 2018). The fast development of the aquaculture worldwide is the depletion of the wild fisheries in the recent years (Ahmed et al., 2019). The aquatic animal farming in some countries has surpassed the wild fisheries yield (Cao et al., 2007). In the coming future, it is expected that the main industry which provide the aquatic products to the human beings will be the aquaculture. In another side, the water pollution will become a serious problem, when the continuous expansion of the aquaculture with their more production and development regarding water pollution with development of the aquaculture (Boyd, 1985; Adler et al., 2000). To control wastewater pollution into the aquaculture and to enhance the aquatic animal survival efficiency, a lot of efforts was devoted. Using traditional practices like filtration, aeration, anaerobic-anoxic-oxic system (A_2O) are the most common techniques to reduce the aquaculture wastewater pollution and also for the removal of nutrients from the wastewater (Altmann et al., 2016; Abyar et al., 2018). The feasible way to prevent the animal's disease and decrease the risks from aquaculture, some medicines and antibiotics are used commonly. In sometimes, the overuse of medicines or antibiotics in aquaculture way causes the problems for food safety and may cause some negative impact on the quality of meat of the aquatic animals (Liu et al., 2017).

7.2 USE OF MICROALGAE AS FEED IN AQUACULTURE

Microalgae used as a mixed diet or mono species showed good nutritional properties and the microalgae which are used as mixed diet in aquaculture includes *Thalassiosira pseudonana*, *Isochrysis* species, *T. suecica*, *C. calsitrans*, *C. mulleri*, *S. costatum* and *P. lutheri*. Based on healthy benefit, microalgae species can fluctuate altogether and that may likewise modify in various culture conditions (Enright et al., 1986; Brown et al., 1997). Different algal species combination provides improved growth in fish and better-balanced nutrition in comparison to the diet which is composed with only a single species (Spolaore et al., 2006). The microalgae which are used in the aquaculture have some specific role like that should have nutrients availability, digestible cell wall, correct cell and their size, lack of toxicity with high nutritional value and has to ease of culturing (Patil et al.,

2007). Microalgae are used in the field of aquaculture because they have some great positive effect on weight gain, some output towards decreased nitrogen into the environment, deposition of protein in muscle, increased triglyceride, increased fish digestibility, carcass quality and starvation tolerance and improved disease resistance (Becker, 2004; Fleurence et al., 2012). Microalgae also play a key role in aquaculture because they are used as a natural pigment source for the culture of ornamental fishes, prawns, and salmonid fishes, etc., with the help of species *Dunaliella salina*, *Spirulina* species, and *Haematococcus pluvialis*. These three species are also used successfully for the high concentration production of proteins, pigments, and lipids like valuable compounds (Abe and Hirano, 1999; Abd El-Baky et al., 2002; El-Baz et al., 2002). Microalgae are considered as an important added substance in taking care of in an aquaculture for the requirements of protein and to lessen the significant expense of the fish supper in the current occasions. Some other experiments are also going on microalgae to decrease the significant expense of the fish meal in the present time. Some other experiments are also going on microalgae to decrease the significant expense of the fish meal with replacement of algal meal. Some species like *Siganus canaliculatus*, *Oncorhynchus mykiss* are replaced as fish meal for feeding purposes (Figure 7.1).

As microalgae are essential food hotspot for the lower tropic fish and zooplankton that are therefore feed fish higher up the evolved way of life and are an important wellspring of key supplements and they are called next-generation microalgal-based feeds. The algal species which are used for microalgal-based feed contain up to 60% carbohydrate, 60% proteins and 70% oils (Draaisma, 2013) which is depending on the used species and conditions of their growth. These different species which is used for feed can produce different valuable pigments, hormones, natural antioxidants, anti-inflammatory, anti-microbial benefits like immune stimulants benefits to the aquatic animals (Garcia-Chavarria and Lara-Flores, 2013; Michalak and Chojnacka, 2015). Cultivated microalgae like *Isochrysis galbana*, *Pavlova lutheri*, *Skeletonema costatum* and *Chaetoceros calcitrans* were used as hatchery and feed for larval finfish, shrimp, bivalve mollusks used as nursery feeds and help to bring broodstock into spawning condition (Muller, 2000; Heasman et al., 2001). They (Muller and Heasman) cultivated mixed microalgae like *Isochrysis*, *Chaetoceros*, *Pavlova*, and *Skeletonema* are also feed to copepods, brine shrimp (*Artemia* species) are fed to juvenile shellfish and finfish, together with crustaceans.

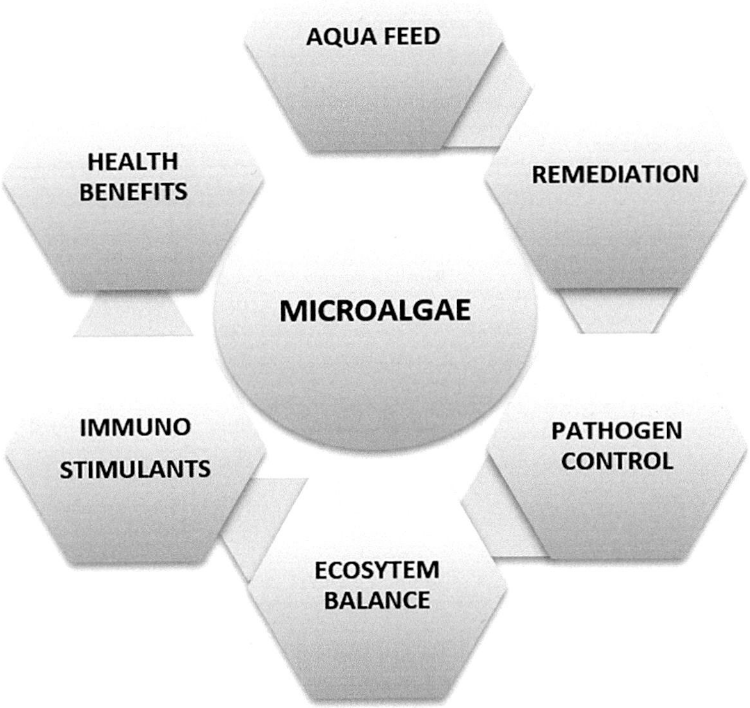

FIGURE 7.1 Application of microalgae in aquaculture.

The use of microalgae as live feed in aquaculture for all development stages of bivalve mollusks like Clams, Mussels, Oysters, and Scallops for the juvenile stages of Crustaceans, Abalone with few species of fishes along with zooplankton which are used in aquaculture food chains (Brown, 2002). Promising replacements for algal live feeds are concentrated algal pastes of high quality, and the pastes were include the microalgae like *Tetraselmis* sp., *Thalassiosira pseudonana* and *Isochrysis* sp. In sandfish larva (*Holothuria scabra*) production and winged pearl oysters (*Pteria penguin*), a similar growth rate has been reported, which is used as the commercial algal concentration and as live algal feed in aquaculture for feeding purpose (Southgate et al., 2016; Militz, 2018). Propelled huge scope algal creation framework coordinated with some great cell gathering, protection, and capacity and effective dispersion systems offers noteworthy potential for cost decrease in the field of aquaculture feed production (Borowitzka, 1997; Heasman et al., 2001). Table 7.1 shows the list of microalgae used in aquaculture.

TABLE 7.1 Microalgae Species Used in Fish Aquaculture

Microalgal Species	Fish Species	References
Skeletonema costatum, Tetraselmis tetrahele, Nannochloropsis oculata	Shrimp larva, Rotifers	Lim (1991)
Spirulina	Red sea bream, Tilapia	Appler (1985); Mustafa et al. (1994); El-Hindawy et al. (2006)
Tetraselmis isochrysis	Shrimp larva, Rotifers	Villegas et al. (1990)
Spirulina	Salmon, Shrimp	Appler (1985); Zeinhom (2004); Regunathan and Wesley (2006)
Scenedesmous	Tilapia, Red Sea bream	Appler (1985); Mustafa et al. (1994); El-Hindawy et al. (2006)
Haematococcus	Salmon, Shrimp	Appler (1985); Zeinhom (2004); Regunathan and Wesley (2006)
Hydrodictyon	Tilapia, Red Sea bream	Appler (1985); Mustafa et al. (1994); El-Hindawy et al. (2006)
Haematococcus	Tilapia, Red Sea bream	Appler (1985); Mustafa et al. (1994); Olvera-Novoa et al. (1998)
Chlorella	Shrimp larva, Rotifers	Villegas et al. (1990)
Chaetoceros, Pavlova, Phaeodactylum, Thalassiosira pseudonana, T. suecica, C. calcitrans, C. muelleri, S. coctatum, P. lutheri	–	Enright et al. (1986); Thompson et al. (1993); Brown (1997); Brown et al. (2002)

7.3 NUTRITIONAL PROPERTIES OF MICROALGAE

Microalgae have also been recognized as the potential source of different novel compounds like various fatty acids, proteins, phenolic compounds, carbohydrates, carotenoids, and lipids (Marella et al., 2020). The variety of microalgae can shift fundamentally based on their dietary benefit and with different types of culture conditions; this may change (Enright et al., 1986; Brown et al., 1997). For larval animals, by the means of directly or indirectly, a mixed microalgae culture could be an excellent package of nutritional point of view (with a combination of zooplankton). *Navicula* sp. and *Nitzschia* sp. like benthic diatoms is used commonly for the purpose of mass refined and afterward settled onto plates for brushing adolescent abalone as a diet. The factors like microalgal size and shape can supply to the dietetic worth. Also, some factors like digestibility (which is related to composition and cell wall structure), composition of biochemical like (enzymes, nutrients, and toxins if present) and some other needs on the microalgae for the animal feed purpose. Some early reports and many studies have demonstrated that the differences in biochemical in gross composition in between the microalgae (Persons et al., 1961) with comparison to the fatty acid (Webb and Chu, 1983) and these studies evaluate the biochemical profile and the nutritional values of the microalgae (Table 7.2). Some general conclusions can be reached when different literature data examining, which included those experiments in which the diets of algal species was supplemented with emulsions or compounded diets (Knauer and Southgate, 1999). When microalgae were cultured through to its stationary phase, a significant change was seen in the microalgal composition; for example, when carbohydrate levels can be double, the nitrate is limiting at the outflow of the protein (Harrison et al., 1990; Brown et al., 1993). Also, when microalgae grown to the late logarithmic growth phase typically contain 30–40% of protein, 05–15% carbohydrate and 10–20% of lipids (Brown et al., 1997; Renaud et al., 1999). For the purpose of paramount increase in juvenile mussels (*Mytilus trossulus*; Kreeger and Langdon, 1993), they provided the protein with high dietary and the same with pacific oysters (*Crassostrea gigas*) (Knuckey et al., 2002).

7.4 ROLE OF MICROALGAE IN AQUACULTURE REMEDIATION

There are a few general methods which is used to remediating the aquaculture wastewater including physical, chemical, and biological method. The method which is generally suited for aquaculture wastewater treatment is biological treatment or method as low cost, high efficiency and absence of secondary pollution (Webb et al., 2012, 2013; Quinta et al., 2015).

TABLE 7.2 Nutritional Properties of Microalgae

Algal Species	Lipid (%)	Protein (%)	Carbohydrate (%)	References
Pavlova sp.	9–14	24–29	6–9	Becker et al. (1994)
Dunaliella sp.	6–8	49–57	4–32	Eddy (1956); Parson et al. (1961)
Ulva lactura	4.36	8.44	35.27	Chakraborty and Sharma (2008)
Polysiphonia mollis	5.79	16.59	25.81	Chakraborty and Sharma (2008)
Porphyridium sp.	–	28–39	50–57	Becker (1994)
Tetraselmis sp.	16–45	52	15	Becker (1994)
Cantella sp.	5.29	8042	28.96	Chakraborty and Sharma (2008)
Rhizoclonium riparium	3.37	21.09	15.34	Chakraborty and Sharma (2008)
Enteromorpha intestinalis	7.13	6.15	30.58	Chakraborty and Sharma (2008)
Spirulina maxima	6–7	60–71	13–16	Becker (1994)
Spirogyra	11–21	6–20	33–64	Becker (1994)
Lola capillaris	4.05	4.87	22.32	Chakraborty and Sharma (2008)
Scenedesmus sp.	12–14	50–56	10–52	Hindak and Probil (1968); Becker (1984, 1994)
Synechococcus sp.	11	63	15	Trubachev et al. (1976)
Chlorella sp.	14–22	12–17	51–58	Renaud (1994); Becker (1994)
Chlamydomonas sp.	14–22	43–56	2.9–17	Becker (1994); Renaud et al. (1994)
Prymnesium sp.	22–38	28–45	25–33	Ricketts (1966)
Anabaena sp.	4–7	48	25–30	Becker (1994)
Euglena sp.	14–20	39–61	14–18	Collyer and Fogg (1955); Becker (1994)
Spirogyra	11–21	6–20	33–64	Becker (1994)
Botryococcus braunii	34.4	39.9	18.5	Tibbetts et al. (2015)
Synechococcus sp.	11	63	15	Becker (1994)
A. plentesis	11	42.33	–	Sydney et al. (2010)
Arthrospira sp.	11.6	58	10.8	–
Tetraselmis chuii	12.3	46.5	25	Tibbetts et al. (2015)

Microalgae are environment friendly, low cost relatively, most efficiently for the purpose of wastewater treatments and proven as promising species towards reducing the pollutant load as well as carbon sequestration (Hii et al., 2011; Marella et al., 2019; Singh et al., 2020). The use of microalgae in the treatment of wastewater is well suited because they are available easily and rapidly growing organism. Furthermore, different species of algae or microalgae are available with their different characteristics (Chaitanawisuti et al., 2011; Su et al., 2012). One of the best quality of algae is that they use phosphorous and nitrogen to eliminate the nutrients which is present in aquaculture wastewater (Quinta et al., 2015). The crucial class of algae which include benthic diatom can obtain in natural environment by the process of separation and purification (Xing et al., 2011). The common species which is used frequently for the treatment of aquaculture wastewater naturally and also for the purpose of feeding and production of different aquatic species like sea cucumber, sea urchin and abalone, etc. (Xing et al., 2011). For the purpose of food and biomass production or production of biofuel, microalgae are usually cultivated (Velichkova et al., 2012, 2013). In aquaculture, one of the most common wastes which are secreted by aquatic animals is nitrogen. Also, ammonium is another waste which is secreted by aquatic animals in a water body which is toxic and mostly unfavorable for that aquatic animal too. The nitrite, nitrate, and are some of the common forms of nitrogen which are present in wastewater (Randall and Tsui, 2002; Zhou et al., 2006). Through active transport, microalgae can be absorbing ammonia with their cells and they utilized it for amino acid synthesis directly. For the growth of microalgae in the aquaculture wastewater, organic carbons like volatiles, fatty acids and saccharides while inorganic carbon like HCO_3 and CO_2 are the main source of carbon. In aquaculture wastewater, the algal-bacterial consortia have been studied widely promoted for assimilation of carbon (Su et al., 2012; Hernandez et al., 2013). In a sea cucumber aquaculture wastewater, *Nitzschia* sp. was used to remove the different types of nutrients. The removal rate of total nitrogen was greatest (54.58%) with a mixture of aquaculture wastewater and artificial seawater with 30% concentration of aquaculture wastewater. The removal rate of nitrate-nitrogen was found maximum (51.08%) with a mixture of aquaculture wastewater with 0% concentration. The removal rate of nitrate nitrogen was absorbed maximum (46.43%) with a mixture of aquaculture wastewater with 80% concentration and F/2 media. The removal rate of ammoniacal nitrogen was found maximum (90% with different concentration of aquaculture and f/2 medium) when *Nitzscia* species were cultured in sea cucumber aquaculture wastewater. The total phosphorus removal rate was ranged between 10.80% and 31.47% when *Nitzschia* species were cultured in an aquaculture wastewater. And total phosphorus was removed maximum in a concentration of 30% aquaculture wastewater and f/2 medium. In the production of *Nitzschia* species with aquaculture wastewater, 100% concentration of wastewater has the greatest nutritional value. The ability of absorbance of nitrogen and phosphorus is higher in aquaculture wastewater which supports nutrient removal and remediation of aquaculture wastewater (Xing et al., 2018).

7.5 USE OF MICROALGAE AS IMMUNOSTIMULANTS

In regard to immunostimulants, some earlier reports explained that the utilization of biomass of microalgae with combination diets may help to increase the physiological condition and immune system of crustaceans and different stages of fish like larval, juvenile, and adult fishes. Immunostimulants, besides identified as immune stimulators are substances (nutrients and drugs that stimulate the immune system by inducing activation or increasing activity of any of its components. In current years, in aquaculture, the role of microalgae has extended further for its use as commercially and very important for the species of aquaculture with a potential of immunostimulant. And in aquaculture for shrimp larva, to improve the survival rate and immune system, microalgae can play an important role. In support of immune-stimulating properties, there are several studies which is highlighted in different microalgae in crustaceans and fishes. The species *Chlorella* is referred as a good additive and it also helps to promote the physiological parameter and growth performance, which is involved in regulating the innate and adaptive immunity of Gibel carp or *Carassius auratus gibelio* (Xu et al., 2014; Zhang et al., 2014).

According to Cerezuela et al. (2012), defense activity could be enhanced by microalgae like *Tetraselmis chuii, P. tricornutum,* and *Nanochloropsis gaditana* in gilthead seabream (*S. aurata)*. Antioxidant factors like super oxidase dismutase and catalase or immunologically strong enhancement are induced by *Dunaliella salina,* which is also, enhanced the rate of survival of many species that was presented as white spot syndrome virus-infected shrimp *P. monodon* (Madhumati and Rengasamy, 2011). *A.*

platensis can be introduced as immunostimulants in diet because they have the ability to increase white blood count. Red blood count, entire protein, hemoglobin, and albumin level significantly in rainbow trout (*O. mykiss*) (Yeganeh et al., 2015). To raise the nonspecific immune activity, *Arthrospira* (5 to 10%) has been reported and also it can enhance both the juvenile and grow out survival of different species when they were presented with some of the unambiguous pathogens (Watanuki et al., 2006; Ibrahen et al., 2013). The rate of survival with increased level of lysozyme in guppy fish (*Poecilia reticulata*) with microalgal diet was given with a composition of microalgae *P. incisa*. These diets helped to reduce the stress level and increase stress resistance (Nath et al., 2012). The immunostimulants which are derived by utilization of microalgae are beneficial in aquaculture for the purpose of aquafeed and also in the industrial aqua feed all over the world. The parameters of immunity like superoxide dismutase (SOD), peroxide (POD) and lysozyme apply a pattern of increasing when fish feed with some increased pattern of dietary with *Chlorella* (Xu et al., 2014). In the presence of the chlorella diet which fed and added as dietary purpose, the results showed that there was increased immune response for Gibel carp. According to Pham et al. (2009), species *Chlorella* can enhance the SOD, which is called the first line for antioxidant defense and it helps the cells to protect tissues and cells against oxidative stress. Lysozyme plays an important role as a defense molecule which is involved in innate type of immunity. And with the help of this innate immunity, organisms can protect themselves against any microbial invasion (Saurabh and Sahoo, 2008). For the purpose of immunity enhancement and vitality enhancement, a benthic diatom likes *Nitzschia* species can be used as a feed generative individual for the cultivation of sea cucumber (Zhang et al., 2015).

7.6 MICROALGAE IN AQUACULTURE FOR DISEASE CONTROL

The cause of significant disease in shrimps are viral pathogens (yellow head disease (YHD) and white spot syndrome disease (WSSD) and from bacterial pathogens early mortality syndrome (EMS)) (Kusumaningrum and Zainuri, 2015; Thitamadee et al., 2016). Infections in fishes are for the most part brought about by bacterial pathogens, the mainly invasive of which are variety of *Vibrio*, *Pseudomonas*, and *Aeromonas*. In recent times, various viral gatherings have additionally been accounted for as causative operators of infection in fish, with Betanodavirus, for example, nervous necrosis virus (NNV),

Megalocyticvirus, for example, infectious spleen and kidney corruption infection (ISKNV), and red sea bream iridovirus (RSIV) (Crane et al., 2011; Subramaniam et al., 2012). There is subsequently a squeezing need to discover compelling techniques to address these nonstop and erratic sicknesses. A superior and more secure methodology for infection control is immunization, and various vaccines have been demonstrated to be profoundly successful against explicit viral or bacterial pathogens of fish. But there are financial, specialized, and administrative difficulties related with immunization. The least complex and generally looked for after strategy is oral deliverance of the antibody by figuring it into the aquaculture feed. This methodology permits mass immunization of fish everything being equal, is non-unpleasant, requires minimal specialized aptitude by the administrator, and permits continue dosing (Dhar et al., 2011; Plant and Lapatra, 2011). As an outcome, oral inoculation frequently brings about just a constrained span of resistance and may require continue dosing. Microalgal species show normal competitor of bacterial exercises, and some likewise contain biomolecules that fill in as immunostimulants (Falaise et al., 2016; Shah et al., 2018). Thus, both unrefined algal concentrates and entire cells have been utilized as a substitution or as an enhancement in shrimp or fish feed to facilitate deal with the issue of bacterial diseases (Austin et al., 1990; Dang et al., 2011; Yaakob et al., 2014). Microalgae homogenate from *Tetraselmis suecica* indicated great antibacterial action *in vitro* against a few *Vibrio* animal types including huge shrimp pathogens, for example, *Vibrio vulnificus*, *Vibrio parahaemolyticus*, *Vibrio anguillarum*, and *Vibrio alginolyticus* (Austin et al., 1990). Other microalgal species generally utilized as aquaculture feed have comparably indicated antibacterial movement *in vitro* adjacent to specific shrimp and fish pathogens. These algae incorporate *Dunaliella tertiolecta, Chaetoceros lauderi, Stichochrysis immobilis, Phaeodactylum tricornutum, Euglena viridis* (Viso et al., 1987; Das and Pradhan, 2010; Falaise et al., 2016).

The pigments of carotenoid like astaxanthin, lutein, and zeaxanthin that are sufficient in microalgae were likewise answered to build the endurance of shrimp and fish, just as of different scavengers, when exposed to sickness disease (Merchie et al., 1998; Babin et al., 2010). C vitamin, which is found in high sums in a few microalgal animal types, was accounted for to help invulnerability in shrimp bringing about diminished mortality from *Vibrio* species (Kanazawa et al., 1995; Kontara et al., 1997). A few studies have detailed the utilization of a feed diet containing microalgae to fight bacterial pathogen in aquaculture and credited this to a mix of antibacterial movement and host resistance enlistment by microalgal-determined mixes (Tayag et al., 2010; Debtanu et al., 2013). Immunostimulants likewise assume an important responsibility regarding improvement for the defense against viral pathogens (Debtanu et al., 2013). Chlorellin, which is a lipophilic antibacterial substance which is used as an auto inhibitor of the algae, reported while *Chlorella vulgaris* extracted in a culture medium. Microalgae can produce antimicrobial or antibacterial synthesis that hinders the development of bacteria or other microbes (Fukami et al., 1997; Amin et al., 2012). The exudates synthesized by microalgae can be an origin of assimilated carbon for microbes (Murray et al., 1986). Some of the studies reported on the microalgal inhibitory effect against pathogenic bacteria like *Vibrio* species (Navinar et al., 1999). A broad diversity of chemical compounds produced by diatoms with various kind of bioactivity like antibacterial activity. *Phaeodactylum tricornutum* and *Skeletonema costatum* produce secondary metabolites that have an effect on the pathogenic bacteria like *Vibrio anguillarum* and *Staphylococcus* (Desbois et al., 2009).

7.7 CONCLUSION

This chapter reports the uses of microalgae in aquaculture and their diverse potential in enhancing the growth of cultured aquatic animals with nutritious food supplements concomitant with disease control, aquaculture wastewater remediation, stress resistance, improved nutritional quality, etc. The future aspects with microalgae in aquaculture industry include higher production and faster growth rate with enriched nutrients towards an eco-friendly system for a better tomorrow.

KEYWORDS

- **aquaculture**
- **disease control**
- **fish feed**
- **immunostimulants**
- **microalgae**
- **remediation**

REFERENCES

Abd El-Baky, H. H., Moawd, A., El-Behairy, A. N., & El-Baroty, G. S., (2002). Chemoprevention of benzopyrene induced carcinogen and lipid peroxidation in mice by lipophilic algae extracts (phycotene). *J. Med. Sci., 2*, 185–193.

Abe, N. K. N., & Hirano, M., (1999). Simultaneous production of β-carotene, vitamin E and vitamin C by the aerial microalga *Trentepohia aurea. J. Appl. Phycol., 11*, 33–36.

Abyar, H., Younesi, H., Bahramifar, N., & Zinatizadeh, A. A., (2018). Biological CNP removal from meat-processing wastewater in an innovative high rate up-flow A_2O bioreactor. *Chemosphere, 213*, 197–204.

Adler, P. R., Harper, J. K., Takeda, F., Wade, E. M., & Summerfelt, S. T., (2000). Economic evaluation of hydroponics and other treatment options for phosphorus removal in aquaculture effluent. *Hortic. Sci., 35*, 993–999.

Ahmed, N., Thompson, S., & Glaser, M., (2019). Global aquaculture productivity, environmental sustainability, and climate change adaptability. *Environ. Manag., 63*, 159–172.

Altmann, J., Rehfeld, D., Trader, K., Sperlich, A., & Jekel, M., (2016). Combination of granular activated carbon adsorption and deep-bed filtration as a single advanced wastewater treatment step for organic micropollutant and phosphorus removal. *Water Res., 92*, 131–139.

Amin, S. A., Parker, M. S., & Ambrust, E. V., (2012). Interactions between diatoms and bacteria. *Microbiol. Mol. Biol. Rev., 76*(3), 667–684.

Appler, H. N., (1985). Evaluation of *Hydrodictyon reticulatum* as protein source in feeds for *Oreochromis niloticus* and *Tilapia zillii*. *J. Fish Biol., 27*(3), 327–334.

Austin, B., & Day, J., (1990). Inhibition of prawn pathogenic *Vibrio* sp. by a commercial spray-dried preparation of *Tetraselmis suecica. Aquaculture, 90*, 389–392.

Babin, A., Biard, C., & Moret, Y., (2010). Dietary supplementation with carotenoids improves immunity without increasing its cost in a crustacean. *Am. Nat., 176*, 234–241.

Becker, E. W., (1984). Biotechnology and exploitation of the green alga *Scenedesmus obliquus* in India. *Biomass, 4*, 1.

Becker, E. W., (1994). *Microalgae: Biotechnology and Microbiology* (p. 293). Cambridge: Cambridge University Press.

Becker, W., (2004). Microalgae for aquaculture: The nutritional value of microalgae for aquaculture. In: Richmond, A., (ed.), *Handbook of Microalgal Culture: Biotechnology and Applied Phycology* (pp. 380–391). Oxford.

Bhattacharjya, R., Marella, T. K., Tiwari, A., Saxena, A., Singh, P. K., & Mishra, B., (2020). Bioprospecting of marine diatoms *Thalassiosira, Skeletonema* and *Chaetoceros* for lipids and other value-added products. *Bioresour. Technol., 318*, 124073.

Bhattacharjya, R., Singh, P. K., Mishra, B., Saxena, A., & Tiwari, A., (2021). Phycoprospecting the nutraceutical potential of *Isochrysis* sp. as a source of aquafeed and other high-value products. *Aquac. Res.*, 1–8.

Borowitzka, M. A., (1997). Microalgae for aquaculture: Opportunities and constraints. *J. Appl. Phycol., 9*, 393.

Boyd, C. E., (1985). Chemical budgets for channel catfish ponds. *Trans. Am. Fish. Soc., 114*, 291–298.

Brown, M. R., (2002). Nutritional value and use of microalgae in aquaculture. In: *Avances en Nutricion Acuieola VI: Memorias del VI Simposium Internacional de Nutricion Acuicola* (Vol. 3, pp. 281–292).

Brown, M. R., Garland, C. D., Jeffrey, S. W., Jameson, I. D., & Leroi, J. M., (1993b). The gross and amino acid compositions of batch and semi-continuous cultures of *Isochrysis* sp. (clone T.ISO), *Pavlova lutheri* and *Nannochloropsis oculata. J. Appl. Phycol., 5*, 285–296.

Brown, M. R., Jeffrey, S. W., Volkman, J. K., & Dunstan, G. A., (1997). Nutritional properties of microalgae for mariculture. *Aquaculture, 151*, 315–331.

Cao, L., Wang, W., Yang, Y., et al., (2007). Environmental impact of aquaculture and countermeasures to aquaculture pollution in China. *Environ. Sci. Pollut. Res. Int., 14*, 452–462.

Cerezuela, R., Guardiola, F. A., Meseguer, J., & Esteban, M. A., (2012). Enrichment of gilthead seabream (*Sparus aurata* L.) diet with microalgae: Effects on the immune system. *Fish Physiol. Biochem., 38*, 1729–1739.

Chaitanawisuti, N., Santhaweesuk, W., & Kritsanapuntu, S., (2011). Performance of the seaweeds *Gracilaria salicornia* and *Caulerpa lentillifera* as biofilters in a hatchery scale recirculating aquaculture system for juvenile spotted Babylon snail (*Babylonia areolata*). *Aquac. Int., 19*, 1139–1150.

Chakraborty, S., & Santra, S. C., (2008). Biochemical composition of eight benthic algae collected from Sunderban. *Indian J. Mar. Sci., 37*(3), 329–332.

Collyer, D. M., &. Fogg, G. E., (1955). Studies on fat accumulation by algae. *J. Exp. Bot., 6*, 256.

Crane, M., & Hyatt, A., (2011). Viruses of fish: An overview of significant pathogens. *Viruses, 3*, 2025–2046.

Dang, V. T., Li, Y., Speck, P., & Benkendorff, K., (2011). Effects of micro and macroalgal diet supplementations on growth and immunity of greenlip abalone (*Haliotis laevigata*). *Aquaculture, 320*, 91–98.

Das, B., & Pradhan, J., (2010). Antibacterial properties of selected freshwater microalgae against pathogenic bacteria. *Indian J. Fish., 57*, 61–66.

Debtanu, B. P. N., Mandal, S. C., & Kumar, V., (2013). Immunostimulants for aquaculture health management. *J. Mar. Sci. Res. Dev., 3*.

Desbois, A. P., Mearns-Sprag, A., & Smith, V. A., (2009). Fatty acid from the diatom *Phaeodactylum tricornutum* is antibacterial against diverse bacteria including multi-resistant *Staphylococcus aureus* (MRSA). *Mar. Biotechnol., 11*, 45–52.

Dhar, A., & Allnutt, F., (2011). Challenges and opportunities in developing oral vaccines against viral diseases of fish. *J. Mar. Sci. Res. Dev., 2*, 1–6.

Draaisma, R. B., (2013). Food commodities from microalgae. *Curr. Opin. Biotechnol., 24*, 169–177.

Eddy, B. P., (1956). The suitability of some algae for mass cultivation for food with special reference to *Dunaliella bioculata. J. Exp. Bot., 7*, 372.

El-Baz, F. K., Aboul-Enein, M. A., El-Baroty, G. S., Youssef, A. M., & El-Baky, H. H. A, (2002). Accumulation of antioxidant vitamins in *Dunaliella salina. J. Biol. Sci., 2*, 220–223.

El-Hindawy, M. M., Abd-Razic, M. A., Gaber, H. A., & Zenhom, M. M., (2006). Effect of various levels of dietary algae *Scenedesmus* spp. on physiological performance and digestibility of Nile tilapia fingerlings. In: *1st Scientific Conference of the Egyptian Aquaculture Society* (pp. 137–149). Sinai: Sharm El-Sheikh.

Enright, C. T., Newkirk, G. F., Craigie, J. S., & Castell, J. D., (1986). Evaluation of phytoplankton as diets for juvenile *Ostrea edulis* L. *J. Exp. Mar. Biol. Ecol., 96*, 1–13.

Falaise, C., François, C., Travers, M. A., et al., (2016). Antimicrobial compounds from eukaryotic microalgae against human pathogens and diseases in aquaculture. *Mar. Drugs, 14*, 159.

FAO, (2010). *The State of World Fisheries and Aquaculture* (p. 179). Rome: Food and Agriculture Organization.

FAO, (2017). FAO and the SDGs. *Indicators: Measuring up to the 2030 Agenda for Sustainable Development* (p. 39). Rome: FAO.

FAO, (2018). *The State of World Fisheries and Aquaculture 2018: Meeting the Sustainable Development Goals*. Rome: FAO. http://www.fao.org/3/i9540en/I9540EN.pdf (accessed on 12 February 2022).

Fleurence, J., Morançais, M., Dumay, J., et al., (2012). What are the prospects for using seaweed in human nutrition and for marine animals raised through aquaculture? *Trends Food Sci. Technol., 27*, 57–61.

Fukami, K., Nishijima, T., & Ishida, Y., (1997). Stimulative and inhibitory effects of bacteria on the growth of microalgae. *Hydrobiologia, 358*, 185–191.

Garcia-Chavarria, M., & Lara-Flores, M., (2013). The use of carotenoid in aquaculture. *Res. J. Fish. Hydrobiol., 8*, 38–49.

Harrison, P. J., Thompson, P. A., & Calderwood, G. S., (1990). Effects of nutrient and light limitation on the biochemical composition of phytoplankton. *J. Appl. Phycol., 2*, 45–56.

Heasman, M. P., Sushames, T. M., Diemar, J. A., O'Connor, W. A., & Foulkes, L. A., (2001). *Production of Microalgal Concentrates for Aquaculture, Part 2: Development and Evaluation of Harvesting, Preservation, Storage and Feeding Technology.* NSW Fisheries.

Hernandez, D., Riaño, B., Coca, M., & García-González, M., (2013). Treatment of agro-industrial wastewater using microalgae–bacteria consortium combined with anaerobic digestion of the produced biomass. *Bioresour. Technol., 135*, 598–603.

Hii, Y. S., Soo, C. L., Chuah, T. S., Mohd-Azmi, A., & Abol-Munafi, B., (2011). Interactive effect of ammonia and nitrate on the nitrogen uptake by *Nannochloropsis* sp. *J. Sustain Sci. Manag., 6*(1), 60–68.

Hindak, F., & Probil, S., (1968). Chemical composition, protein digestibility and heat of combustion of filamentous green algae. *Plant Biol., 10*, 234.

Ibrahem, M., Mohamed, M. F., & Ibrahim, M. A., (2013). The role of *Spirulina platensis* (*Arthrospira platensis*) in growth and immunity of Nile tilapia (*Oreochromis niloticus*) and its resistance to bacterial infection. *J. Agr. Sci., 5*, 109–117.

Kanazawa, A., (1995). Recent developments in shrimp nutrition and feed industry. In: *INDAQUA '95 Exposition of Indian Aquaculture* (pp. 28, 29). Madras. The Marine Products Export Development Agency: Cochin, India.

Knauer, J., & Southgate, P. C., (1999). A review of the nutritional requirements of bivalves and the development of alternative and artificial diets for bivalve aquaculture. *Rev. Fish. Sci., 7*, 241–280.

Knuckey, R. M., Brown, M. R., Barrett, S. M., & Hallegraeff, G. M., (2002). Isolation of new nanoplanktonic diatom strains and their evaluation as diets for the juvenile Pacific oyster (*Crassostrea gigas*). *Aquaculture, 211*, 253–274.

Kontara, E., Merchie, G., Lavens, P., et al., (1997). Improved production of post-larval white shrimp through supplementation of L-ascorbyl-2-polyphosphate in their diet. *Aquac. Int., 5*, 127–136.

Kreeger, D. A., & Langdon, C. J., (1993). Effect of dietary protein content on growth of juvenile mussels, *Mytilus trossulus* (Gould 1850). *Biol. Bull., 185*, 123–139.

Kusumaningrum, H. P., & Zainuri, M., (2015). Detection of bacteria and fungi associated with *Penaeus monodon* postlarvae mortality. *Procedia Environ. Sci., 23*, 329–337.

Lim, L. C., (1991). An overview of live feeds productions systems in Singapore. In Harvey, B. J., (ed.), *Rotifer and Microalgae Culture Systems* (pp. 203–221).

Liu, H., Lu, Q., Wang, Q., et al., (2017). Isolation of a bacterial strain, *Acinetobacter* sp. from centrate wastewater and study of its cooperation with algae in nutrients removal. *Bioresour. Technol., 235*, 59–69.

Madhumathi, M., & Rengasamy, R., (2011). Antioxidant status of *Penaeus monodon* fed with *Dunaliella salina* supplemented diet and resistance against WSSV. *Int. J. Eng. Sci. Tech., 3*, 7249–7259.

Marella, T. K., Datta, A., Patil, M. D., Dixit, S., & Tiwari, A., (2019). Biodiesel production through algal cultivation in urban wastewater using algal floway. *Bioresour. Technol., 280*, 222–228.

Marella, T. K., Lopez-Pacheco, I. Y., Parra-Saldivar, R., Dixit, S., & Tiwari, A., (2020). Wealth from waste: Diatoms as tool for phytoremediation of wastewater and for obtaining value from the biomass. *Sci. Total Environ., 724*, 137960.

Merchie, G., Kontara, E., Lavens, P., Robles, R., Kurmaly, K., & Sorgeloos, P., (1998). Effect of vitamin C and astaxanthin on stress and disease resistance of postlarval tiger shrimp, *Penaeus monodon* (Fabricius). *Aquac. Res., 29*, 579–585.

Michalak, I., & Chojnacka, K., (2015). Algae as production systems of bioactive compounds. *Eng. Life Sci., 15*, 160–176.

Militz, T. A., (2018). Successful large-scale hatchery culture of sandfish (*Holothuria scabra*) using micro-algae concentrates as a larval food source. *Aquac. Rep., 9*, 25–30.

Muller-Feuga, A., (2000). The role of microalgae in aquaculture: Situation and trends. *J. Appl. Phycol., 12*, 527–534.

Murray, R. E., Cooksey, K. E., & Priscu, J. C., (1986). Stimulation of bacterial DNA synthesis by algal exudates in attached algal-bacterial interaction. *Appl. Environ. Microbiol., 52*(5), 1177–1182.

Mustafa, M. G., Umino, T., & Nakagawa, H., (1994). The effect of spirulina feeding on muscle protein deposition in red seabream, *Pagrus major*. *J. Appl. Ichthyol., 10*, 141–145.

Nath, P., Khozin-Goldberg, I., Cohen, Z., Boussiba, S., & Zilberg, D., (2012). Dietary supplementation with the microalgae *Parietochloris incisa* increases survival and stress resistance in guppy (*Poecilia reticulata*) fry. *Aquac. Nutr., 18*, 167–180.

Naviner, M., Bergé, J. P., Durand, P., & Le Bris, H., (1999). Antibacterial activity of the marine diatom *Skeletonema costatum* against aquacultural pathogens. *Aquaculture, 174*, 15–24.

Olvera-Novoa, M. A., Daminguez-Cen, L. J., Olivera-Castillo, L., & Martinez-Palacios, A. C., (1998). Effect of the use of the microalgae *Spirulina maxima* as fish meal replacement in diets for tilapia, *Oreochromis mossambicus* (Peters) fry. *Aquac. Res., 29*, 709–715.
Parson, T. R., Stephens, K., & Strickland, J. D. H., (1961). On the chemical composition of eleven species of marine phytoplankters. *Journal of the Fisheries Research Board Canada, 18*, 1001–1016.
Patil, V., Källqvist, T., Olsen, E., Vogt, G., & Gislerød, H. R., (2007). Fatty acid composition of 12 microalgae for possible use in aquaculture feed. *Aquacult. Int., 15*, 1–9.
Pham, T. M., Fujino, Y., Ando, M., et al., (2009). Relationship between serum levels of superoxide dismutase activity and subsequent risk of lung cancer mortality: Findings from a nested case-control study within the Japan collaborative cohort study. *Asian Pac. J. Cancer P, 10*, 75–79.
Plant, K. P., & Lapatra, S. E., (2011). Advances in fish vaccine delivery. *Dev. Comp. Immunol., 35*, 1256–1262.
Quinta, R., Santos, R., Thomas, D. N., & Le, V. L., (2015). Growth and nitrogen uptake by *Salicornia europaea* and *Aster tripolium* in nutrient conditions typical of aquaculture wastewater. *Chemosphere, 120*, 414–421.
Randall, D., & Tsui, T., (2002). Ammonia toxicity in fish. *Mar. Pollut. Bull., 45*, 17–23.
Regunathan, C., & Wesley, S. G., (2006). Pigment deficiency correction in shrimp broodstock using *Spirulina* as a carotenoid source. *Aquac. Nutr., 12*(6), 425–432.
Renaud, S. M., Thinh, L. V., & Parry, D. L., (1999). The gross composition and fatty acid composition of 18 species of tropical Australian microalgae for possible use in mariculture. *Aquaculture, 170*, 147–159.
Ricketts, T. R., (1966). On the chemical composition of some unicellular algae. *Phytochemistry, 5*, 67.
Saurabh, S., & Sahoo, P. K., (2008). Lysozyme: An important defense molecule of fish innates immune system. *Aquac. Res., 39*, 223–239.
Saxena, A., Mishra, B., & Tiwari, A., (2021). Development of diatom entrapped alginate beads and application of immobilized cells in aquaculture. *Environ. Technol. Innov., 23*(53), 101736.
Shah, M. R., Lutzu, G. A., Alam, A., et al., (2018). Microalgae in aquafeeds for a sustainable aquaculture industry. *J. Appl. Phycol., 30*, 197–213.
Singh, P. K., Bhattacharjya, R., Saxena, A., Mishra, B., & Tiwari, A., (2020). Utilization of wastewater as nutrient media and biomass valorization in marine Chrysophytes-*Chaetoceros* and *Isochrysis*. *Energy Convers. Manag.: X*, 100062.
Sirakov, I., Velichkova, K., Stoyanova, S., & Staykov, Y., (2015). The importance of microalgae for the aquaculture industry. *Int. J. Fish. Aquat. Stud., 2*, 81–84.
Southgate, P. C., Beer, A. C., & Ngaluafe, P., (2016). Hatchery culture of the winged pearl oyster, *Pteria penguin*, without living micro-algae. *Aquaculture, 451*, 121–124.
Spolaore, P. C., Joannis-Cassan, E. D., & Isambert, A., (2006). Commercial application of microalgae. *J. Biosci. Bioeng., 101*, 87–96.
Su, Q., Xing, R. L., & Wang, H. Y., (2012). The uptake kinetics of nitrogen and phosphorus by *Nitzschia* sp. *Adv. Mat. Res.*, 549–553.
Su, Y., Mennerich, A., & Urban, B., (2012). Synergistic cooperation between wastewater-born algae and activated sludge for wastewater treatment: Influence of algae and sludge inoculation ratios. *Bioresour. Technol., 105*, 67–73.

Subramaniam, K., Shariff, M., Omar, A. R., & Hair-Bejo, M., (2012). Megalocytivirus infection in fish. *Rev. Aquac., 4*, 221–233.

Sydney, E. B., Sturm, W., De Carvalho, J. C., et al., (2010). Potential carbon dioxide fixation by industrially important microalgae. *Bioresour. Technol., 101*(15), 5892–5896.

Tayag, C. M., Lin, Y. C., Li, C. C., Liou, C. H., & Chen, J. C., (2010). Administration of the hot-water extract of *Spirulina platensis* enhanced the immune response of white shrimp *Litopenaeus vannamei* and its resistance against *Vibrio alginolyticus*. *Fish Shellfish Immun., 28*, 764–773.

Thitamadee, S., Prachumwat, A., Srisala, J., et al., (2016). Review of current disease threats for cultivated penaeid shrimp in Asia. *Aquaculture, 452*, 69–87.

Thompson, P. A., Guo, M. X., & Harrison, P. J., (1993). The influence of irradiance on the biochemical composition of three phytoplankton species and their nutritional value for larvae of the Pacific oyster (*Crassostrea gigas*). *Mar. Biol., 117*, 259–268.

Tibbetts, S. M., Milley, J. E., & Lall, S. P., (2014). Chemical composition and nutritional properties of freshwater and marine microalgal biomass cultured in photobioreactors. *J. Appl. Phycol., 27*(3), 1109–1119.

Tilman, D., Balzer, C., Hill, J., & Befort, B. L., (2011). Global food demand and the sustainable intensification of agriculture. *Proc. Natl Acad. Sci. USA, 108*(50), 20260–20264.

Trubachev, N. I., Gitelzon, I. I., Kalacheva, G. S., Barashkov, V. A., Belyanin, V. N., & Andreeva, R. I., (1976). Biochemical composition of several blue-green algae and *Chlorella*. *Prikladnaia Biohimiia i Mikrobiologiia, 12*, 196–202.

Velichkova, K., Sirakov, I., & Georgiev, G., (2013). Cultivation of *Scenedesmus dimorphus* strain for biofuel production. *J. Agric. Sci. Technol., 5*(2), 181–185.

Verdegem, M. C. J., (2013). Nutrient discharge from aquaculture operations in function of system design and production environment. *Rev. Aquacult., 4*, 1–14.

Villegas, C. T., Millamena, O., & Escritor, F., (1990). Food value of (*Brachionus plicatilis*) fed three selected algal species as live food for milkfish, *Chanos chanos* Forsskal, fry production. *Aquac. Res., 21*, 213–219.

Viso, A., Pesando, D., & Baby, C., (1987). Antibacterial and antifungal properties of some marine diatoms in culture. *Bot. Mar., 30*, 41–46.

Watanuki, H., Ota, K., Tassakka, A. R., Sakai, M., Kato, T., & Sakai, M., (2006). Immunostimulant effects of dietary *Spirulina platensis* on carp, *Cyprinus carpio*. *Aquaculture, 258*, 157–163.

Webb, J. M., Quinta, R., Papadimitriou, S., et al., (2012). Halophyte filter beds for treatment of saline wastewater from aquaculture. *Water Res., 46*, 5102–5114.

Webb, J. M., Quinta, R., Papadimitriou, S., et al., (2013). The effect of halophyte planting density on the efficiency of constructed wetlands for the treatment of wastewater from marine aquaculture. *Ecol. Eng., 61*, 145–153.

Webb, K. L., & Chu, F. E., (1983). Phytoplankton as a food source for bivalve larvae. In: Pruder, G. D., Langdon, C. J., & Conklin, D. E., (eds.), *Proceedings of the Second International Conference on Aquaculture Nutrition: Biochemical and Physiological Approaches to Shellfish Nutrition* (pp. 272–291). Louisiana State University, Baton Rouge, LA.

Xing, R. L., Ma, W. W., Shao, Y. W., et al., (2018). Growth and potential purification ability of *Nitzschia* sp. benthic diatoms in sea cucumber aquaculture wastewater. *Aquac. Res.*, 1–9.

Xing, R. L., Qun, S. U., Wang, C. H., & Sun, L. Q., (2011). Growth and purification ability of *Nitzschia* sp. in wastewater. *Mar. Envir. Sci., 30*, 72–75.

Xu, W., Gao, Z., Qi, Z., Qiu, M., Peng, J. Q., & Shao, R., (2014). Effect of dietary *Chlorella* on the growth performance and physiological parameters of gibel carp, *Carassius auratus gibelio*. *Turk. J. Fish Aquat. Sci., 14*, 53–57.

Yaakob, Z., Ali, E., Zainal, A., Mohamad, M., & Takriff, M. S., (2014). An overview: Biomolecules from microalgae for animal feed and aquaculture. *J. Biol. Res. – Thessalon., 21*, 6.

Yeganeh, S., Teimouri, M., & Amirkolaie, A. K., (2015). Dietary effects of *Spirulina platensis* on hematological and serum biochemical parameters of rainbow trout (*Oncorhynchus mykiss*). *Res. Vet. Sci., 101*, 84–88.

Zeinhom, M. M., (2004). *Nutritional and Physiological Studies on Fish* (pp. 127–136). PhD Thesis. Faculty of Agriculture, Zagazig University, Egypt.

Zhang, L. B., Song, X. Y., Hamel, J. F., & Mercier, A., (2015). Aquaculture, stock enhancement and restocking. In Yang, H. S., Mercler, A., & Hamel, J. F., (eds.), *The Sea Cucumber Apostichopus japonicus: History, Biology and Aquaculture* (pp. 289–318).

Zhang, Q., Qiu, M., Xu, W., Gao, Z., Shao, R., & Qi, Z., (2014). Effects of dietary administration of *Chlorella* on the immune status of gibel carp, *Carassius auratus gibelio*. *Ital. J. Anim. Sci., 13*, 3168.

Zhou, P., Su, C., Li, B., & Qian, Y., (2006). Treatment of high-strength pharmaceutical wastewater and removal of antibiotics in anaerobic and aerobic biological treatment processes. *J. Environ. Eng., 132*, 129–136.

CHAPTER 8

Microalgal Food Biotechnology: Prospects and Applications

ANEELA NAWAZ,[1] USMAN ALI CHAUDHRY,[2] MALIK BADSHAH,[1] and SAMIULLAH KHAN[1]

[1]*Department of Microbiology, Faculty of Biological Sciences, Quaid-i-Azam University, Islamabad – 45320, Pakistan*

[2]*Infection Control and Disease Prevention Center, Ministry of Health, Tabuk, Kingdom of Saudi Arabia*

ABSTRACT

Microalgae are the potential source of sustainable products that are being used in the food industry, animal feed, pharmaceutical, cosmeceuticals, nutraceuticals, and in the production of biofuel. Microalgae being the richest source of protein can potentially meet the population's need for protein energy. Microalgae produce a diverse group of bioactive compounds that are beneficial for human health. Bioactive compounds of microalgae have been attributed to the anticancerous, antioxidant, anticoagulant, immunomodulatory, and hepatoprotective agents. Microalgae produce high valued compounds such as carotenoids, phycobilin pigments, single-cell proteins, polysaccharides, long-chain fatty acids, and exceptionally produce arachidonic acids (ARAs), docosahexaenoic, and eicosapentaenoic that are beneficial to human health. However, the exploitation of microalgae as a food source has drawbacks because of underdeveloped technologies and processes for the processing of microalgae. Systematic improvement in the technology that is applied to the cultivation and processing of microalgae

Microalgal Biotechnology: Bioprospecting Microalgae for Functional Metabolites towards Commercial and Sustainable Applications. Jeyabalan Sangeetha, PhD, Svetlana Codreanu, PhD, & Devarajan Thangadurai, PhD (Eds.)

can make the processing of algae cost-effective. This chapter describes the chemical composition of microalgae with their potential health benefits in food products.

8.1 INTRODUCTION

The term algae refer to the polyphyletic group of diverse species that ranges from unicellular to multicellular organisms and have successfully adapted to different kinds of environmental conditions (Savoie and Saunders, 2019). Microalgae are eukaryotic organisms that can be heterotrophic or autotrophic. Autotrophic microalgae use inorganic nutrients and solar light to produce valuable organic compounds while heterotrophic microalgae require small organic molecules as carbon and energy source (Lubarsky et al., 2010). By providing desired conditions to microalgae, a bulk of biomass containing valuable compounds that include active polysaccharides, proteins, fatty acids, vitamins, and minerals can be produced (Michalak and Chojnacka, 2015). For centuries microalgae had been used by the indigenous population for food purposes, but the cultivation of microalgae had been started a few decades ago. Till now, 30,000 species of microalgae have been reported among the few 1,000 species are kept in the collection, and only a few 100 were investigated for their chemical composition (Ariede et al., 2017). The industrial production of microalgae has been flourished for the last 50 years, but very few species of microalgae have been cultivated for industrial application. Microalgae because of the production of the wide range of biologically active and high-valued compounds have been applied in the food, feed, cosmetics, and pharmaceutical industry as well as in ecological applications (Spolaore et al., 2006).

Hunger is the most lethal problem that is affecting one out of nine individuals globally and is causing deaths globally because of protein malnutrition (Elia, 2000). This problem will get worsen with an increase in population and increase in demand for protein while protein supplies are getting limited. The population of the world is projected to reach 10 billion by 2050 and currently produced food must be doubled to fulfill the demand of the growing population and still there is no solution to the problem of increased demand for food that can be deployed (Tomlinson, 2013). Globally plants are used as the source of protein for humans and animals. In Europe, animal-based protein is mostly used, but health issues and animals' welfare has boosted the use of plants as a protein source (Day, 2013). Most of the agricultural lands have been maximally exploited but the agriculture yield is badly affected by climate changes. Some regions are showing an increase in yield while some are showing decline and the overall balance is negative (Olesen and Bindi, 2002). Expansion of agricultural lands to urban areas is posing a serious challenge to the future of agriculture (Tscharntke et al., 2012). Drastic changes need to be implemented in current agro-technology to cope with the problems of rising temperature, climate change, and loss of cultivatable lands that are affecting agricultural yields. Increasing the yield of agricultural products, farming, and fishing will not fulfill the future demands of food. To deal with the increasing demand for food, there is a need for a scalable and sustainable option. The sustainable solution for the increasing demand for food is the utilization of microalgae for food purposes because microalgae are water-efficient when compared to plants. They can be grown on non-arable lands and even on marine water and saline soil so the area that is not suitable for crops cultivations could be used for large-scale cultivation of microalgae (Smetana et al., 2019). The nutritional compositions of microalgae make it an alternative to conventional crops. Today's food industry is in the search of cheap, healthy, and convenient foods. Natural ingredients like antioxidants and polyunsaturated fatty acids (PUFA) are important for the biochemical and physiological functioning of the human body and can also reduce the incidences of chronic diseases (Lordan et al., 2011). The introduction of natural substances in usual food provides health benefits and show long-term effects, unlike traditional therapeutics that show short-term effects. Microalgae because of their nutritional value and balanced chemical composition are considered as the promising source of balanced food. The use of biomass and metabolites of microalgae in the development of food products is becoming innovative as well as an interesting approach in food biotechnology. Microalgal biotechnology that is similar to conventional agriculture has been given great importance for the last few decades because of high productivities, higher photosynthetic efficiency, and can be grown in environments that are considered unsuitable for agriculture. Microalgae also pose huge advantages over plants like easy manipulation of genetic material, easy scaling of the processes, higher content of protein, high yield of biomass, and ability to be grown on even saltwater. Microalgae can also reduce the global heating of the environment and climatic changes using the fixation of atmospheric carbon dioxide (CO_2) and by recycling the carbon that is fixed in organic products (Kumar et al., 2010). Microalgae cultivation cannot be affected by seasonal variations and can effectively be used as feed to aquaculture to support aquatic life. In the past, yeast and bacteria had been exploited for the production of food products but microalgae are preferred because microalgae use sunlight and fix the atmospheric carbon and do not need a carbon source to be provided for fermentation.

In different cultures, algae had been used as a food source for thousands of years, and the oldest known use of algae is reported from the archeological record of Chile (14,000 years ago). Algae as food had been used globally for the last few centuries and records from Spanish conquistadors showed that *Spirulina* from Lake Texcoco had been harvested by Aztecs. Data showed that microalgae (*Nostoc*) had been first used by the Chinese during famine dated back 2000 years ago for their survival (Sánchez et al., 2003). Microalgae can be incorporated into food as nutritional supplements and as natural colorants. The use of microalgae in foods can also promote health by protecting against renal failure, hypertension, and control blood glucose level, and promotes the growth of *Lactobacilli* in the human gut. The most biotechnologically relevant microalgae are *Dunaliella salina*, *Chlorella vulgaris*, *Haematococcus pluvialis*, *Aphanizomenon flos-aquae* (AFA), *Spirulina maxima*, and *Arthrospira* are commercialized for human nutritional supplements and animal feed additives because of various health benefits that include anti-tumor effects, promoting activities of the immune system and for promotions of animals growth because of the abundance of proteins, vitamins, and active polysaccharides (Kovač et al., 2013). China and India produce *Arthrospira* in large amounts (Khan et al., 2005). *Chlorella* is used in food for enhancing food flavors and as a coloring agent and *D. salina* is used as the source of β-carotene (Herrero et al., 2006). Cognis Nutrition and Health is the world's largest producer of *Dunaliella* and sells it in the form of powder as a functional food and dietary supplement (Shahidi, 2011). *A. flos-aqua* is a commercial strain used in combination with food products or alone and has been reported for the promotion of good health (Spolaore et al., 2006). Microalgae had long been used as a direct or indirect feed source in aquaculture feed that promotes hatchery and provides excellent nutrients to juveniles of shrimp, shellfish, and farmed fishes. The species of microalgae that had been used in aquaculture feed are *Chaetoceros gracilis*, *Chaetoceros muelleri*,

Nannochloropsis oculata and *Isochrysis galbana*. The chapter deals with the application of microalgae in food for the enhancement of its nutritive value, the biochemical composition of microalgae concerning its nutraceutical values, and the safety of microalgae for their applications in the food industry and animal feed industry.

8.2 UTILIZATION OF MICROALGAE AS NUTRITIONAL PRODUCTS AND FOOD

The estimation of the global market of microalgae is difficult because of the diverse and large number of sets of products as well as individual products that are produced at a small scale. The estimate of products of macroalgae in the form of carrageenans and alginates showed a market value of 6.7 billion US dollars per annum. In Asian cuisine, many types of seaweed were used traditionally, like in Japanese sushi Nori is used, which is produced from the dried leaves of *Porphyra* with a market value of approximately 1 billion US dollars per annum. The market of microalgae as food ingredients is also very large, and the market value is more than 1.25 billion US dollars per annum, but this amount is of the dried biomass, not of the processed products that are obtained as food supplements and additives from microalgae (Benemann et al., 2018).

8.3 NUTRITIONAL COMPOSITION OF MICROALGAE

The crucial factor for considering algae as a food source is its nutritional content as well as biochemical composition. The nutritional composition varies among species as well as the same species depending on the environmental conditions. The important nutritional components of algae are proteins, lipids, vitamins, and minerals that have a positive impact on human health. Tocopherols are present widely in photosynthetic and non-photosynthetic tissues of algae and higher plants. *Euglena* contains a higher amount of tocopherols when compared to molds, other algae, and yeast (Takeyama et al., 1997). Studies showed that the growth rate of bivalves is dependent on the amount and type of sterols present in the diet. The polyhydroxy sterols obtained from marine algae showed various biological activities that include anticancer and cytotoxic effects.

8.3.1 PROTEIN AND AMINO ACID COMPOSITION

Protein is the most crucial element in the human diet as it provides the nitrogen required by the human body. Essential amino acids cannot be synthesized by the human body, such as histidine, leucine, isoleucine, methionine, lysine, phenylalanine, threonine, tryptophan, and valine and these amino acids must be provided in the human diet. Moreover, the most important factor of hunger in the world is malnutrition of protein-energy that is causing health problems, and to solve this problem balanced protein diet is needed. Microalgae have been reported for the high percentage of protein content, the biomass of *Arthrospira platensis* contains 70% protein content and the species of microalgae that are recognized as safe contain 40% of protein content (Avila-Leon et al., 2012) when compared to plants sources like a pea that contains 2.8%, rice that contains 10% and soybean that contain 38%, animal sources like egg contains 13% and milk contain 4% of protein. Microalgae not only have high protein content but the content of essential amino acid in the microalgal protein is also very high and the amount of essential amino acids present in a protein determines its quality, in general, the proteins of animal and plants origin are of low quality (Lourenço et al., 2002). Proteins obtained from a certain group of plants have been found deficient in certain essential amino acids like corn is deficient in lysine and tryptophan, legumes are deficient in methionine and cereals are deficient in lysine. Unlike most plants, the protein of *E. gracilis* (that is recognized as safe) contains all the essential amino acids. The main source of protein in today's time are plants in both food and feed, but to meet the increasing requirement cultivation area needs to be expanded, cropping frequency should be changed, but these practices can generate serious problems such as deforestation, biodiversity loss, and degradation of land. The animal-based proteins require cost-effective and appropriate proteins in the feed from plants. To consider these problems, microalgae are a suitable alternative to plants and animal-based proteins. Microalgae have also the benefit of growing on non-arable land and in minimal requirement of water.

Proteins rich microalgae like *Arthrospira*, *Aphanizomenon*, and *Nostoc* have been used thousands of years ago in food (García et al., 2017). Aztecs used to consume *Arthrospira* cakes. In 1952 at the Algae Mass Culture Symposium, microalgae were suggested for their application in food. The first facilities for the commercial production of *Chlorella* were started in Japan (Spolaore et al., 2006). In the United States of America, the regulatory status of algal products and food additives is given to Food and Drug Administration which has assigned GRAS (Generally Recognized as Safe) status to these products. In Europe, the first assessment of a new product is made by a member of the state and if a member state does not object then European Commission authorizes it and if there is any objection made by state member then, in that case, European Food Safety Authority assesses novel food for safety (Lehto et al., 2017).

8.3.2 PHYCOBILIPROTEINS (PBPs)

The phycobiliproteins (PBPs) that are deep-colored water-soluble pigment acting as major components of the photosynthetic system playing role in the harvesting of sunlight and are present in Cyanobacteria, Rhodophyta, and Cryptomonads besides carotenoids and chlorophyll. Phycobiliprotein is formed by the covalently linking of tetrapyrrole chromophoric prosthetic group (phycobilins) to the protein backbone. PBPs have a wide range of applications like they are used as fluorescent markers in clinical diagnosis, labeling of antibodies, and coloring agents in food products and cosmetics. Several studies revealed various pharmacological properties like hepatoprotective, neuroprotective, antioxidant, and anti-inflammatory activity of PBPs (Kannaujiya et al., 2017).

8.3.3 LIPIDS AND FATTY ACIDS

Lipids are the essential molecules that act as a precursor of many essential molecules and appropriate intake of lipid is important in the diet. Due to the absence of enzymes in humans and animals responsible for the synthesis of PUFA containing more than 18 carbon atoms, essential fatty acids must be intake in food. Some algae can accumulate up to 70% of lipids of their dry mass, for example, *Auxenochlorella protothecoides* (Rismani-Yazdi et al., 2015). There are some lipids that are essential like essential amino acids in the human body such as linoleic acid, γ-linolenic acid (GLA), docosahexaenoic acid (DHA), eicosapentaenoic acid (EPA) and α-linolenic acid. Some lipids have been reported with their positive impacts on health like DHA and EPA (Swanson et al., 2012). DHA is vital structural fatty acid playing an important role in the development of the eye and brain of infants (Innis, 2008). Seafood and cold-water fish were used in the diet to fulfill the body's requirement of fatty acids. PUFA help in lowering blood cholesterol and triglycerides and ultimately reduce the risk of atherosclerosis and heart diseases. Fish contain a high amount of omega-3 fatty acids because, in their food, they eat planktons and algae. Algae can produce PUFA, and certain species of microalgae can accumulate 30–40% of EPA of total fatty acids, for example, *Phaeodactylum tricornutum* and some species of microalgae can accumulate 50% of DHA of total fats, for-example, *Schizochytrium* species (Adarme-Vega et al., 2012). Microalgae can therefore be the source of healthy fatty acids alternative to non-sustainable fish oil, having an odor, presence of toxins in fish, unpleasant taste, mixed fatty acids, and of non-vegetarian nature. PUFA plays a vital role in cellular metabolism like help in the regulation of fluidity of plasma membrane, in the transport of electrons and oxygen in the cells, and help in the thermal adaptation of cells and tissue. DHA and EPA produce by *Schizochytrium* strains have been used as dietary supplements in health foods, beverages, animal feed, breakfast cereals, yogurt, cheese, and are also used as a food supplement in the food of nursing and pregnant women and of cardiovascular patients. EPA is produced by *Chlorella minutissima* in high quantity and its production can further be enhanced by lowering the temperature and increasing salinity (Wen and Chen, 2003). *S. platensis* is considered the richest source of GLA (Dunstan et al., 1992). The composition of fatty acids of several cyanobacteria containing 16 and 18 carbon long chains is considered a significant constituent of food. GLAs that occur rarely in nature can be obtained from *Spirulina* and *Nostoc* (Pradhan et al., 2014).

8.3.4 POLYSACCHARIDES

Arthrospira, *Chlorella*, and *Nannochloropsis* had been reported as the best source of polysaccharides and oligosaccharides (Bernaerts et al., 2018). Polysaccharides and oligosaccharides provide potential health benefits by acting as prebiotics. In the food industry, polysaccharides are used as solidifying or thickening agents. Polysaccharides are produced by most of the algal species, but few of the species are used for commercial production because of their easy optimization and fast growth. Highly sulfated polysaccharides of algal origin are used for the stimulation of the human immune system (Leiro et al., 2007).

8.3.5 CAROTENOIDS

Carotenoids are lipophilic molecules with isoprenoids structures that are present in microalgae, non-photosynthetic organisms, and in higher plants, and till now, 400 types of carotenoids have been identified. The widely commercialized types of carotenoids are β-carotene and astaxanthin. Most of the carotenoids are considered anti-cancerous and anti-inflammatory and they also protect organisms from oxidative stress (Pallela et al., 2010).

8.3.5.1 β-CAROTENE

β-carotene, the active provitamin A is considered as most important carotenoid and is used in health food products and as an additive, and in the preparation of multi-vitamins. As compared to the synthetic form of β-carotene its natural form can easily be absorbed by the human body. β-carotene has been intensively used in cosmetics, feed, food, and pharmaceutical products, because of its increasing demand. *D. salina* contains 10–14% of the dry weight of algal biomass. β-carotene production can be enhanced by providing optimal conditions such as increasing temperature, providing nutrients supplements, increasing salinity and light intensity. The 9-cis isomer of β-carotene is important in quenching free radicals while trans is much more active than cis in forming vitamin A (Levin and Mokady, 1994). A large amount of β-carotene is produced in Australia while China, Israel, and the USA are also producing it commercially.

In animals and humans, β-carotene appeared as a protective agent against atherosclerosis (Kritharides and Stocker, 2002). *Dunaliella* rich in β-carotene, when consumed by humans, inhibited the oxidation of high-density lipoproteins and reduced the level of low-density lipoproteins, cholesterols, and triglycerides (Shaish et al., 2006). Epidemiological studies showed that the food rich in β-carotene maintain normal serum level and decreases the chances of degenerative diseases and several types of cancers and oral intake of β-carotene obtained from *Dunaliella* can protect from the erythema induced by UV radiations (Stahl and Sies, 2005).

8.3.5.2 ASTAXANTHIN

The second most important type of carotenoid is astaxanthin, a keto-carotenoid produced by microalgae mostly by *Haematococcus pluvialis*. Astaxanthin can also be produced by *Chlorella zofingiensis* but production is lower than *Haematococcus pluvialis (Hong et al., 2015)*. The production of astaxanthin by *Chlorella zofingiensis* can be increased by giving light and nitrogen stress (Liu et al., 2014). During stress conditions, the production of hydroxyl radicals is increased that enhances the production of astaxanthin. Astaxanthin is used in nutraceuticals, the feeds industry, the food industry, and aquaculture as a source of pigmentation. Astaxanthin is sold in capsule form in nutraceuticals. This metabolite has strong anti-oxidant activity.

Astaxanthin appeared as a UV protectant, protects peroxidation of lipids, inhibits oxidation of low-density lipoproteins, and inhibits oxidation of plasma membrane by its strong free radicals scavenging capacity (Yuan et al., 2011). Studies have shown that astaxanthin protects against diabetes, neurodegenerative diseases, inflammatory diseases, cancers, metabolic diseases, eye diseases, and diabetic neuropathy (Ambati et al., 2014). In mice and rats, astaxanthin appeared to be slowing down colon cancer (Palozza et al., 2009). In response to T-cells dependent stimuli, astaxanthin appeared as enhancing the production of antibodies in human blood cells (Jyonouchi et al., 1993). Astaxanthin had been reported as a protectant of photoreceptors in rats, and it can cross the blood-brain barrier easily without forming crystals in the eyes. The antioxidant potential of astaxanthin protects the cells from oxidative stress and ultimately prevents apoptosis and necrosis (Fang et al., 2017). Astaxanthin also reversed oxidative stress in neutrophils of rats in which diabetes was induced by alloxan. Few studies reported the protective effect of astaxanthin against ulcers caused by *Helicobacter pylori*, preventing obesity and heart diseases (Lee et al., 2020).

8.3.5.3 CANTHAXANTHIN

Canthaxanthin is produced by *Coelastrella striolata* and *C. zofingiensis* under nitrogen deprivation and salt stress (Abe et al., 2007). It is used to color the skin of chicken, egg yolk and also be used as a drying agent in food. It has a neuroprotective effect, anti-inflammatory effect, antioxidative effect and can increase the level of vitamin E in the liver.

8.3.5.4 LUTEIN

Muriellopsis sp. and *Scenedesmus almeriensis* produce and accumulate a higher amount of lutein at optimal temperature. Production of lutein in *Chlamydomonas zofingiensis* and *Auxenochlorella* (*Chlorella*) *protothecoides* increases with the variation in pH (D'Alessandro and Antoniosi Filho, 2016). Lutein is the most important component of food, human serum, the lens of the eye, and macula lutea (Roberts et al., 2009). Commercial production of lutein from microalgae has not been started yet but pilot-scale production has been set up for the production of lutein from *Muriellopsis* sp. and *Scenedesmus* by outdoor cultivation. Lutein is sold as a food additive for the prevention of metabolic and inflammatory diseases in humans.

8.3.6 CHLOROPHYLL

All algae contain one or more types of chlorophyll that account for 0.5–1.5% of total dried algal biomass. Chlorophyll is used as a natural colorant in food and pharmaceutical products. The derivatives of chlorophyll can promote health, heal wounds, can protect the production of crystals of calcium oxalates in the human body, and can protect from colorectal cancers (Gouveia et al., 2010).

8.3.7 OTHER NUTRIENTS

Vitamins and minerals are crucial in the human diet and are not synthesized in animals; plants produce them and are taken by animals and humans in their diet. Like green vegetables, algae are also considered the richest source of vitamins and minerals. The great source of Vitamin E, B1, B9, and A is *Dunaliella tertiolecta.* The consumption of microalgae can provide vitamins and protects from aging, oxidative stress, cancers, and cardiovascular diseases (CDVs) (Wang et al., 2020).

8.4 MICROALGAE AS FOOD AND DIETARY SUPPLEMENTS

Microalgae because of their high nutritional value are consumed as food in several countries. In Mexico and Chad, the natives have harvested *Spirulina* for their food purposes since ancient times. In Chad *Spirulina* has been used for the preparation of dry cake that is named dihe and it was harvested from the Kossorom Lake that is an alkaline lake. In Mexico, *Spirulina* has been used for the production of dry cake and is collected from Texcoco Lake (da Silva Vaz et al., 2016). In present days, the trade of dihe is more than 100,0000.8 US Dollars and *Spirulina* is contributing a lot to the economy of Chad. In China, India, Thailand, and United States, *Spirulina* is cultured on a large scale using open ponds for a dietary supplement. The estimated annual production of *Spirulina* is 3,000 to 4,000 metric tons due to its high protein content, minerals, vitamins, and GLA is considered a nutritious food. The known implications of *Spirulina* in therapeutics are in cancers, anemia, arthritis, diabetes, and CDVs. *Spirulina*, because of its nutritional qualities is incorporated in functional foods to enhance the nutritional value and in the therapeutic management of certain chronic diseases like heart diseases, diabetes, and hypertension. The well-known compounds of *Spirulina* with antioxidant potential are vitamin E and phycocyanin (PC) (Mani et al., 2007). *Nostoc* is another cyanobacterium that has been consumed by the Chinese, 2,000 years ago as food. This alga has been used because of its pigments, low fat, and high protein content. The main species of *Nostoc* with economic value is *Nostoc flagelliforme* that contains a high amount of pigments such as chlorophyll, PC, allophycocyanin (APC), myxoxanthophyll, and echinenone. Since 400 years ago, *Nostoc flagelliforme* had been used in traditional medicines in China for the treatment of diarrhea, hepatitis, and hypertension. The proteins of *Nostoc flagelliforme* contain 19 amino acids and eight of them are essential ones that account for 35.8–38% of total protein content. *Nostoc sphaeroides* is collected from paddy fields in China for food purposes. *Chlorella* is produced on a commercial scale and is sold in powder and pill form, the first pilot-scale plant of mass culture of *Chlorella* was tested in Boston, and later, the cultivation plants of mass culture were tested in Japan, Czechoslavakia, and Israel. Nihon Chorella Inc. in 1961 has established the first commercial production plant of *Chlorella* and later Taiwan, Indonesia, and Malaysia have started their production plants. In Asia, by 1980, 46 commercial-scale plants for the production of *Chlorella* were established with 1,000 kilograms of production, monthly. In Taiwan more than 70 companies are producing *Chlorella* with the sale of more than 38 billion US dollars, annually. In the early 1970s, Malaysia has discontinued its *Chlorella* plants and now is importing them from Taiwan and Japan. *Chlorella* is nutritionally rich and contains 51–58% of proteins of their dry weight, vitamins, carotenoids, and β-glucan. β-glucan helps in reducing blood lipids and is an immunostimulatory and antioxidant. The growth factors extracted from *Chlorella* are used for the stimulation of the growth of lactic acid bacteria. Now a days the active research on microalgae is on increasing the capacity of algae for the production of minerals, nutrients, and trace elements. Food and Drug Administration has recognized some species of microalgae safe because of safe production and consumption history (Novoveská et al., 2019).

Many products and ingredients derived from microalgae are in the commercial market for human consumption because the biomass of microalgae is a source of protein, lipids, carbohydrates, and vitamins. The proteins from the microalgae are considered sustainable compared to plant and animal-based proteins (Palanisamy et al., 2019). Microalgae are also a source of biologically active secondary metabolites that are providing therapeutic benefits in curing inflammation and protect against neurodegenerative diseases (Montero-Lobato et al., 2018). The biomass of microalgae is marketed in powder form that is used as a dietary supplement for improving health and is also be added in blended beverages. Powder of microalgae appeared as safe by clinical trials and showed promising outcomes like; some species of *Chlorella* and *Spirulina* showed antioxidant potential and reduce the lipid content of the blood.

The public prefers natural food additives over synthetic ones, and products of microalgae are considered as appealing alternatives because they are derived from nature and are highly bioactive. The ingredients of microalgae because of free radicals scavenging potential are preferred in the fortification of food for health promotion. Synthetic pigments are being replaced with natural ones because of their adverse health effect, toxicities and low nutritional value (Oplatowska-Stachowiak and Elliott, 2017). Microalgae have been reported for the production of fat-soluble pigments that include carotenoids and chlorophyll and water-soluble pigments that include phycobilins. Bioactive pigments such as β-carotene and astaxanthin are stored at a significant level in some species of *Dunaliella* and *Haematococcus*. Astaxanthin is the bright red colored pigment used to color seafood. Humans lack the capability of production of astaxanthin and are taking from the seafood diet (Guerin et al., 2003). Astaxanthin is also produced synthetically but the production from the microalgae is preferred. PC produced by *Spirulina* is a blue pigment and is used as natural coloring agents in jellies, drinks, candies, chewing gums, and popsicles. Phytosterols are compounds extracted from plants and used as food supplements and for lowering the blood cholesterol level. The sterols produced by microalgae are extremely diverse and are similar to the sterols synthesized by animals, plants, and fungi. Some of the sterols that are uncharacterized like; triterpenoids and some are novel like brevisterol and gymnodinosterol. Stigmasterols produced by microalgae are used in food for lowering cholesterol. The production of phytosterols by microalgae is equal or greater than the production of phytosterols by plants that's why algae are considered as an alternative source of cheap and sustainable phytosterols (Luo et al., 2015).

PUFA like DHA and EPA are very important for human health and are essential for neural development, these fatty acids are used in fortified beverages, infant food, and are also sold as dietary supplements (Whelan and Rust, 2006). Presently, the natural source of DHA and EPA are sea animals. Microalgae, the primary producers can produce and store DHA up to 50% of dry mass and appeared as an alternative and promoting source. Unlike DHA, the utilization of microalgae for the production of EPA is less progressed. In many microalgae, a high quantity of EPA is produced and stored, but the production of EPA from microalgae has not been seen as economical, to make its production economical, processing technology needs to be enhanced to achieve the market viability of EPA. The industry of microalgae is currently contributing to the food and nutrition industry, but the greater utilization is still facing problems because the optimization of production of specific metabolites required excessive screening as well as biological evaluation and also the understanding of the effect of seasonal and geographical variation on the production of desired metabolites (Mimouni et al., 2012). Phenomics is an emerging technology that could be used for finding multiple traits for making the optimization process easy. To compete with the metabolites of other sources and synthetic ones, cheap bioprocessing is required to make the product economically viable. To cater to the increasing demand for food phenomics and bioprocessing technologies are being used as a tool to reduce the production and processing cost of microalgae.

The utilization of novel algae strains for food products need to be first approved as safe by Food and Drug Administration. Familiarity and complete knowledge must be gained before achieving the confidence of consumers. To overcome the negative perceptions of consumers about bio-products, clinical trials should be done for ensuring the safety of food. Adoption of products by the consumers is based on safety, taste, sustainability, and nutritional benefits.

Foods also provide health benefits to the consumers in addition to energy and growth. Proteins and peptides are considered as the main compound that provides biological functionalities. Peptides are made up of 20–30 amino acids that remain inactive in primary structure until released in the gastrointestinal tract by digestive enzymes or when the digestion of food is done by bioprocesses like fermentation, ripening, cooking as well as during storage. The specific protein fragments have shown a positive impact on health. Some of the peptides of microalgae were attributed as hepatoprotective, antihypertensive, anti-carcinogenic, antioxidant, anti-coagulant, and immune-modulator. Very rare knowledge is present on the health benefits of peptides, but the use of peptides of microalgae in functional food is gaining huge interest.

In most industrial applications, the antioxidant compounds from microalgae are gaining interest. During the storage of food, oxidation of proteins and lipids occur which leads to the production of toxic compounds. During the oxidation of proteins and lipids, the small molecules are produced that give food an unacceptable taste and ultimately affect consumer acceptability. Moreover, these compounds can also be toxic and can lead to chronic diseases like neurological disorders, coronary heart disease, cancer, arteriosclerosis, and diabetes. The solution to prevent the oxidation of food components is by adding antioxidants, protecting food items from light, evacuating air, and reducing the contents of pro-oxidants in food (Shashirekha et al., 2015). Since 1970, the chemical antioxidants that are used in foods preservation to avoid oxidation are butylated hydroxytoluene (BHT) and butylated hydroxyanisole, utilization of these chemical antioxidants was regulated by law but the adverse effect of these synthetic antioxidants are reported (Shahidi and Ambigaipalan, 2005). Now consumers are diverting their preferences on natural preservatives and to fulfill the demands of consumers, food industries are now adopting technologies to use natural antioxidants for the preservation of food from microalgae.

8.5 INCORPORATION OF MICROALGAE IN FOOD FOR PROMOTION OF HEALTH

In regularly consumed food, bioactive compounds can be added to fulfill the human body's requirements. In food with different purposes, whole cells of microalgae, as well as extracted compounds, are being added while the addition of peptides from microalgae has not been reported to be used in food, till now. The addition of certain species of microalgae in food emulsion enhances the antioxidant properties and techno-functional properties of food (Yousefi and Jafari, 2019). The addition of microalgae in jelly food enhances texture, improves its structure, and provides omega 3 fatty acids and antioxidants properties (Sidari and Tofalo, 2019). To enhance the number of bioactive compounds in dairy products, microalgae can also be added. The addition of some species of *Arthrospira* in fermented milk and yogurt stimulates the growth of probiotics and the viability of probiotics also increases. Desired bacteria in dairy products are promoted by the vitamins, trace elements, and other bioactive compounds of microalgae. A synergistic relationship has been found between probiotics and microalgae in dairy products; microalgae produce exopolysaccharides (EPS) that promotes the growth of bacteria in yogurt (Beheshtipour et al., 2013). The successful incorporation of *Chlorella* in cheese and yogurt has been made.

Microalgae-based products for human consumption in the form of cookies and biscuits are also prepared because of acceptable taste, appearance, texture, and convenience consumption. In cookies, *Chlorella vulgaris* and *Isochrysis galbana* have been incorporated as a nutritional supplement, a good source of antioxidants, and as a coloring agent and to provide omega-3 fatty acids. For enhancing the protein and fiber content of cookies *A. platensis* has been incorporated. In the previous study conducted by Batista et al. (2013), *A. platensis*, *C. vulgaris*, *Tetraselmis suecica*, and *Phaeodactylum tricornutum* are added in cookies for enhancing the nutritional content. The addition of *Haematococcus pluvialis* in cookies enhanced the nutritional value and health benefits of cookies (Batista et al., 2013). The incorporation of these microalgae in cookies has increased antioxidant capacity. Bread is also consumed in large quantities worldwide, several years ago, it was reported by several authors that the incorporation of microalgae in bread enhances its nutritive value. To enhance the protein content of bread *Arthrospira* and its extract were used. *Dunaliella* was also suggested to increase the protein content of bread. Protein quantity and the bread quality of gluten-free bread were improved by the addition of *Arthospira* compared with non-supplemented bread (Tertychnaya et al., 2020). Nutritional enhancements of snacks were also enhanced by the addition of *Arthrospira*. Pasta is a widely accepted food, the addition of *Arthrospira maxima* and *C. vulgaris* in fresh spaghetti enhanced the taste and nutritional content. The incorporation of *I. galbana* and *Diacronema vlkianumin* in pasta increased the content of omega 3 fatty acids and antioxidants potential of pasta. The powder of *Dunaliella salina* was used to increase the nutritional value of pasta but the addition of microalgae in low quantity (below 3%) increased the number of minerals in a significant amount. The utilization of microalgae in some food is limited because of some of the characteristics of microalgae, for example, *Arthrospira*, and *Chlorella* in addition to enhancing antioxidants contents also change the flavor and food color, that consumer found undesirable (Fradique et al., 2010).

The applicability of different food additives is determined by the techno-functional properties of the additives. Some microalgae have been reported for the properties like fat absorption, emulsification, gelation, and foaming. The protein and the protein hydrolysate of some microalgae are reported for the absorption of fats from foods. The application of microalgae into emulsion has shown a reduction in its oil content and protection from oxidation. Vegetable proteins are used for the stability of emulsion, but algal proteins could be alternative to these plant-based proteins, as algal protein can preserve emulsion without affecting its stability and enhancing the resistance of emulsion to oxidation. Incorporation of microalgae in vegetarian dessert as coloring agent protected the thermal degradation of pigments of vegetables during processing. The introduction of microalgae in the gel has also improved the rheological and structural properties of gel but improvement in these properties depends on the species as well as lipid, protein, and salt content of microalgae. The addition of *Chlorella* in processed cheese has increased its hardness and springiness and has reduced the cohesiveness and meltability of cheese. The firmness of cookies and bread was increased by the incorporation of microalgae. The addition of microalgae in dough affects the texture and structure of dough because of lipids incorporation and changes in the absorption of water due to complex ingredients of microalgae (Tang et al., 2020). The addition of microalgae has not been found to affect the textural and cooking properties of pasta, but the mechanical strength of raw pasta has been decreased by the addition of *Arthrospira*.

8.6 SAFE STATUS OF ALGAE

Food and Drug Administration has given safe status to algae and algae-derived products. The consumption of whole algae or any chemical substance of algae is considered safe for human consumption. The safe status is either gained by documented consumption by humans use for years and/or by providing scientific evidence that the substance poses no harm to the consumer. If the organism is recognized as safe then the product obtained from it requires little or no purification but if the organism has not been given safe status, in that case, a high degree of purification is required to ensure the safety of the product and the product cost gets increased because of expenses investments on a high level of purification (Day et al., 2009).

The species of microalgae that are recognized as safe by the Food and Drug Administration Authority are *Arthrospira platensis*, *Chlamydomonas reinhardtii*, *Auxenochlorella protothecoides*, *Chlorella vulgaris*, *Dunaliella bardawil*, and *Euglena gracilis*. The safe status of the species of microalgae is limited because obtaining safe status is costly (expensive safety tests) and also time-consuming. United States jurisdiction applies GRAS (Generally Recognized as Safe) designation (Torres Tiji et al., 2020) while in Europe the regulation about animals and human foods are overseen by European Food Safety Authority. Food that was consumed before May 1997 with precautions is considered safe while genetically modified food is considered as a novel food by European Food Safety Authority. Before marketing these novel foods, safety assessments are made (Robinson et al., 2013). In Canada, Health Canada is an organization that has been given charge of stipulating and supervising the safety of food. In Canada, novel foods are the foods that are new or changes are made in the composition of any existing foods. In India, Japan, China, and Canada, *Arthrospira platensis* is considered safe for human consumptions (Niccolai et al., 2019).

8.7 MARKETABILITY OF PRODUCTS OF MICROALGAE

Drawbacks are associated with the large-scale commercialization of microalgae products. The bulk products in the market that are extracted from seaweeds or algae are allocated to aquaculture and are processed for the production of high-value products. Multiple barriers are there for getting approval for new products. The proven market demand and the value of microalgae-based products are difficult to estimate and ensure because the investors before making investments seek an opportunity of having long-term market demand, so for the competitive market, research on the cultivation of microalgae and the development of sustainable processes is required (Jong et al., 2012). If the comparison of microalgae proteins is made with crop proteins, the optimization of algae for the maximum production of protein is easy as compared to crops based because crops take a year to grow. By increasing the cultivation area the biorefining cost of biomass reduces. Implementation of changes in the production processes of microalgae is difficult. The incorporation of microalgae in food is dependent on the composition, which is dependent on the species and cultivation conditions. The production cost of biomass of microalgae is also affected by the harvesting technologies, facility size, and cultivation system as well as the dewatering process. For improving the economic scale of the cultivation and bioprocessing of microalgae, systematic improvements are required in technology readiness level to enhance the development and maturity of technology.

Biorefining of microalgae costs 20–40% of total production cost and in underdeveloped technologies this cost further increases to 50–60%. The cost of biorefining is reduced by the proper optimization of processes involved in the disruption of cells, extraction, and cell fractionation among other processes. With the technology level and economic scale production of microalgae proteins, the sustainability of microalgae proteins is improved as the substitutes of meat. Compared to pork and meat high moisture extruded *Chlorella* appeared as more sustainable.

The cost of metabolites such as PC, astaxanthin, and β-carotene is between hundreds to thousands of euros per kilogram depending on the level of purity and is becoming an appealing business. The cost of whole cells of microalgae is below 40 euro per kilogram, marketed as food supplements. The economic feasibility of microalgae-based biofunctional compounds could be improved by introducing new cultivation and processing technologies. The business of microalgae biomass and the products is expanding and large enterprises are taking interest globally. The development of sustainable biorefinery for the recovery of microalgae-based products for food applications could improve the economic conditions (Lian et al., 2015).

8.8 NEW TRENDS IN THE APPLICATIONS OF MICROALGAE IN FOOD ITEMS

The commercial production and the utilization of microalgae in food items is reality and are adapted all over the world. Microalgae in combination with other foods or the form of the mixture are marketed in pills, capsules, powder, or liquid form for the supplementation of food. Microalgae are also incorporated in pasta, snacks, bread, biscuits, yogurt, candies, and soft drinks for improving health benefits and modulating immune responses. Nowadays, consumers are demanding natural food supplements for the promotion of health. The combination of microalgae with functional foods is convenient and provides health benefits with attractiveness to consumers (Sidari and Tofalo, 2019). Food production and distribution companies of the USA, China, Germany, Japan, France, and Thailand have started the addition of microalgae in functional foods. In the exploitation of microalgae in food biotechnology, food safety regulations are becoming the main constraints, but the successful approval of few species by the safety regulation authorities as novel food is promoting the utilization of microalgae (Buono et al., 2014). In Portugal novel and healthy foods rich in PUFA and carotenoids were prepared from algal biomass after assessment of toxicological studies of the microalgae incorporated in these foods (Sousa et al., 2009). In Europe, traditional foods like biscuits, pasta, cereals, desserts, and mayonnaise can be used as vehicles to the nutraceuticals obtained from microalgae, but Europeans have strong cultural motivations, and they show resistance to innovations in foods. The addition of natural substances in usual foods has long-term health effects compared to traditional medicines.

8.9 MICROALGAE IN FEED

Algae, with the potential of emerging biological resources, are used as feedstock in some of the well-established industries. Microalgae have been used in the feed of aquaculture by the aquaculture industry for decades (Muller-Feuga, 2000). The balanced nutritional composition of microalgae and rapid growth rate is making it the best aquaculture feed and is used indirectly as an algal meal, that is, a residual material extracted after the extraction of oil (Guedes and Malcata, 2012). Traditionally fishmeal that is obtained by cooking, drying, and grinding of fish is used as aquaculture feed because of its high protein content and low cost of production, fishmeal because of deterioration of ecosystem and collapse of local fisheries is considered as unsuitable. Alternatives to fishmeal are legumes, cottonseed meal, soy meal, and algae. Algal fishmeal is considered the best alternative to fishmeal because the production of the algal meal at a large-scale cost less compared to small scale production and low requirement of freshwater and land (Ansari et al., 2020). Large-scale business incorporating multiple products is helping the production of algae-based fish feed because the extraction of PUFA from algae produces residual biomass that can be used as aquafeed. Residual biomass of *Nannochloropsis* has shown great potential for the feed of common carp, white leg shrimp, and Atlantic salmon (Sørensen et al., 2017). Algal aquafeed is a viable and emerging feedstock if it is obtained by a business model that is integrated with the biorefinery of multiple products (Subhadra, 2010). The use of algae in fish feed is not surprising because it is the natural food consumed by fish in their natural habitat and the species of algae used in fish feed belong to the genera *Isochrysis*, *Pavlova*, *Nannochloropsis*, *Tetraselmis*, and *Chlorella* (Priyadarshani and Rath, 2012).

At the current time, 30% of the total microalgae biomass is incorporated in the animal feed while 50–60% of *Arthrospira* is used as a feed supplement (Madeira et al., 2017). Nutritional and toxicological evaluation of microalgae proved it suitable for the feed of cats, dogs, bulls, horses, cows, and ornamental birds. The large profile of vitamins, minerals, and fatty acids of microalgae appeared to affect the physiology of animals positively by improving weight gain and fertility and have also affected the appearance of animals like lustrous and healthy skin (Spolaore et al., 2006). In rations of poultry 5–10% of algae are used as partial replacement of plant-based proteins.

8.10 CONCLUSION

In recent years, research on the use of microalgae in food and animals feed has increased. The biochemical composition of microalgae is making it a good alternative to animals and plants-based food. Moreover, the issues associated with agriculture like scarcity of fertile land, water resources, and uncontrollable environmental factors can also be overcome by the use of microalgae as an alternative source. The health benefits posed by microalgae are making its application acceptable as food additives.

KEYWORDS

- **bioactive compounds**
- **biorefining**
- **dietary supplements**
- **food technology**
- **health benefits**
- **microalgae**
- **pigments**
- **protein-energy**

REFERENCES

Adarme, V. T. C., Lim, D. K., Timmins, M., Vernen, F., Li, Y., & Schenk, P. M., (2012). Microalgal biofactories: A promising approach towards sustainable omega-3 fatty acid production. *Microbial Cell Factories, 11*(1), 96.

Ambati, R. R., Phang, S. M., Ravi, S., & Aswathanarayana, R. G., (2014). Astaxanthin: Sources, extraction, stability, biological activities and its commercial applications: A review. *Marine Drugs, 12*(1), 128–152.

Ansari, F. A., Nasr, M., Guldhe, A., Gupta, S. K., Rawat, I., & Bux, F., (2020). Techno-economic feasibility of algal aquaculture via fish and biodiesel production pathways: A commercial-scale application. *Science of the Total Environment, 704*, 135–259.

Ariede, M. B., Candido, T. M., Jacome, A. L. M., Velasco, M. V. R., De Carvalho, J. C. M., & Baby, A. R., (2017). Cosmetic attributes of algae - A review. *Algal Research, 25*, 483–487.

Avila-Leon, I., Chuei, M. M., Sato, S., & De Carvalho, J., (2012). *Arthrospira platensis* biomass with high protein content cultivated in continuous process using urea as nitrogen source. *Journal of Applied Microbiology, 112*(6), 1086–1094.

Batista, A. P., Gouveia, L., Bandarra, N. M., Franco, J. M., & Raymundo, A., (2013). Comparison of microalgal biomass profiles as novel functional ingredient for food products. *Algal Research, 2*(2), 164–173.

Benemann, J. R., Woertz, I., & Lundquist, T., (2018). Autotrophic microalgae biomass production: From niche markets to commodities. *Industrial Biotechnology, 14*(1), 3–10.

Bernaerts, T. M., Gheysen, L., Kyomugasho, C., et al., (2018). Comparison of microalgal biomasses as functional food ingredients: Focus on the composition of cell wall-related polysaccharides. *Algal Research, 32*, 150–161.

D'Alessandro, E. B., & Antoniosi, F. N. R., (2016). Concepts and studies on lipid and pigments of microalgae: A review. *Renewable and Sustainable Energy Reviews, 58*, 832–841.

Da Silva, V. B., Moreira, J. B., De Morais, M. G., & Costa, J. A. V., (2016). Microalgae as a new source of bioactive compounds in food supplements. *Current Opinion in Food Science, 7*, 73–77.

Day, A. G., Brinkmann, D., Franklin, S., et al., (2009). Safety evaluation of a high-lipid algal biomass from *Chlorella protothecoides*. *Regulatory Toxicology and Pharmacology, 55*(2), 166–180.

Day, L., (2013). Proteins from land plants–potential resources for human nutrition and food security. *Trends in Food Science and Technology, 32*(1), 25–42.

Dunstan, G., Volkman, J., Jeffrey, S., & Barrett, S., (1992). Biochemical composition of microalgae from the green algal classes *Chlorophyceae* and *Prasinophyceae*. 2. Lipid classes and fatty acids. *Journal of Experimental Marine Biology and Ecology, 161*(1), 115–134.

Elia, M., (2000). Hunger disease. *Clinical Nutrition, 19*(6), 379–386.

Fang, Q., Guo, S., Zhou, H., Han, R., Wu, P., & Han, C., (2017). Astaxanthin protects against early burn-wound progression in rats by attenuating oxidative stress-induced inflammation and mitochondria-related apoptosis. *Scientific Reports, 7*(1), 1–13.

Fradique, M., Batista, A. P., Nunes, M. C., Gouveia, L., Bandarra, N. M., & Raymundo, A., (2010). Incorporation of *Chlorella vulgaris* and *Spirulina maxima* biomass in pasta products. Part 1: Preparation and evaluation. *Journal of the Science of Food and Agriculture, 90*(10), 1656–1664.

García, J. L., De Vicente, M., & Galán, B., (2017). Microalgae, old sustainable food and fashion nutraceuticals. *Microbial Biotechnology, 10*(5), 1017–1024.

Gouveia, L., Marques, A. E., Sousa, J. M., Moura, P., & Bandarra, N. M., (2010). Microalgae–source of natural bioactive molecules as functional ingredients. *Food Sci Technol Bull Funct Foods, 7*(2), 21–37.

Guedes, A. C., & Malcata, F. X., (2012). Nutritional value and uses of microalgae in aquaculture. *Aquaculture, 10*(1516), 59–78.

Guerin, M., Huntley, M. E., & Olaizola, M., (2003). Haematococcus astaxanthin: Applications for human health and nutrition. *Trends in Biotechnology, 21*(5), 210–216.

Herrero, M., Jaime, L., Martín-Álvarez, P. J., Cifuentes, A., & Ibáñez, E., (2006). Optimization of the extraction of antioxidants from *Dunaliella salina* microalga by pressurized liquids. *Journal of Agricultural and Food Chemistry, 54*(15), 5597–5603.

Hong, M. E., Hwang, S. K., Chang, W. S., Kim, B. W., Lee, J., & Sim, S. J., (2015). Enhanced autotrophic astaxanthin production from *Haematococcus pluvialis* under high temperature via heat stress-driven Haber–Weiss reaction. *Applied Microbiology and Biotechnology, 99*(12), 5203–5215.

Innis, S. M., (2008). Dietary omega 3 fatty acids and the developing brain. *Brain Research, 1237*, 35–43.

Jyonouchi, H., Zhang, L., & Tomita, Y., (1993). Studies of immunomodulating actions of carotenoids. II. Astaxanthin enhances in vitro antibody production to T-dependent antigens without facilitating polyclonal B-cell activation. *Nutr. Cancer 19*(3), 269-280. doi: 10.1080/01635589309514258.

Khan, Z., Bhadouria, P., & Bisen, P., (2005). Nutritional and therapeutic potential of *Spirulina*. *Current Pharmaceutical Biotechnology, 6*(5), 373–379.

Kovač, D. J., Simeunović, J. B., Babić, O. B., Mišan, A. Č., & Milovanović, I. L., (2013). Algae in food and feed. *Food and Feed Research, 40*(1), 21–32.

Kritharides, L., & Stocker, R., (2002). The use of antioxidant supplements in coronary heart disease. *Atherosclerosis, 164*(2), 211–219.

Kumar, A., Ergas, S., Yuan, X., et al., (2010). Enhanced CO_2 fixation and biofuel production via microalgae: Recent developments and future directions. *Trends in Biotechnology, 28*(7), 371–380.

Lee, H., Lim, J. W., & Kim, H., (2020). Effect of astaxanthin on activation of autophagy and inhibition of apoptosis in *Helicobacter pylori*-infected gastric epithelial cell line AGS. *Nutrients, 12*(6), 1750.

Lehto, S., Buchweitz, M., Klimm, A., Straßburger, R., Bechtold, C., & Ulberth, F., (2017). Comparison of food color regulations in the EU and the US: A review of current provisions. *Food Additives and Contaminants: Part A, 34*(3), 335–355.

Leiro, J. M., Castro, R., Arranz, J. A., & Lamas, J., (2007). Immunomodulating activities of acidic sulphated polysaccharides obtained from the seaweed *Ulva rigida* C. Agardh. *International Immunopharmacology, 7*(7), 879–888.

Levin, G., & Mokady, S., (1994). Antioxidant activity of 9-cis compared to all-trans β-carotene *in vitro*. *Free Radical Biology and Medicine, 17*(1), 77–82.

Liang, Y., Kashdan, T., Sterner, C., Dombrowski, L., Petrick, I., Kröger, M., & Höfer, R., (2015). Algal biorefineries. *Industrial Biorefineries and White Biotechnology* (pp. 35–90), Elsevier.

Lordan, S., Ross, R. P., & Stanton, C., (2011). Marine bioactives as functional food ingredients: Potential to reduce the incidence of chronic diseases. *Marine Drugs, 9*(6), 1056–1100.

Lourenço, S. O., Barbarino, E., De-Paula, J. C., Pereira, L. O. D. S., & Marquez, U. M. L., (2002). Amino acid composition, protein content and calculation of nitrogen-to-protein conversion factors for 19 tropical seaweeds. *Phycological Research, 50*(3), 233–241.

Lubarsky, H. V., Hubas, C., Chocholek, M., et al., (2010). The stabilization potential of individual and mixed assemblages of natural bacteria and microalgae. *PLoS One, 5*(11), e13794.

Mani, U. V., Iyer, U. M., Dhruv, S. A., Mani, I. U., & Sharma, K. S., (2007). Therapeutic utility of *Spirulina*. *Spirulina in Human Nutrition and Health*, 71–99.

Michalak, I., & Chojnacka, K., (2015). Algae as production systems of bioactive compounds. *Engineering in Life Sciences, 15*(2), 160–176.

Montero, L. Z., Vázquez, M., Navarro, F., et al., (2018). Chemically induced production of anti-inflammatory molecules in microalgae. *Marine Drugs, 16*(12), 478.

Muller, F. A., (2000). The role of microalgae in aquaculture: Situation and trends. *Journal of Applied Phycology, 12*(3–5), 527–534.

Niccolai, A., Venturi, M., Galli, V., et al., (2019). Development of new microalgae-based sourdough "crostini": Functional effects of *Arthrospira platensis* (*Spirulina*) addition. *Scientific Reports, 9*(1), 1–12.

Novoveská, L., Ross, M. E., Stanley, M. S., Pradelles, R., Wasiolek, V., & Sassi, J. F., (2019). Microalgal carotenoids: A review of production, current markets, regulations, and future direction. *Marine Drugs, 17*(11), 640.

Olesen, J. E., & Bindi, M., (2002). Consequences of climate change for European agricultural productivity, land use and policy. *European Journal of Agronomy, 16*(4), 239–262.

Oplatowska-Stachowiak, M., & Elliott, C. T., (2017). Food colors: Existing and emerging food safety concerns. *Critical Reviews in Food Science and Nutrition, 57*(3), 524–548.

Palanisamy, M., Töpfl, S., Berger, R. G., & Hertel, C., (2019). Physico-chemical and nutritional properties of meat analogues based on *Spirulina*/lupin protein mixtures. *European Food Research and Technology, 245*(9), 1889–1898.

Pallela, R., Na-Young, Y., & Kim, S. K., (2010). Anti-photoaging and photoprotective compounds derived from marine organisms. *Marine Drugs, 8*(4), 1189–1202.

Palozza, P., Torelli, C., Boninsegna, A., et al., (2009). Growth-inhibitory effects of the astaxanthin-rich alga *Haematococcus pluvialis* in human colon cancer cells. *Cancer Letters, 283*(1), 108–117.

Pradhan, J., Das, S., & Das, B. K., (2014). Antibacterial activity of freshwater microalgae: A review. *African Journal of Pharmacy and Pharmacology, 8*(32), 809–818.

Priyadarshani, I., & Rath, B., (2012). Commercial and industrial applications of microalgae – A review. *Journal of Algal Biomass Utilization, 3*(4), 89–100.

Rismani-Yazdi, H., Hampel, K. H., Lane, C. D., et al., (2015). High-productivity lipid production using mixed trophic state cultivation of *Auxenochlorella* (*Chlorella*) *protothecoides*. *Bioprocess and Biosystems Engineering, 38*(4), 639–650.

Roberts, R. L., Green, J., & Lewis, B., (2009). Lutein and zeaxanthin in eye and skin health. *Clinics in Dermatology, 27*(2), 195–201.

Robinson, C., Holland, N., Leloup, D., & Muilerman, H., (2013). Conflicts of interest at the European food safety authority erode public confidence. *J. Epidemiol Community Health, 67*(9), 717–720.

Sánchez, M., Bernal-Castillo, J., Rozo, C., & Rodríguez, I., (2003). *Spirulina* (*Arthrospira*): An edible microorganism: A review. *Universitas Scientiarum, 8*(1), 7–24.

Savoie, A. M., & Saunders, G. W., (2019). A molecular assessment of species diversity and generic boundaries in the red algal tribes polysiphonieae and streblocladieae (*Rhodomelaceae, Rhodophyta*) in Canada. *European Journal of Phycology, 54*(1), 1–25.

Shahidi, F., (2011). Global trends in marine nutraceuticals an expanding body of scientific research indicates that the marine environment is a unique resource of functional food ingredients with health-promoting properties. *Global Trends, 65*(12).

Shaish, A., Harari, A., Hananshvili, L., et al., (2006). 9-cis β-carotene-rich powder of the alga *Dunaliella bardawil* increases plasma HDL-cholesterol in fibrate-treated patients. *Atherosclerosis, 189*(1), 215–221.

Shashirekha, M., Mallikarjuna, S., & Rajarathnam, S., (2015). Status of bioactive compounds in foods, with focus on fruits and vegetables. *Critical Reviews in Food Science and Nutrition, 55*(10), 1324–1339.

Sidari, R., & Tofalo, R., (2019). A comprehensive overview on microalgal-fortified/based food and beverages. *Food Reviews International, 35*(8), 778–805.

Smetana, S., Schmitt, E., & Mathys, A., (2019). Sustainable use of *Hermetia illucens* insect biomass for feed and food: Attributional and consequential life cycle assessment. *Resources, Conservation and Recycling, 144*, 285–296.

Sørensen, M., Gong, Y., Bjarnason, F., et al., (2017). *Nannochloropsis oceania*-derived defatted meal as an alternative to fishmeal in Atlantic salmon feeds. *PLoS One, 12*(7), e0179907.

Sousa, I., Gouveia, L., Batista, A. P., Raymundo, A., & Bandarra, N. M., (2008). Microalgae in novel food products. *Food Chemistry Research Developments*, 75–112.

Spolaore, P., Joannis-Cassan, C., Duran, E., & Isambert, A., (2006). Commercial applications of microalgae. *Journal of Bioscience and Bioengineering, 101*(2), 87–96.

Takeyama, H., Kanamaru, A., Yoshino, Y., Kakuta, H., Kawamura, Y., & Matsunaga, T., (1997). Production of antioxidant vitamins, β-carotene, vitamin C, and vitamin E, by two-step culture of *Euglena gracilis Z*. *Biotechnology and Bioengineering, 53*(2), 185–190.

Tertychnaya, T., Manzhesov, V., Andrianov, E., & Yakovleva, S., (2020). New aspects of application of microalgae *Dunaliella salina* in the formula of enriched bread. *Paper Presented at the IOP Conference Series: Earth and Environmental Science.*

Tomlinson, I., (2013). Doubling food production to feed the 9 billion: A critical perspective on a key discourse of food security in the UK. *Journal of Rural Studies, 29*, 81–90.

Torres-Tiji, Y., Fields, F. J., & Mayfield, S. P., (2020). Microalgae as a future food source. *Biotechnology Advances, 41*, 107536.

Tscharntke, T., Clough, Y., Wanger, T. C., et al., (2012). Global food security, biodiversity conservation and the future of agricultural intensification. *Biological Conservation, 151*(1), 53–59.

Wang, A., Yan, K., Chu, D., et al., (2020). Microalgae as a mainstream food ingredient: Demand and supply perspective. *Microalgae Biotechnology for Food, Health and High Value Products* (pp. 29–79), Springer.

Wen, Z. Y., & Chen, F., (2003). Heterotrophic production of eicosapentaenoic acid by microalgae. *Biotechnology Advances, 21*(4), 273–294.

Whelan, J., & Rust, C., (2006). Innovative dietary sources of n-3 fatty acids. *Annu. Rev. Nutr., 26*, 75–103.

Yousefi, M., & Jafari, S. M., (2019). Recent advances in the application of different hydrocolloids in dairy products to improve their techno-functional properties. *Trends in Food Science and Technology, 88*, 468–483.

Yuan, J. P., Peng, J., Yin, K., & Wang, J. H., (2011). Potential health-promoting effects of astaxanthin: A high-value carotenoid mostly from microalgae. *Molecular Nutrition and Food Research, 55*(1), 150–165.

CHAPTER 9

Biofuel Production from Microalgae: Current Trends and Future Perspectives

PAVITHRA SURESH, AAVANY BALASUBRAMANIAN, and JYOTHISH JAYAKUMAR

Anna University Regional Campus, Coimbatore – 641046, Tamil Nadu, India

ABSTRACT

Algal biofuel production has renewed its interest in various fields of research and entrepreneurship in recent years. In this field, microalgae are the most agreeable biofuel feedstock. Microalgae are recognized for their high potential toward faster growth rate and relative biofuel production. The major steps in algal biofuel production comprise cultivation, harvesting, and conversion of unusable fuel. The cultivation of microalgae for fuel production comprises of two types of system: photoautotrophic algal culture system that includes light and carbon dioxide (CO_2) as source for carbon and energy and heterotrophic algal culture system where microalgae grow in darkness. The commonly used harvesting methods are by gravity settlement or centrifugation. The biomass used for oil extraction through implementing a solvent extraction method and is further processed to biodiesel. In the current research on algal biofuels, newer steps are under development against the present restricting features of algal biofuel production from commercializing purposes. The integration of algal biology with culture cultivation engineering will also enhance the present scenarios and drawbacks in the algal production system and will

Microalgal Biotechnology: Bioprospecting Microalgae for Functional Metabolites towards Commercial and Sustainable Applications. Jeyabalan Sangeetha, PhD, Svetlana Codreanu, PhD, & Devarajan Thangadurai, PhD (Eds.)

take biofuels to another level of commercialization. The chapter emphasizes on the current methods used for the algal biofuel production and various advanced techniques used by the industries.

9.1 INTRODUCTION

The utilization of fossil fuels globally and the non-renewable nature of fossil fuels has opened avenues in the field of research for the identification of renewable alternate fuels. Bioenergy-based fuels are deemed to be an alternative for the conventional fossil fuels with biomass as an efficient and energy-rich source. Biomass is a part of the renewable energy sources available in alternate to fossil fuels. Bioenergy-based biofuels have the most positive environmental impact due to lower emission of carbon dioxide (CO_2) and sulfur content. Biofuels are cost-effective and has drastic economic effects in the pricing of fuels (Demirbas, 2008). Biofuel comprises energy-rich compounds generated directly by biological processes or due to the chemical conversion of biomass from living sources.

The harnessing of biofuels is from organisms that are photosynthetic, i.e., bacteria which are photosynthetic, micro-algae, macro-algae, and vascular plants. The major form of biofuels synthesized is either in the form of gas, liquid, or solid. Biofuels are classified as primary and secondary biofuels and the biofuel-based products are a resultant of biochemical, physical, and thermochemical processes; primary biofuels are direct products resulting from activities such as burning wood or cellulosic plant material and dry animal waste. Secondary biofuels are indirect products produced by the usage of animal or plant material, the secondary biofuels are classified based on the sources, and it is diversified into three generations. First-generation biofuels are ethanol-based and are from starch-rich food crops and fats from animal waste.

The second generation of biofuels was ethanol-based, cellulosic biomass was the source, and initially, biodiesel production was being carried out using soybean and jatropha seeds which are abundant in oil content. Third-generation biofuels are comprised of energy from microbial sources such as cyanobacteria, microalgae, which are a promising way forward to handle global energy demands. Biofuel is a less polluting, highly accessible, reliable, and sustainable fuel, which is renewable.

Liquid biofuels are an alternative to the currently available fossil fuels categorized under (a) bioalcohols; (b) vegetable oils; (c) biodiesels; (d) bio-crude and (e) biosynthetic oils. The widely available liquid biofuels are biodiesel and bioethanol that includes ETBE partially made along with bioethanol, these fuels contribute around 90% of the biofuel market globally. The biofuels require defined physical and chemical properties in meeting the application of the fuel along with its combustion rate, energy density, and transportation condition.

All biofuels have to showcase the necessary chemical and physical properties, to meet the demand of engine application such as stability and predictable combustion at high pressures as well as the demands of transportation such as safety and energy density.

Biofuel is produced from various sources that include plants, microbes, etc. Vegetable oils such as rapeseed oil, soybean oil, palm oil, and sunflower oil are the main resources of biodiesel. Bioethanol production majorly depends on wheat, maize, potatoes, and sugar beet, the first generation of biofuel developed attained worldwide acceptance and extracted from food and oil crops such as maize, sugar cane, sugar beet and rapeseed. The usage of conventional methods and techniques for biofuel extraction yielded biofuels from animal fats and vegetable oils (Rossi and Lambrou, 2009). The usage of food as a source for biofuel generated controversy in the food sector.

The second-generation biofuel mainly focused on plant biomass, creating a huge impact worldwide. The raw material availability was cheaper and the production was effective. Later the usage of plants in abundant quantity resulted in severe problems. Plant biomass is abundantly available and a promising resource present the over usage of this creates enormous problems around worldwide. This also created a negative impact on fuel production (Naik and Goud, 2010).

The analysis of algal biomass are popular rather than the food and terrestrial plant biomass. Technically and commercially algal biomass usage has procured world acceptance. The fuel produced from algal biomass is cost-effective compared to commercially available fuels. To grow this biomass does not require any additional land, requires minimum water usage, and enables quality improvement in the air (Wang and Li, 2008). The use of algae has a suitable alternative because algae are considered the most versatile biomass source and an important easily renewable crop. The major reasons to consider algal biomass for biofuel production is: (i) higher photosynthetic rate; (ii) higher biomass productivity; (iii) high growth rate than other terrestrial plants; (iv) good CO_2 fixation and O_2 production; (v) grown easily in a liquid medium; and (vi) can withstand temperature variations (Chisti, 2008).

Algae are one of the oldest and primitive life forms of plants that come under the thallophyte division of the plant kingdom (Falkowski and Raven, 2013). Algal biomass has the potential of producing carbohydrates, proteins, and lipids. Algae have the capability of producing a larger amount of lipids in comparison to terrestrial plants due to their high growth rate (Metting, 1996). Therefore, the crude petroleum products can be substituted by natural oils (lipids) produced by mass culturing of algal biomass for sustainable fuel production.

Algae are generally classified into two classes namely autotropic and heterotrophic. For biofuel production, autotrophic species of algae are mostly considered. The algae usually utilize CO_2, salts, and light sources for development and its sustainability in the environment. The heterotrophic species of algae requires external nutrients as well as organic materials for their sustainability (Kadir et al., 2018). The algae have the capability of producing three major biochemical compounds namely carbohydrates, proteins, and lipids. The amount of lipids produced by autotrophic algae is 20 times higher in comparison to the rates at which terrestrial plants crops such as corns, soybean, coconut is producing (Tickell and Tickell, 2003). A major problem faced is the lipid yield and growth rate of algae differ in different species. It was reported that lipid content produced by some algae which have higher lipid content (Table 9.1) and has the capability of producing a larger concentration of lipids. (Antoni et al., 2007).

TABLE 9.1 Lipid Content of Some Algal Species

Species	Lipids (%)
Chlorella vulgaris	14–40/56
Chlorella vulgaris EPS-31	14.83
Chlamydomonas sp. JSC4 residue	10.93
Dunaliella salina	18–25

Source: Adapted from: Borowitzka (1999); Balat and Balat (2010).

In the view of Table 9.1, it is evident that algae have the capability of producing a convenient amount of lipids using this biofuel of high oxidation can be developed (Dote et al., 1994). Depending on the species, algae produce different kinds of lipids. Not all kinds of algae produce biofuel. Autotropic algal organism is a rich source of oil production.

The heterotrophic organism cannot produce a sufficient amount of oil, this is because they majorly depend on the organic compound supplied to them for growth and development. Variation in the supply leads to the production of an undesired amount of oil (Guschina and Harwood, 2006). The biofuel extract from the algae has physical and fuel properties similar to available automobile fuel. The major challenge is to identify the strain which has a high growth rate as well as lipid production rate (Miao and Wu, 2006).

The major advantages of using algae as biofuel are that the growth rate is high and biomass can be harvested yearly basis. They have high lipid production capability. Lipids usually produced by algae have a high saturation level which makes them suitable for biofuel production. Algae require less freshwater for growing when compared to terrestrial plants. For the cultivation purpose, it requires only non-arable land in brackish water. Thus reducing the environmental effects caused by the cultivation of other plants. They do not require any kind of herbicides or pesticides which adversely affect the environment. The requirement of land for the growing of algae is estimated to be around only 2% in contact to the amount of land utilized by oil-bearing crops. The algae can remove the phosphates and nitrates present in the wastewater when utilized as a source for algal growth (Wood et al., 2017).

The cost regarding the cultivation of algae is comparatively very low compared to terrestrial oil-producing crops. A major advantage of algal biofuel is that it is a carbon-neutral fuel. The fuel is developed through the photosynthetic fixation of atmospheric CO_2. These biofuels have properties similar to the commercially available crude oil, which includes density, viscosity, cold flow, and heating value. Algae during the production of biofuels provide by-products such as pigments, carbohydrates, proteins that can be utilized for commercial as well as the pharmaceutical purpose (Daroch et al., 2013). Pyrolysis is used as an effective method for the production of feedstock. Algae having a higher amount of cellular lipids are utilized for oil production.

9.2 MAJOR SOURCES OF ALGAL BIOFUEL

9.2.1 NATURAL OILS

The biodiesel industry in the European countries utilizes rapeseed oil and canola oil for commercial biodiesel production. In the United States, soybean oil is an important source for biodiesel production; annually the quantity of soybean oil produced is higher than all known sources of fats and oil produced. A variety of sources is available in the feedstock, which includes fats, oilseed, and recycled oil (Hossain et al., 2017).

The European Union utilizes majorly about 77% rapeseed oil for biodiesel production; soybean oil and palm oil that is utilized for biodiesel production are imported into Europe. The contributory shares of 9% soybean oil, 12% of palm oil and 2% of waste materials show the various sources' contribution in commercial production of biodiesel in the European Union.

Vegetable oils are a part of renewable energy material currently employed for the production of biodiesel. Transesterification of vegetable oils produces biodiesel, i.e., a methyl or ethyl ester fatty acid produced from used vegetable oils (edible and nonedible). The major sources of nonedible plants used for the biofuel production are *Calophyllum inophyllum* (nagchampa), *Jatropha curcas* (ratanjyot), *Hevea brasiliensis* (rubber) and *Pongamia pinnata* (karanj), etc. (Kakati et al., 2017).

Rice bran oil is a good feedstock for biodiesel production due to its nature and availability as an agricultural waste, which is currently utilized for inexpensive biodiesel production. The biodiesel production process using rice bran is more effective as most of the bioactive compounds retain. The processing and purification condition includes the usage of an acid catalyst in place of a base catalyst to avoid damage to the bioactive compounds. Low-temperature (vacuum distillation) purification is carried out due to the heat liable nature of the bioactive compounds. The biodiesel purification process also includes the recovery of bioactive compounds shall be carried out using the appropriate process design. The active utilization and sale of the high-value bioactive compounds along with the sale of biodiesel improves the process economics of the cost involved in biodiesel production in comparison to the prices of commercial fossil fuels (Sinha et al., 2008).

9.2.2 PLANT SOURCES FOR BIODIESEL PRODUCTION

Research focusing on biodiesel production using sweet sorghum, *Jatropha*, and *Pongamia* ensure the utilization of plant-based resources in the arid regions are used. The livelihood and food security of farmers are ensured and the increased use of biofuels results in high farmer produce and a reduction in fossil fuel usage. The crops do not require larger quantities of water and have a high tolerance to stress and with a low capital investment resulting in the cultivation of such plants in drylands.

The Ministry of Rural Development under the Indian government, by the suggestions of the Planning Commission, has taken a National Mission for the utilization of biodiesel; the mission encompasses the cultivation and harvesting of physic nut and *Jatropha curcas*. The seed consists of 30% to 40% oil, which after transesterification shall be mixed with diesel.

In India, molasses is the major source of raw material for biofuel production; the increased demand of the automobile sector requires a higher quantity of produce. Sweet sorghum, which is equivalent with sugarcane, is easily grown in dry conditions with less water usage for cultivation and lower production cost makes an ideal alternative for sugarcane-based molasses.

9.2.3 MICROBE BASED

The forms of biofuel obtained from the organisms and the method of production is mentioned in Table 9.2 (Antoni et al., 2007).

9.2.4 ANIMAL FATS

One of the sources for biodiesel production is fats and tallow derived from animals. Animal fats have a greater advantage over plant crops because they are priced favorable for conversion into biodiesel. The presence of saturated fats in this animal source the biodiesel produced from this particularly turns to gel and hence is incapable of extensive applications.

TABLE 9.2 Biofuel from Microbes and Method of Production

Fuel	Method Produced	Organism Involved	Status
Hydrogen	Bio-photolysis	Algae and cyanobacteria	Microbiological hydrogen production yet to be developed as an economically viable technology
	"Dark" fermentation	Acidogenic bacteria	
	Photo-fermentation	Photosynthetic bacteria	
Biogas	Uncontrolled secondary fermentation	Hydrolytic and the methanogenic bacteria	Usage of bacteria directly in car combustion engines
Bioethanol	Fermentation	*Saccharomyces cerevisiae*	Water and gasoline phase separation is difficult
	Fermentation of glucose syrups	Yeast	Commercially proven method
	Bacterial ethanol fermentation	*Zymomonas mobilis*	Continuous ethanol fermentation in FBR
	Bioethanol production with cheese whey as substrate	*Kluyveromyces fragilis*	Commercial plants
	Fermentation using cellulosic biomass	Metabolically engineered *Clostridium thermocellum*	Feasible and strain development for ethanol production is a promising prospect
	Conversion of Syngas (a CO/H_2 mixture from gasified biomass) to ethanol	*Clostridium ljungdahlii*	A commercial plant is expected to be established (Henstra et al., 2007; http://www.brienergy.com)
	Fermentation of molasses and starch	–	Large scale production has various limitations
Biodiesel	Micro diesel	Metabolically engineered *Escherichia coli* cells, *Zymomonas mobilis*, *Acinetobacter baylyi*	Large scale bacterial cultures are not available

Source: Adapted from: Antoni et al. (2007).

9.2.5 MICROALGAE – SOURCE OF BIOFUEL

Algae are autotrophic organisms that exhibit growth in various ranges of temperature, humidity, light intensities, and pH. The population is classified into micro (phytoplankton) and macro (seaweed) depending on the size of the species. Microalgae grow rapidly that during the peak of their growth phase its doubling time is about 3.5 hours. The organism shows a lipid content from 20% to 80% that accounts for its ability to produce a higher quantity of oil than that of the already available renewable sources of biofuel. Biofuel is produced from the oil content in the microalgae by transesterification.

Algae have become one of the prime importance as it is a renewable energy source. It produces extremely valuable byproducts that include lipids, carbohydrates, proteins that involve in the production of biofuel and other useful materials. Hu et al. (2008) projected a yield of 200 barrels of oil per hectare of land using photosynthetic algae, which is 100 times greater than that for soybeans, a common source of biodiesel. Algae-based biofuel production has several potential advantages (Greenwell et al., 2009):

- Availability of higher variations of algal sources to produce biofuels and byproducts
- Increased growth rate
- Habitats of algal cultivation are numerous
- Human and animal waste can be used as a major supply of nutrients (silicon, hydrogen, nitrogen, sulfate) for algae
- Algae can be cultivated in unsuitable lands for agriculture
- Algae can sequester carbon dioxide (CO_2) from industrial sources
- Although algal biomass may be rich in energy, its growth in suspensions with dry solids (Zamalloa et al., 2011) poses greater challenges in achieving a commercially viable energy balance in algal process operations

Certain other challenges of algae harvesting include:

- The size of micro-algal cells that is below 30 μm (Grima et al., 2003)

- The density of the algal cells to the growth medium is a bigger challenge
- Formations of dispersed algal suspensions in the growth phase is due to the negative charge of the algal surface (Moraine et al., 1979; Edzwald, 1993; Packer, 2009)
- The rapidity of algal growth rates needs frequent and suitable harvesting compared to it of terrestrial plants

9.3 MICROALGAE-BASED BIOFUEL PRODUCTION PATHWAYS

Microalgae-derived biomass is utilized to harvest biofuels by the process of thermochemical conversion (Raheem et al., 2018), transesterification process, biochemical conversion process and process which involves the photosynthetic based microbial fuel cell (Naik et al., 2010). Thermochemical conversion technologies utilize heat to convert algal biomass into intermediate products, which are then converted into biofuels via chemical and biological routes. Biochemical conversion technologies involve the utilization of micro-organisms or enzymes for the breakdown of water molecules in the pretreated biomass to obtain broken-down sugars. Biochemical processing of biomass to biofuels is more selective and greener but is of higher production costs as compared to thermochemical processing (Brennan and Owende, 2010). The identification and development of a conversion process is one of the key challenges for the viability of algal biofuels economically and or a sustainable production.

9.4 INDUSTRIAL PRODUCTION OF BIOFUELS BASED ON MICROALGAE

In today's date, fossil fuels that we use in our day-to-day life are on the verge of exhaustion, and the increasing use of fossil fuel is also causing an increase in pollution globally. In this situation, biofuel production using microalgae is referred to as a major choice. Microalgae are tiny factories of the cell which has an extensive property to act as ecofriendly fuel. The production mechanism includes three types such as photoautotrophic, heterotrophic, and mixotrophic production. These mechanisms lead to natural growth (Shen et al., 2009).

Photoautotrophic production is the general photosynthesis process, heterotrophic production is the utilization of organic substances, e.g., glucose to stimulate growth. Several algal strains combines, i.e., autotrophic and heterotrophic hence they are an example of mixotrophic production. Generally, photoautotrophic production is considered as the only method which is practically and economically feasible for the large-scale production of algae biomass.

9.4.1 BIOREACTOR

Bioreactors are the manufactured vessels or systems inside which a biologically active environment is created and maintained to sustain microbial life, and hence active substances can be derived from these organisms. These processes can maintain under both aerobic and/or anaerobic conditions. Bioreactors used are to study the scaling up of biofuel production from the lab scale to the industrial scale (Verma et al., 2010).

Bioreactors are generally cylindrical, their size ranges from liters to cubic meters and is mostly made up of stainless steel, glass, or plastic having a different size. Regarding the mode of operation, reactors are classified as batch, fed-batch, and continuous (Mata et al., 2010). Designing a bioreactor is relatively a complex task for biochemical engineers. The main factor to be considered is the optimal condition under which the microorganism or cell can perform its desired function successfully (Balat and Balat, 2010). The environmental condition includes:

- The flow rate of gases (oxygen, nitrogen, and carbon dioxide)
- Variation in temperature and pH
- Dissolved oxygen concentration
- Monitoring and controlling agitation speed

Most of the industries are using a reactor having sensors and a controlling system. A controlled temperature is maintained in the reactor using a heat exchanger. To maintain the increasing temperature, the bioreactors are usually jacketed with the help of external coil in large industries.

9.4.1.1 GROWTH OF MICROALGAE IN BIOREACTOR

In the present day of increasing demand for fuels, microalgae are considered the only alternate source of biodiesel for the global demand of transport fuels. The algal biomass is considered as feed for the production of biofuels (hydrogen, methane, and bioethanol). The production cost of microalgal biomass as biofuel is considered to be an expensive process than growing a crop (Amorim et al., 2020).

Microalgae growth requires the natural resources such as light, CO_2, water, and nutrients. Nutrients are micro and macro essentials. The selection of reactor is a major factor which influences the productivity of the yield from the microalgae biomass.

9.4.1.2 POND PRODUCTION SYSTEM

The culturing of algae using an open pond system has been used since the 1950s (Borowitzka, 1999). The system for algal production has been categorized into a natural water source, artificial ponds, or containers depending upon the source and availability of water (Jiménez et al., 2003). In an industry setup open tanks such as raceway, shallow tanks big or circular are used.

A commonly used artificial type of pond system is the raceway ponds system. They are mainly made up of closed-loop and oval-shaped recirculation channels, generally between 0.2- and 0.5-meters depth. Mixing and circulation are two main factors required to stabilize the growth and productivity of algae. Raceway ponds are generally built-in concrete that is compact earth lined with white plastic. In a continuous cycle of production, algae broth and nutrients are introduced by using a paddlewheel and circulated in the loop to the harvest extraction point. These wheels are mainly used to reduce sedimentation. The surface air will provide the required amount of CO_2.

The absorption of CO_2 is enhanced by the already installed submerged aerator. While comparing raceway ponds and closed tank systems, the raceway pond system is considered one of the cheapest systems for the large-scale production of microalgae. This system requires low energy and feasible to maintain and is easy to clean. The productivity in the raceway pond system is higher than that of an unmixed pond. Generally, small outdoor raceway ponds are capable of a larger yield. According to the current situation, 98% of algae are produced commercially using the open pond system. The open pond system can be extended up to large acres while individual ponds can be converted into an indoor system by transforming it into a layered system.

The most commonly cultured algal species using raceway pond type of open culture system are *Spirulina*, *Haematococcus*, and *Dunaliella* (Terry and Raymond, 1985). Circular ponds are commonly used in the Asia region for the production of *Chlorella* species. These ponds have a rotating agitator. These systems were the oldest method implemented for large-scale production of the algal mass, nowadays these techniques of the system are used for wastewater treatment. This type of system shows its efficiency till the pond size of about 10,000 m^2 and a limitation of uneven mixing in large ponds.

A thin layer, inclined ponds consists of shallow trays over which thin layers of algae flow to the bottom, where the culture is collected and again returned to the top. In this system, the velocity of water is typically close to 30 cm/s. Mild mixing using paddlewheels is used. Higher velocities demand an excessive amount of energy.

An open system that is unmixed is not the actual algal production system because in this type of system, the maximized production cannot be achieved and produced biomass is rarely harvested. The construction of the open tank is low cost and they are easy to operate; however, they have certain limitations like:

- Difficulty to control contamination and only highly selective species are not affected by contamination by other microalgae and microorganisms
- The other challenge for open pond systems is that they are easily affected by environmental frictions like air pollution, heavy metal accumulation, etc.
- Carbon dioxide pumped into the ponds is escaped into the atmosphere
- Cell culture is directly influenced by environmental variations;
- Maintenance of the cell density is low due to the shadowing of the cells (Amaro et al., 2011)
- Factors like the intensity of light, temperature, pH, and dissolved oxygen concentration are the other factors that limit the growth parameters (Harun et al., 2010)

Open photobioreactors have a lower yield than closed systems due to loss by evaporation temperature fluctuations, nutrient limitation, light limitation and inefficient homogenization (Brennan and Owende, 2010).

9.4.1.3 CLOSED PHOTOBIOREACTOR SYSTEM

The major reason behind the designing of a closed photobioreactor was to overcome the problems associated with the open pond production system. For example, the risk of pollution and contamination by an organism like fungi or bacteria are the main problem faced by the industries like pharmaceuticals and cosmetics (Ugwu et al., 2008).

Photobioreactor permits the microbial culturing of single species for a prolonged duration with less contamination. Photobioreactors can be planted indoor with the provision of artificial light or natural light vial light collection and distribution system. To reduce the expense outdoor plants can be installed with direct sunlight as the source of light. Photobioreactors can be operated in batch or continuous mode.

The photoreactors are engineered in such a way that they may have tilts at different angles to enhance the reflection of light which is very much needed for the production process. Materials such as plastic or glass sheets, as collapsible or rigid tubes, with no toxicity and high transparency, high mechanical strength, high durability, chemical stability, and low cost are generally used.

Closed photobioreactor systems have high efficiency, a shorter harvesting period, and less contamination. This system also allows for the good selection of species used for cultivation and has a high surface-to-volume ratio (25–125/m) compared to the open pond production system. Mixing mechanisms require high energy consumption and it is the highest cost factor for the system.

Light is implemented inside the bioreactor using optical fibers, fluorescent light, or sun. Photobioreactors have two zones, photo-limited central dark zone close to the surface of the bioreactor. CO_2 enriches air is spared to the reactor, which creates a turbulent flow that further helps in the movement of cells from the light to dark zones. The movement of cells from light to dark zone depends on turbulence, cell concentration, optical properties of culture, and diameter of the tube. Dissolved oxygen levels, as well as CO_2 levels, must be regulated in a photobioreactor in the case of algal culture.

Overheating of photo limiting zones in the inner zones, photoinhibition in the peripheral zones, cell structure damage due to hydrodynamic stress and growth on the reactor wall and cost are some of the challenges faced by photobioreactors. The scale-up of the bioreactor will lead to an increase in the percentage of dark zones and a decrease in algal growth.

One of the popular types of photobioreactor systems is a tubular system, helical photobioreactor system, plastic bag system, well system, pyramid photobioreactor system, airlift photobioreactor system, annular photobioreactor system, column photobioreactor system, bubble-column photobioreactor system, vertical column photobioreactor system, Flat-plate photobioreactor system, stirred-tank photobioreactor system, rectangular tanks, and immobilized bioreactor.

9.4.1.4 FLAT-PLATE PHOTOBIOREACTOR

Flat-plate photobioreactors are closed photobioreactor systems (Samson and Leduy, 1985). The system has been designed to have a large surface area exposed to illuminations and has received much research attention. These reactors were generally made of transparent materials which allow for the capture of maximum solar energy (Richmond et al., 2003). A thin layer of dense culture flows across the system which absorbs the radiation for a few millimeters thickness. In the system, the highest density of photoautotrophic cells was observed to be >80 gL^{-1} (Hu et al., 1998). Because of less accumulation of dissolved oxygen and greater photosynthetic efficiency flat plate bioreactors are used for mass culturing of algae (Richmond, 2000).

9.4.1.5 HELICAL-TYPE PHOTOBIOREACTOR

These types of bioreactors consist of a small diameter coiled transparent and flexible tube with a degassing unit. The cultures are driven along the tube due to the pressure exerted by the centrifugal pump.

9.4.1.6 TUBULAR BIOREACTOR

Tubular photobioreactor consists of straight transparent tubes of plastic or glass. The sunlight is captured by the tubular arrays or by the solar collectors. The diameter of the tube is limited to <0.1 m so that the light can penetrate through the surface and can reach deep into the densely cultured broth which is very necessary to ensure high biomass production. In this system, the microalgal culture is cycled continuously from reservoir to solar collector and again back to the reservoir.

To maximize the capture of solar light, the solar panels are oriented from north to south. Mainly panels are placed parallel and arranged flat above the ground. Sometimes the tubes can also be made up of flexible plastic and can be coiled around the supporting frame to make up a helical coil tubular photobioreactor. Tubular photobioreactors are deemed to be more suitable for outdoor mass culture since they expose a large surface area to direct sunlight.

Tubular photobioreactor has certain design limitations on the length of tubes which is dependent on the potential of oxygen accumulation, CO_2 depletion, and pH (Eriksen, 2008) variation in the systems because of which they cannot be scaled up indefinitely hence large-scale production plants are based on the integration of multiple reactor units. Tubular photobioreactors are the largest closed photobioreactors, e.g., the 25 m^3 plant at Mera Pharmaceuticals, Hawaii (Olaizola, 2000) and the 700 m^3 plant in Klötze, Germany (Pulz, 2001).

9.4.1.7 BUBBLE-COLUMN BIOREACTOR

Bubble-column bioreactors offer the most efficient mixing, the highest volumetric mass transfer rates and the best controllable growth condition. In the bubble-column bioreactor, the tubes are placed vertically. In this type of bioreactor air is pumped through the bottom portion of each tube and bubbles are formed. The output of a column is connected as the input of the next tubular column.

Low capital cost, high surface area to volume ratio, lack of moving parts, satisfactory heat and mass transfer, residual gas mixture and efficient release of oxygen are some of the advantages of bubble-column bioreactor. Bubbling of the gas mixture from the sparger helps in the mixing and CO_2 mass transfer. The photosynthetic efficiency of the culture depends directly on the gas flow rate. The gas flow rate will regulate the light and dark cycle as the liquid is circulated from the central dark zone to the external photic zone at a higher gas flow rate (Kumar et al., 2011).

9.4.1.8 HYBRID PRODUCTION SYSTEMS

This method of hybrid two-stage cultivation combines distinct growth stages in photobioreactors and open ponds. The first stage is in a photobioreactor were under controllable conditions and low risk of contamination by other organisms or environmental factors and under favorable continuous cell divisions. The second production stage is aimed at exposing the cells to nutrient stress which will enhance the synthesis of desired lipid product. This stage is suited to the open pond system, as the environmental stress that stimulates the production can occur naturally through the transfer of the culture from the photobioreactor to the open pond.

This two-stage production system was used by Huntley and Redalje (2007) for the production of both oil and astaxanthin from *Haematococcus pluvialis* and they produced microbial oil at a rate >10 toe ha^{-1} per annum as an annual average with a maximum rate of 24 toe ha^{-1} per annum. On their demonstration with another species with higher oil content and photosynthetic efficiency under similar conditions, the rate of up to 76 toe ha^{-1} was achieved and feasible.

9.4.2 HETEROTROPHIC PRODUCTION

Heterotrophic production systems have been successfully used for the production of algal biomass and metabolites. In this production system, microalgae are grown on organic carbon substrates such as glucose in stirred tank bioreactor or the fermenters. This system provides a high degree of growth control and also lower harvesting cost due to the higher cell densities achieved (Chen and Chen, 2006). The setup cost for this production system is very low but the operation and maintenance cost is very high (Chisti, 2007). Li et al. (2007) have proved the feasibility for large-scale biodiesel production from *Chlorella protothecoides*.

9.4.3 MIXOTROPHIC PRODUCTION

Number of the algae has dual capability of performing either autotrophic or heterotrophic metabolism process. The ability of mixotrophic are to generate organic substrates, which cell growth is strictly not dependent on photosynthesis and the cells can also survive in the absence of light. The cells can survive either in light, organic carbon substrate (Chen et al., 1996). The microalgae that displays the mixotrophic metabolism process for growth are the cyanobacteria *Spirulina platensis* and *Chlamydomonas reinhardtii* (Andrade and Costa, 2007).

Chojnacka and Noworyta (2004) compared the growth of *Spirulina* species in photoautotrophic, heterotrophic, and mixed trophic cultures. They observed that the mixed trophic cultures reduced the photoinhibition and improved growth rates over both autotrophic and heterotrophic cultures.

9.5 CURRENT TRENDS IN ALGAL BIOFUEL PRODUCTION

9.5.1 BIOFUEL FROM WASTEWATER

Microalgae promote the treatment of wastewater, while the production of biofuel with wastewater-harvested algae is one of the current methods of biofuel production (Benemann et al., 1977). Microalgae generate significant concentration of lipids depending on the species and its growth condition. The type of lipids that accumulate includes saturated fatty acids, polyunsaturated fatty acids (PUFA), glycolipids or triacylglycerol (TAG) (Chisti, 2007; Hu et al., 2008; Griffiths and Harrison, 2009). The high concentrations of lipid in algal cells when induced by environmental stress are frequently coupled with low biomass (Rodolfi et al., 2009; Pittman et al., 2010). Hence, the improvement of higher lipid concentrations along with higher biomass is of particular interest (McGinn et al., 2011).

The dual advantage of higher biomass and higher lipid productivity suggests the potential use of high nutrient resource, wastewater for cost-effective production of biofuel. However, there are limitations that include the need for:

- Effective and efficient algae harvesting techniques
- Analysis of life cycle of wastewater-derived algal biofuel
- Pond scale demonstration of high biomass and lipid productivity

Therefore, the cultivation of microalgae for wastewater treatment along with biofuel generation is an attractive option in terms of reducing the cost of energy, GHG emissions, and freshwater resource costs.

9.5.2 BIOFUEL PRODUCTION USING PHOTOBIOREACTORS

Photobioreactors are solar receivers that allow the optimum environmental conditions for the growth of the algal cells along with required light intensity and CO_2. This promotes an increase in the biomass of algal cells while the light energy received if not utilized completely in the process of photosynthesis causes thermal energy resulting in an increase of the temperature of culture causing efficient culture production failure even on usage of highly tolerant species.

Microalgae cultivation for energy production implies harvesting of solar energy over large areas along with other growth conditions, light and CO_2. The efficiency of this process is photo conversion efficiency (PCE) and depends on the efficiency of photosynthesis and cell anabolism.

The photobioreactors are safe and reliable and are of various types, among which only a few 100 tons are produced in closed photobioreactors. The only reason for higher preference of open pond bioreactors to closed photobioreactor for cultivation is that the former is cost-effective. Hence in an industrial plant, a strategy that combines the use of PBR for production of active inoculum and ponds for bulk cultivation. Often the highest concentrations of lipids that are reported tend to be either from photo-bioreactor-grown cells or batch culture-grown cells in the laboratory (Griffiths and Harrison, 2009).

Cost being a major issue in the production of biofuel, an increase in the productivity of biomass might change the current status of the biofuel economy. The current cost of algal biomass production exceeds by 20 times that is required for economic fuel production (about US$ 0.25 per kg). Hence application of photobioreactors promotes an increase in the yield of the biofuel further triggering a reduction in the cost of the biofuel (Amer et al., 2011).

9.5.3 ENZYMATIC ALGAL HYDROLYSIS METHOD FOR BIOFUEL PRODUCTION

Microalgae hydrolysis promotes the rupture of the cell wall, thereby releasing the organic substances present in the cell. This is one of the most economical pre-treatment steps in the production of biofuel from the cell-derived organic substances. The hydrolysis of microalgae is catalyzed by chemical (alkali/acid), physical (pyrolysis, grinding, and milling) or biological (enzyme) methods.

Physical hydrolysis of microalgae is time consuming and involves in loss of salient features of the extracted organic substances. The chemical method has its advantages for biomass as the physical hydrolysis requires higher energy consumption and is not commercially viable. The usage of acid and alkali has been implemented in the large-scale production where the acid pretreatment process which uses hydrochloric acid and sulfuric acid gives better yield in the conversion of cellulosic material in comparison to the alkaline treatment using lime and sodium hydroxide. The comparison of both the alkaline pretreated system and acid pretreatment shows that the acid pretreated system gives a yield of 58 wt.% (g ethanol/g microalgae) while the alkali pretreated system gives 38 wt.% (g ethanol/g microalgae). The usage of diluted hydrolysis is the most common among the chemical hydrolysis.

Microalgae have been reported to have low to no lignin composition that makes it a cellulose-based material. Hence, the biological hydrolysis of algal cellulose is carried out by highly specific cellulase enzyme where the products are reducing sugars. The utilization of microbes or enzymes to effectively breakdown the biomass has resulted in a low hydrolysis rate and it prolongs the process completion time. Hydrolysis is influenced by factors including substrate surface area, cellulose crystallinity, cell wall thickness, mass transfer, porosity, and hemicellulose or lignin contents. The advantages of enzymatic hydrolysis include requirement of mild conditions of pH and temperature; hydrolysis of the cellulose content completely.

Various other areas explored in enzymatic hydrolysis include method of extraction, solvent used for extraction enzyme immobilization and separation of products. Immobilization of the enzyme influences the stability of the enzyme that increases the efficiency of the enzyme. Enzymatic engineering has proven to increase the yield of the product by reducing triglycerides into fatty acid methyl esters and glycerol, which is temperature resistant, stable, and compatible with solvent used, resulting in higher yield.

The concern surrounding the production of biofuels using enzymatic reaction depends on the efficiency of the enzyme that can be improved by immobilization of the enzyme, thereby resulting in the formation of biofuels along with byproducts. Immobilization is the key for higher percentage conversion of biodiesel that requires a high oil content to obtain higher product yield. The enzymatic hydrolysis is less prone to contamination resulting in the higher yield of biofuel in comparison to chemical reaction method (Noraini et al., 2014).

9.5.4 BIOFUEL PRODUCTION USING GENETICALLY ENGINEERED ALGAE

Microalgal genomics has advanced significantly in the recent decades with the establishment of expressed sequence tag (EST) database which consist of sequences of microalgae mitochondria and chloroplast genomes. The target sequences which were developed for algae, are all based on the target sequences of *Chlamydomonas reinhardtii* and most of the tools used for the expression of gene knockdown, transgenes was specific, and newer target genes are being developed for diatoms and other species of algae for the industrial applications.

The large-scale sequencing technology has fastened the sequencing of microalgal genome that are of interest for research and Industrial scale applications using genetic manipulation. The homologous recombination technique used for the transformation of chloroplast is widely used for biofuel production involving *C. reinhardtii.* The techniques such as transgene expression along with protein localization of chloroplast are required for the functioning of the metabolic gene of interest for biofuel production.

The microalgae are considered one of the promising sources for the production of biofuel, and in the current scenario with techniques such as high throughput analytical techniques and gene sequencing allows analyzing and altering the metabolic pathways with precision. The increasing advances in metabolic engineering allow the increased production of carbon compounds such as TAGs and starch. The usage of such advanced techniques in the production of biofuels using microalgae has opened unprecedented avenues for establishing microalgae as a huge source for large-scale production that will not compete with the existing food production process or the requirement of fresh water and space for microalgae cultivation (Radakovits et al., 2010).

9.6 FUTURE PERSPECTIVES

Due to rise in demand for the fossil fuel and non-renewable sources, the application of microalgae as a major source of biofuel is advantageous. Though over the decades there are numerous ongoing research on microalgae, several areas remain unaccomplished. Theoretically and at a pilot scale, a few approaches have gained attention whilst in industries the conventional methods of biofuel production are being followed owing to the economic factors involved in the plant. By enhancing our knowledge over various areas of biofuel production, algal biofuel can be promoted to be one of the richest sources of biofuel in the near future, which would also serve utilization of various value-added products, applications in environmental approaches. Though interests in algal biofuels are decreasing in recent years, deep insights into the mechanism of production of biofuels, enhancement of fuel production using metabolic engineering would enhance the applications of algal byproducts leading to maintain our environmental resources from diminishing.

9.7 CONCLUSION

This chapter deals with the analysis of biofuels with their status and production efficiency in the current scenario; the first and second-generation biofuel production and its outcomes have been discussed along with the shortfalls and the necessary steps forward. The major sources for the production of biofuels were elaborated in the chapter along with its shortfall and its alternatives have been described. The subsection major sources discuss the usage of algae for the development of biofuels in alternate to the existing fossil fuels.

The industrial production of algal biofuels are discussed elaborately with each industrial case process defined elaborately, the production of algal-based fuels in bioreactors are discussed along with the parameters for cultivation and its extraction. Algae cultivation in open pond system is separately discussed to address the methods of growth and requirements for construction of open pond system. The yield and commercial production are discussed to address the industrial requirements for the extraction of biofuels from microalgae. The usage of bubble column reactors is described to understand the air supply requirements for the growth of algae in bubble column reactors. The photobioreactors for the commercial production of algae is discussed in detail as it is one of the critical methods to grow algae as algae being photoautotrophs, and the increased production of algae can be achieved in the usage of photobioreactors.

The current trends section in the chapter discusses the recent advancements in the usage of algae as a commercial source for biofuel production. The usage of genetic engineering-based production methods enhances the yield as genetically modified algae is highly productive in the production of biofuels. The usage of algal-based biofuels has greater economic importance, and the carbon footprint due to the existing fossil fuels is drastically reduced. The area of research in the usage of algae as a source for biofuel is enormous and requires further investigations to improve the yield and cost-effective methods.

KEYWORDS

- **autotrophic**
- **biodiesel**
- **biofuel**
- **commercialization**
- **genetically engineered**
- **heterotrophic**
- **hydrolysis**
- **lipids**
- **microbes**
- **mixotrophic**
- **photobioreactor**
- **wastewater**

REFERENCES

Amaro, H. M., Guedes, A. C., & Malcata, F. X., (2011). Advances and perspectives in using microalgae to produce biodiesel. *Applied Energy, 88*(10), 3402–3410.

Amer, L., Adhikari, B., & Pellegrino, J., (2011). Techno-economic analysis of five microalgae-to-biofuels processes of varying complexity. *Bioresour. Technol., 102*(20), 9350–9359.

Amorim, M. L., Soares, J., Coimbra, J. S. D. R., Leite, M. D. O., Albino, L. F. T., & Martins, M. A., (2020). Microalgae proteins: Production, separation, isolation, quantification, and application in food and feed. *Critical Reviews in Food Science and Nutrition, 26*, 1–27.

Andrade, M. R., & Costa, J. A., (2007). Mixotrophic cultivation of microalga *Spirulina platensis* using molasses as organic substrate. *Aquaculture, 264*(1–4), 130–134.

Antoni, D., Zverlov, V. V., & Schwarz, W. H., (2007). Biofuels from microbes. *Applied Microbiology and Biotechnology, 77*(1), 23–35.

Balat, M., & Balat, H., (2010). Progress in biodiesel processing. *Applied Energy, 87*(6), 1815–1835.

Benemann, J. R., Weissman, J. C., Koopman, B. L., & Oswald, W. J., (1977). Energy production by microbial photosynthesis. *Nature, 268*(5615), 19–23.

Borowitzka, M. A., (1999). Commercial production of microalgae: ponds, tanks, tubes and fermenters. *Journal of Biotechnology, 70*(1–3), 313–321.

Brennan, L., & Owende, P., (2010). Biofuels from microalgae – a review of technologies for production, processing, and extractions of biofuels and co-products. *Renewable and Sustainable Energy Reviews, 14*(2), 557–577.

Chen, F., Zhang, Y., & Guo, S., (1996). Growth and phycocyanin formation of *Spirulina platensis* in photoheterotrophic culture. *Biotechnology Letters, 18*(5), 603–608.

Chen, G. Q., & Chen, F., (2006). Growing phototrophic cells without light. *Biotechnology Letters, 28*(9), 607–616.

Chisti, Y., (2007). Biodiesel from microalgae. *Biotechnology Advances, 25*(3), 294–306.

Chisti, Y., (2008). Biodiesel from microalgae beats bioethanol. *Trends in Biotechnology, 26*(3), 126–131.

Chojnacka, K., & Noworyta, A., (2004). Evaluation of *Spirulina* sp. growth in photoautotrophic, heterotrophic and mixotrophic cultures. *Enzyme and Microbial Technology, 34*(5), 461–465.

Daroch, M., Geng, S., & Wang, G., (2013). Recent advances in liquid biofuel production from algal feedstocks. *Applied Energy, 102*, 1371–1381.

Demirbas, A., (2008). Biofuels sources, biofuel policy, biofuel economy and global biofuel projections. *Energy Conversion and Management, 49*(8), 2106–2116.

Dote, Y., Sawayama, S., Inoue, S., Minowa, T., & Yokoyama, S. Y., (1994). Recovery of liquid fuel from hydrocarbon-rich microalgae by thermochemical liquefaction. *Fuel, 73*(12), 1855–1857.

Edzwald, J. K., (1993). Algae, bubbles, coagulants, and dissolved air flotation. *Water Science and Technology, 27*(10), 67–81.

Eriksen, N. T., (2008). The technology of microalgal culturing. *Biotechnology Letters, 30*(9), 1525–1530.

Falkowski, P. G., & Raven, J. A., (2013). *Aquatic Photosynthesis*. Princeton University Press.

Greenwell, H. C., Laurens, L. M. L., Shields, R. J., Lovitt, R. W., & Flynn, K. J., (2009). Placing microalgae on the biofuels priority list: A review of the technological challenges. *JR Soc. Interface, 7*(46), 703–726.

Griffiths, M. J., & Harrison, S. T., (2009). Lipid productivity as a key characteristic for choosing algal species for biodiesel production. *Journal of Applied Phycology, 21*(5), 493–507.

Grima, E. M., Belarbi, E. H., Fernández, F. A., Medina, A. R., & Chisti, Y., (2003). Recovery of microalgal biomass and metabolites: Process options and economics. *Biotechnology Advances, 20*(7, 8), 491–515.

Guschina, I. A., & Harwood, J. L., (2006). Lipids and lipid metabolism in eukaryotic algae. *Progress in Lipid Research, 45*(2), 160–186.

Harun, R., Singh, M., Forde, G. M., & Danquah, M. K., (2010). Bioprocess engineering of microalgae to produce a variety of consumer products. *Renewable and Sustainable Energy Reviews, 14*(3), 1037–1047.

Henstra, A. M., Sipma, J., Rinzema, A., & Stams, A. J., (2007). Microbiology of synthesis gas fermentation for biofuel production. *Current Opinion in Biotechnology, 18*(3), 200–206.

Hossain, N., Zaini, J. H., & Mahlia, T. M. I., (2017). A review of bioethanol production from plant-based waste biomass by yeast fermentation. *International Journal of Technology, 8*(1), 5–18.

Hu, Q., Kurano, N., Kawachi, M., Iwasaki, I., & Miyachi, S., (1998). Ultrahigh-cell-density culture of a marine green alga *Chlorococcum littorale* in a flat-plate photobioreactor. *Applied Microbiology and Biotechnology, 49*(6), 655–662.

Hu, Q., Sommerfeld, M., Jarvis, E., et al., (2008). Microalgal triacylglycerols as feedstocks for biofuel production: Perspectives and advances. *Plant Journal, 54*(4), 621–639.

Huntley, M. E., & Redalje, D. G., (2007). CO_2 mitigation and renewable oil from photosynthetic microbes: A new appraisal. *Mitigation and Adaptation Strategies for Global Change, 12*(4), 573–608.

Jiménez, C., Cossío, B. R., Labella, D., & Niell, F. X., (2003). The feasibility of industrial production of *Spirulina* (*Arthrospira*) in Southern Spain. *Aquaculture, 217*(1–4), 179–190.

Kadir, W. N. A., Lam, M. K., Uemura, Y., Lim, J. W., & Lee, K. T., (2018). Harvesting and pre-treatment of microalgae cultivated in wastewater for biodiesel production: A review. *Energy Conversion and Management, 171*, 1416–1429.

Kakati, J., Gogoi, T. K., & Pakshirajan, K., (2017). Production of biodiesel from Amari (*Amoora wallichii* King) tree seeds using optimum process parameters and its characterization. *Energy Conversion and Management, 135*, 281–290.

Kumar, K., Dasgupta, C. N., Nayak, B., Lindblad, P., & Das, D., (2011). Development of suitable photobioreactors for CO_2 sequestration addressing global warming using green algae and cyanobacteria. *Bioresource Technology, 102*(8), 4945–4953.

Li, X., Xu, H., & Wu, Q., (2007). Large-scale biodiesel production from microalga *Chlorella protothecoides* through heterotrophic cultivation in bioreactors. *Biotechnology and Bioengineering, 98*(4), 764–771.

Mata, T. M., Martins, A. A., & Caetano, N. S., (2010). Microalgae for biodiesel production and other applications: A review. *Renewable and Sustainable Energy Reviews, 14*(1), 217–232.

McGinn, P. J., Dickinson, K. E., Bhatti, S., Frigon, J. C., Guiot, S. R., & O'Leary, S. J., (2011). Integration of microalgae cultivation with industrial waste remediation for biofuel and bioenergy production: Opportunities and limitations. *Photosynthesis Research, 109*(1–3), 231–247.

Metting, F. B., (1996). Biodiversity and application of microalgae. *Journal of Industrial Microbiology, 17*(5, 6), 477–489.

Miao, X., & Wu, Q., (2006). Biodiesel production from heterotrophic microalgal oil. *Bioresource Technology, 97*(6), 841–846.
Moraine, R., Shelef, G., Meydan, A., & Levi, A., (1979). Algal single cell protein from wastewater treatment and renovation process. *Biotechnology and Bioengineering, 21*(7), 1191–1207.
Naik, S. N., Goud, V. V., Rout, P. K., & Dalai, A. K., (2010). Production of first and second-generation biofuels: A comprehensive review. *Renewable and Sustainable Energy Reviews, 14*(2), 578–597.
Noraini, M. Y., Ong, H. C., Badrul, M. J., & Chong, W. T., (2014). A review on potential enzymatic reaction for biofuel production from algae. *Renewable and Sustainable Energy Reviews, 39,* 24–34.
Olaizola, M., (2000). Commercial production of astaxanthin from *Haematococcus pluvialis* using 25,000-liter outdoor photobioreactors. *Journal of Applied Phycology, 12*(3), 499–506.
Packer, M., (2009). Algal capture of carbon dioxide; biomass generation as a tool for greenhouse gas mitigation with reference to New Zealand energy strategy and policy. *Energy Policy, 37*(9), 3428–3437.
Pittman, J. K., Dean, A. P., & Osundeko, O., (2010). The potential of sustainable algal biofuel production using wastewater resources. *Bioresource Technology, 102*(1), 17–25.
Pulz, O., (2001). Photobioreactors: Production systems for phototrophic microorganisms. *Applied Microbiology and Biotechnology, 57*(3), 287–293.
Radakovits, R., Jinkerson, R. E., Darzins, A., & Posewitz, M. C., (2010). Genetic engineering of algae for enhanced biofuel production. *Eukaryotic Cell, 9*(4), 486–501.
Raheem, A., Prinsen, P., Vuppaladadiyam, A. K., Zhao, M., & Luque, R., (2018). A review on sustainable microalgae-based biofuel and bioenergy production: Recent developments. *Journal of Cleaner Production, 181*, 42–59.
Richmond, A., (2000). Microalgal biotechnology at the turn of the millennium: A personal view. *Journal of Applied Phycology, 12*(3–5), 441–451.
Richmond, A., Cheng-Wu, Z., & Zarmi, Y., (2003). Efficient use of strong light for high photosynthetic productivity: Interrelationships between the optical path, the optimal population density and cell-growth inhibition. *Biomolecular Engineering, 20*(4–6), 229–236.
Rodolfi, L., Chini, Z. G., Bassi, N., et al., (2009). Microalgae for oil: strain selection, induction of lipid synthesis and outdoor mass cultivation in a low-cost photobioreactor. *Biotechnology and Bioengineering, 102*(1), 100–112.
Rossi, A., & Lambrou, Y., (2009). *Making Sustainable Biofuels Work for Smallholder Farmers and Rural Households* (pp. 6–10). Food and Agriculture Organization of the United Nations, Rome.
Samson, R., & Leduy, A., (1985). Multistage continuous cultivation of blue-green alga *Spirulina maxima* in the flat tank photobioreactors with recycle. *The Canadian Journal of Chemical Engineering, 63*(1), 105–112.
Shen, Y., Yuan, W., Pei, Z. J., Wu, Q., & Mao, E., (2009). Microalgae mass production methods. *Trans. ASABE, 52*(4), 1275–1287.
Sinha, S., Agarwal, A. K., & Garg, S., (2008). Biodiesel development from rice bran oil: Transesterification process optimization and fuel characterization. *Energy Conversion and Management, 49*(5), 1248–1257.

Terry, K. L., & Raymond, L. P., (1985). System design for the autotrophic production of microalgae. *Enzyme and Microbial Technology, 7*(10), 474–487.

Tickell, J., & Tickell, K., (2003). *From the Fryer to the Fuel Tank: The Complete Guide to Using Vegetable Oil as an Alternative Fuel*, p. 53. California: Biodiesel America.

Ugwu, C. U., Aoyagi, H., & Uchiyama, H., (2008). Photobioreactors for mass cultivation of algae. *Bioresource Technology, 99*(10), 4021–4028.

Verma, N. M., Mehrotra, S., Shukla, A., & Mishra, B. N., (2010). Prospective of biodiesel production utilizing microalgae as the cell factories: A comprehensive discussion. *African Journal of Biotechnology, 9*(10), 1402–1411.

Wang, Y., & Li, J., (2008). Molecular basis of plant architecture. *Annual Review of Plant Biology, 59*, 253–279.

Wood, D., Capuzzo, E., Kirby, D., Mooney-McAuley, K., & Kerrison, P., (2017). UK macroalgae aquaculture: What are the key environmental and licensing considerations? *Marine Policy, 83*, 29–39.

Zamalloa, C., Vulsteke, E., Albrecht, J., & Verstraete, W., (2011). The techno-economic potential of renewable energy through the anaerobic digestion of microalgae. *Bioresource Technology, 102*(2), 1149–1158.

CHAPTER 10

Recent Developments in Biodiesel Production from Heterotrophic Microalgae: Insights and Future Prospects

GOURI RAUT,[1] MAHESH KHOT,[2] and SRIJAY KAMAT[3]

[1]*Bioenergy Division, Agharkar Research Institute, Pune – 411004, Maharashtra, India*

[2]*Laboratorio de Recursos Renovables, Centro de Biotecnología, Universidad de Concepción, Concepción – 4030000, Chile*

[3]*Department of Biotechnology, Goa University, Goa – 403206, India*

ABSTRACT

Heterotrophic microalgae have biotechnological advantages as sustainable resources of high-value compounds, ranging from therapeutic proteins to fatty acid-based fuels. The dark cultivation of lipid-accumulating microalgae on organic compounds as simple as glucose is being investigated as a promising alternative to biodiesel and other biofuels. A wide range of literature is available on the production and application of microalgal lipids. In this chapter, critical aspects of algal lipids in producing biodiesel are described systematically. The production of triglycerides as feedstock for biodiesel is summarized for different types of microalgae grown on cheap, low-cost renewable carbon sources. The choice of feedstock and fuel properties of the end product are two critical factors in the biomass-to-biodiesel production process which are in turn determined by the strain

Microalgal Biotechnology: Bioprospecting Microalgae for Functional Metabolites towards Commercial and Sustainable Applications. Jeyabalan Sangeetha, PhD, Svetlana Codreanu, PhD, & Devarajan Thangadurai, PhD (Eds.)

involved and growth substrate. Strategies to improve productivity, e.g., optimization of bioprocess, direct (one-step) transesterification, and genetic engineering are discussed. The importance of physicochemical fuel properties is also highlighted.

10.1 INTRODUCTION

The search for alternative and renewable fuel sources is the need of the hour owing to growing human consumption and the limited availability of these resources. As a result of rapid industrialization, fossil fuels like petrodiesel are being increasingly utilized at an alarming rate. Amongst the transportation fuels, biodiesel is of particular interest because of its renewable nature and sustainable production (Katre et al., 2018; Khot et al., 2018). Recent statistics indicate around 1% rise in the annual production of biodiesel (Raut et al., 2019). Worldwide, out of the total biofuel production of 143 billion liters, 29% was biodiesel. The largest biodiesel producers were the United States and Brazil with 16 and 11% of the global production (REN21, 2018).

Biodiesel has many advantages like renewability, higher flash point, and miscibility in all ratios with conventional diesel and overall improvement in engine efficiency without the requirement of any engine modification to incorporate the use of this biofuel. Additionally, biodiesel reduces sulfur dioxide and greenhouse gases (GHGs) emissions, which can be the leading cause of acid rain and is environment-friendly (Suresh et al., 2018). The general characteristics of biodiesel are represented in Figure 10.1.

10.1.1 GENERATIONS OF BIODIESEL

The first generation of biodiesel includes food crops like soybeans, palm, corn, etc., for biodiesel production. The main problem in using edible sources for biodiesel production is that they give rise to the food vs. fuel debate due to their competition with the food resources. The first-generation biodiesel suffers from a significant disadvantage viz., requirement of arable land and raw materials. Second generation biodiesel refers to the use of non-food sources like non-edible plants (Jatropha, Karanja, etc.), agriculture, and forest residues. The second-generation sources become non-sustainable with respect to competition with available land.

Microalgae grown in photobioreactors and raceways are the source of the third generation of biodiesel. Recently, the fourth generation of biodiesel is increasingly in focus wherein, genetically modified microalgae with engineered biosynthetic metabolic pathways are used for hydrocarbon production that can serve as fuels (Lü et al., 2011). Figure 10.2 depicts the different generations of biodiesel and its sources (Lü et al., 2011). Present efforts are dedicated towards technology for the second, third, and fourth-generation biodiesel.

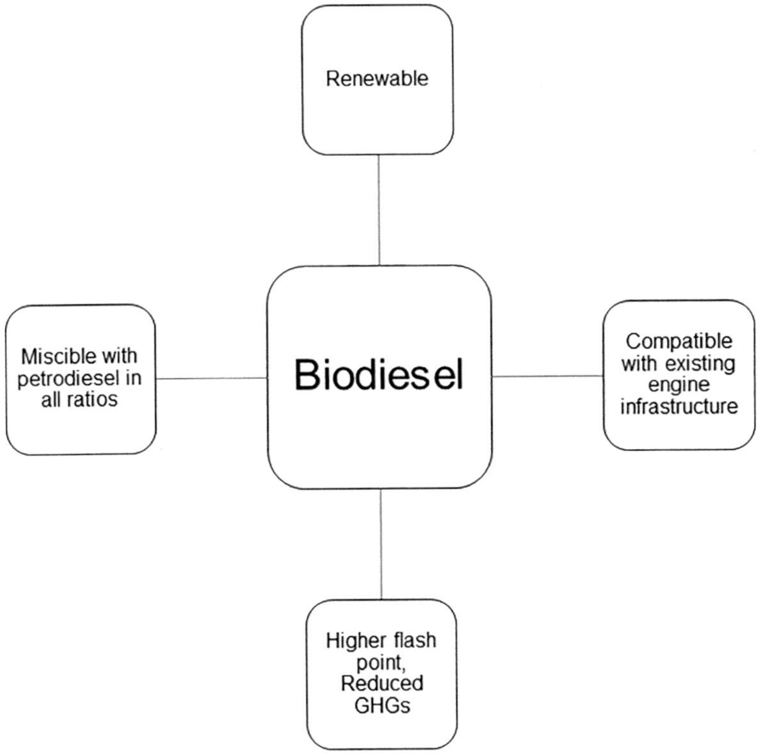

FIGURE 10.1 Characteristics of biodiesel.

10.1.2 DIFFERENT SOURCES OF BIODIESEL

Vegetable oils, animal fats, and microbes are the different sources for biodiesel, which chemically is a mixture of fatty acid methyl esters

(FAMEs). Vegetable oils and animal fats represent the first generation of biodiesel and suffer from many disadvantages as mentioned in section 10.1.1. Hence, the onus is now on microbial oils, also called single cell oils or SCOs. Typically, SCOs from oleaginous bacteria, fungi, and yeasts constitute the second-generation biodiesel feedstock. Oleaginous microbes are those that can accumulate 20% or more of their cellular dry weight as lipids or triacylglycerol (TAG).

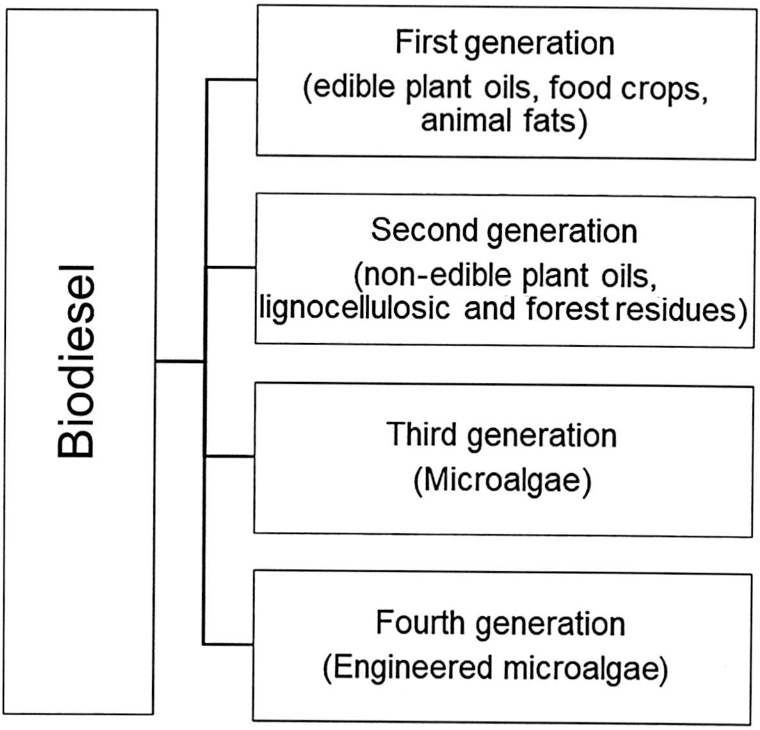

FIGURE 10.2 Different generations of biodiesel and its sources.

Bacteria are capable of lipid accumulation, but they generally accumulate lipids that are not warranted for biodiesel. The biodiesel cetane index that indicates the relation between the density of FAME with cetane number is an important parameter. In a study, the biodiesel cetane index of biodiesel from *Rhodococcus opacus* PD630, an oleaginous bacterium was found to be relatively low, with a value of 41 (Wahlen et al., 2013).

Fungi and yeasts are being used for biodiesel production (Katre et al., 2012; Khot et al., 2012) but to date, there are hardly any reports on commercial production of biodiesel from these sources. They are chemo-heterotrophs, that is, they need organic compounds for their requirement of energy and carbon. Research now is directed towards the third and fourth generation of biodiesel which involves the use of microalgae.

Microalgae have different modes of growth – autotrophic, heterotrophic, and mixotrophic and can switch this mode based on the availability of light and carbon sources. In the autotrophic mode, light is used as an energy source and CO_2 as the carbon source; in heterotrophic mode, microalgae use organic compounds as both energy and carbon sources and in mixotrophic mode, both organic compounds and CO_2 are used as carbon sources. Amongst these, the heterotrophic mode offers several advantages such as higher lipid yield, control better over the process, the requirement of light is minimum, and high biomass yields (Devi et al., 2012).

10.2 MICROALGAE AS A SOURCE FOR BIODIESEL

Microalgae are a promising feedstock for biodiesel production not only in terms of their lipid contents and compositions but also in terms of the fuel properties of the biodiesel obtained from them (Chen et al., 2018). Oils obtained from microalgae have been found to have suitable levels of saturated fatty acids, free fatty acids, water, phosphorus, and sulfur that make them ideally suited for the production of biodiesel. The assessment of biodiesel fuel properties from a wide range of microalgal species has also revealed that the final biodiesel conforms to American (ASTM D6751) and European (EN14214) fuel standards (Kumar and Sharma, 2016; Deshmukh et al., 2019).

One of the main attractive features of microbial biodiesel is that they do not compete with for resources with agriculture and therefore completely circumvent the "food vs. fuel" debate (Gujjala et al., 2017). Marine microalgae have the added advantage of being amenable to cultivation in saline or brackish coastal waters (Maeda et al., 2018). Being unicellular organisms, microalgae also support higher growth rates than terrestrial plants, utilize wastewater for growth, and provide feedstocks for various biofuels (Kumar and Sharma, 2015). The development of intensive culture systems that can support the production and recovery of multiple products from the rich reserves of lipids, pigments, proteins, and carbohydrates of microalgal biomass is essential to realize the industrial-scale production of microalgal biodiesel (Gifuni et al., 2018).

Most microalgal species are obligate photoheterotrophs, but some species can grow in the absence of light and utilize organic carbon and nitrogen sources similar to heterotrophic bacteria and fungi (Perez-Garcia et al., 2011). More recent studies have indicated that microalgae can be cultivated in four modes of cultivation as follows: photoautotrophic, heterotrophic, mixotrophic, and photoheterotrophic (Vasistha et al., 2019). Microalgal biomass can also be grown to high cell densities using heterotrophic conditions. The high growth rates of microalgae increase lipid yield and biodiesel production compared to autotrophic cultivation of microalgae, oil crops, or plants (Zhu, 2015; Tan et al., 2018). Several microalgae such as *Chlorella*, *Scenedesmus*, *Neochloris*, and *Tetraselmis* have been explored for biodiesel production using heterotrophic cultivation strategies to produce lipids, polyunsaturated fatty acids (PUFA), and pigments (Nagarajan et al., 2018). A recent screening of 37 photoautotrophic microalgal strains found five microalgal strains previously unreported for heterotrophic growth (Pavel and Cepák, 2019). *Botryosphaerella sudetica*, *Bracteacoccus* sp., *Dictyosphaerium* sp., *Lemmermannia* sp., *Parachlorella kessleri* were the new heterotrophic microalgal strains, while *Scenedesmus* sp., *Dictyosphaerium chlorelloides*, and *Tribonema aequale* were investigated for their production of carotenoids, exopolysaccharides (EPS), and eicosapentaenoic acid (EPA), respectively.

Azari et al. (2018) compared the carbon footprints of autotrophic and heterotrophic modes of microalgal cultivation and performed a life cycle assessment of the two modes of cultivation. The authors stated that although heterotrophic cultivation of microalgae supports higher production rates, oil contents, faster growth rates, and high cell densities compared to autotrophic cultivation, the initial investment was also high. However, they did concede that wastewater and agriculture residues could be used to alleviate the cost of using pure sugars as medium components.

New cultivation techniques are also being studied for the production of biodiesel from microalgal feedstocks. Mixotrophic mode of microalgal cultivation has been shown to be the most optimal mode to achieve high biomass and lipid yields for *Chlorella vulgaris* (Shen et al., 2019), *Scenedesmus obliquus* (Shen et al., 2018), and *Tribonema* sp. (Wang et al., 2017). However, it is not commonly used for commercial cultivation purposes (Nagarajan et al., 2018). Dual species cultivation of microalgal genera (*Chlorella* and

Spirulina) and yeast genera (*Candida*, *Saccharomyces*, *Yarrowia*, and *Rhodotorula*) are also being explored for the production of biofuel feedstocks and are already found to produce better lipid yield and productivities than mono-cultures (Ananthi et al., 2018; Liu et al., 2018; Qin et al., 2018).

TABLE 10.1 Heterotrophic Microalgae Grown on Different Substrates Used for Biodiesel Production

Microalga	Phylum	Substrate Used	Total Fatty Acid Productivity (mg/L/day)	References
Chlorella protothecoides	Chlorophyta	Sugarcane bagasse hydrolysate	173.66	Chen et al. (2019)
Chlorella vulgaris NIES-227	Chlorophyta	Acetate	72.60	Shen et al. (2019)
Ettlia sp. YC001	Chlorophyta	Fructose	1,180	Kim et al. (2019)
Tetraselmis suecica	Chlorophyta	Gray mullet aquaculture wastewater	49.80	Andreotti et al. (2019)
Dunaliella tertiolecta	Chlorophyta	Gray mullet aquaculture wastewater	11.10	Andreotti et al. (2019)
Chlorella sorokiniana CY-1	Chlorophyta	Glucose and glycerol supplementation to palm oil mill effluent	14.41–18.08	Cheah et al. (2018)
Micractinium sp. ME05	Chlorophyta	Sugarcane vinasse	1,870	Engin et al. (2018)
Chlamydomonas reinhardtii	Chlorophyta	Acetate	28.60	Fan and Zheng (2017)

10.3 BIODIESEL PRODUCTION BY MICROALGAE ON VARIOUS SOURCES

The use of food waste and wastewater from industrial, agricultural, dairy, and municipal sources would be ideal nutrient streams for heterotrophic microalgal cultivation compared to pure carbon sources like glucose (Ende and Noke, 2018; Ray et al., 2018). An evaluation of 19 different waste streams and 6 different algal consortia showed that autoclaved whey effluent was the best nutrient source resulting in biomass of 14.32 g L^{-1} of which 13.23% (1.91 g of lipid L^{-1}) was lipid (Jordaan et al., 2017). A few recent examples of microalgal strains which have been cultivated heterotrophically and evaluated for biodiesel production are summarized in Table 10.1.

10.4 DIRECT TRANSESTERIFICATION OF MICROALGAE FOR BIODIESEL PRODUCTION

In addition to the upstream challenges in large-scale cultivation of oleaginous microalgae, the development of efficient and economic lipid recovery process for biodiesel conversion is equally important for the successful scale-up of the downstream processes (Halim et al., 2012). The extraction of lipids from oleaginous microalgae is a challenging task given the rigid cell wall and localization of lipids in intracellular globules (Lee et al., 2012). The algal cell wall is a complex framework of biomolecules including a polysaccharide, uronic acid, protein, glycolipid, and minerals. This contributes a higher tensile strength to the algal cell wall compared to the plant wall (Carpita, 1985). This prompted the development of a range of physical and mechanical methods including bead mill, homogenizer, microfluidizers, and microwave treatment to efficiently disrupt the cell wall facilitating the complete lipid recovery from rigid microalgal cells (Xue et al., 2018). Cell disruption enhances lipid extraction from microalgae. Physical methods involve extreme conditions of high temperature and pressure. Thus, they are energy-consuming while the mechanical methods are unsuitable because of the microalgal cell size (3 to 10 μm) making cells pass unchanged. This has complicated the scale-up of physicomechanical cell lysis methods in algal bioprocesses. Relatively, passive methods extract lipids through perforation and permeabilization of the cell wall while avoiding an energy-intensive lysis step (Günerken et al., 2015). Such passive methods include *in situ* transesterification, saponification-acidification, supercritical fluids, switchable solvents, enzymatic treatment, extraction using organic solvents and chemical agents (Nagappan et al., 2019). Many of these passive methods have been optimized for maximal recovery of microalgal lipids without the necessity of energy-intensive steps like mechanical cell disruption (Ehimen et al., 2010; Du et al., 2015, 2016, 2018).

The integration of oil extraction and biodiesel conversion as direct (*in situ*) transesterification, has been well studied which is simply the conversion of algal lipids present in biomass to biodiesel in one step (Hidalgo et al., 2013). This process includes both the esterification of the free fatty acid and the transesterification of triglyceride from algal cells. It simplifies the biodiesel production process and improves the yield of the fatty ester compared with conventional extraction because of the elimination of a lipid extraction step that incurs oil loss. The direct transesterification reaction involves the addition of alcohols, catalysts, and biomass, and sometimes co-solvents (Baumgartner et al., 2013). Several studies have been conducted to produce FAMEs by direct transesterification of both dry and wet biomass obtained from freshwater and marine microalgae. *Chlorella* sp. and *Nannochloropsis* sp. are among the most common species of microalgae being studied for direct transesterification to biodiesel. In most studies, sulfuric acid has been used as a catalyst because of higher levels of free fatty acids in most microalgal lipids. The use of base catalyst generally results in soap formation due to neutralization of free fatty acids or due to saponification of TAGs in the presence of moisture, leading to reduced FAME yields. This also gives rise to difficulty in downstream processing for FAME separation and purification (Kakkad et al., 2015).

Methanol acts both as an extraction solvent and an esterification reagent. In general, the direct transesterification process demands a high level of methanol and sulfuric acid compared to the conventional two-step biodiesel production. However, reduced volumes of methanol and sulfuric acid are necessary to avoid the need for a large reactor and reactor corrosion by sulfuric acid. The use of an additional solvent such as pentane, hexane, diethyl ether, or chloroform assists in efficient lipid recovery by enhancing the lipid contact with the esterification reagent (Cao et al., 2013) and by improving the diffusion of the cellular lipids across the cell walls. This is facilitated by increasing the selectivity and solubility of the extraction media, thereby providing greater availability of the oils for the transesterification process (Ehimen et al., 2012). Longer chain alcohols such as ethanol, isopropanol, and butanol are now being introduced as potential candidates to replace methanol being toxic and non-renewable. Similar yields were reported for both fatty acid alkyl esters (Lemões et al., 2016). The characteristic features and prerequisites of an *in situ* trans-esterification reaction using microalgal biomass include the following:

- Either homogeneous alkali, acid, or heterogeneous catalyst.
- High mixing rate (150–500 rpm).
- Methanol volume required is greater for reactions involving a heterogeneous catalyst due to phase transfer limitation compared to the one with either homogeneous alkali or acid catalyst. Co-solvent addition can reduce the methanol requirement.
- Catalyst is not required in the presence of supercritical water or alcohol.
- Room temperature can bring about reactions, particularly with an alkali catalyst.
- FAME yield from microalgal biomass via direct transesterification depends on several factors namely microalgae species, reaction time, temperature, catalyst/oil molar ratio, methanol/oil molar ratio, agitation rate, moisture content of the feedstock/reactants, catalyst type, co-solvent (Salam et al., 2016).

The application of ultrasound or microwave can improve the yield of FAMEs in direct transesterification. Ultrasound can enhance the mass transfer characteristics, and, at the same time, reduces the overall reaction time (Hidalgo et al., 2013).

Water is known to inhibit the direct transesterification reaction of wet algal biomass and its mechanism has been well studied by Sathish et al. (2014). The biodiesel molecules can be hydrolyzed back to methanol and free fatty acids by water as FAME formation is a reversible reaction. Besides water content from biomass can shield lipids from the extracted solvent and prevent access of lipid molecules into reaction. Moreover, the acid catalyst can be deactivated owing to water competing for available protons in the reaction.

Direct transesterification of microalgal biomass to produce biodiesel has been demonstrated to be technically feasible for a range of marine and freshwater species. However, two major setbacks include the relatively higher molar ratio of methanol to oil (100:1 to 1000:1) and energy-intensive drying of the microalgal biomass which significantly hinders the commercial production of large-scale algae biofuels. Despite the disadvantages of water in transesterification, several wet oil extraction technologies have been developed recently, and these methods can be combined with a direct process to overcome water inhibition.

The possible cost savings due to the increased water tolerance of reactive extraction of microalgae lipids to FAME are considerable (Lardon et al., 2009), but they must be weighed against the costs of alcohol regeneration by distillation.

Direct transesterification eliminates lipid loss by converting all available cellular lipids into FAME and offers concurrent production of valuable co-products such as ethyl levulinate, ethyl formate, diethyl ether, and glycerol carbonate. Co-products constitute an important factor for cutting down elevated production costs, making microalgae-based biodiesel production more feasible (Sivaramakrishnan and Incharoensakdi, 2018).

10.5 EVALUATION OF FUEL PROPERTIES

Biodiesel composition reflects the fatty acid profile of the feedstock used and is a mixture of different fatty esters. Each ester component contributes to the properties of the resulting fuel (Knothe, 2005, 2009). Each feedstock used for biodiesel production differs from the others in terms of chain length of fatty acid components with different proportions of saturated, monounsaturated, and PUFA which are known to affect biodiesel fuel properties. For example, FAMEs prepared from saturated fatty acids impart excellent oxidative stability but poor cold flow properties whereas polyunsaturated FAMEs display the opposite effect (Khot, 2016).

A set of well-defined standards is used in the biodiesel industry to certify FAMEs as biodiesel fuels for commercial sale (ASTM D6751, EN14214). The standards, set by ASTM International take into consideration all aspects of the fuel which are important for commercial resale and may not necessarily relate to the performance of the fuel in an engine, e.g., density, viscosity, and biodiesel cetane index are the physical properties of the fuels most related to combustion. The measured fuel property database of microalgae-derived biodiesel is limited. Table 10.1 shows the measured fuel properties of biodiesel derived from different microalgae species. The density of the biodiesel fuels derived from different microalgal species is within the range specified by ASTM for biodiesel (0.86–0.90 g cm^{-3}). The biodiesel standard (ASTM D6751) specifies that the kinematic viscosity for biodiesel fuels must fall within the range of 1.9–6.0 $mm^2 s^{-1}$. All algal fuels tested were within the acceptable range for kinematic viscosity. The energy density, expressed as the heating value (kJ g^{-1} or MJ kg^{-1}) is an important parameter for biodiesel fuel quality. Fuels with a higher heating value can accomplish more work than an equal amount of lower energy density fuels. The algal biodiesel heating value is comparable to or greater than soybean biodiesel (Table 10.2).

Cetane number for diesel fuel serves as an important ignition quality indicator that is equivalent to the octane number of petrol. The higher the CN, the shorter the ignition delay time and vice versa (Knothe, 2005, 2008). The cetane number of algal biodiesel fuels has been observed to lie above the ASTM minimum value of 47 in most cases but slightly lower than edible oil-derived biodiesel (soybean), whereas sulfur content is higher. *Chlorella protothecoides* produced biodiesel with better fuel properties than other species, which also had better oxidative stability (~12 h). Most species of microalgae display poor oxidative stability for the biodiesel produced from their lipids on account of high PUFA content. But overall, microalgae biodiesel properties were found comparable with other vegetable oil sources (Deshmukh et al., 2019).

Since large-scale cultivation of microalgae is challenging, scope remains for studies associated with engine performance and evaluation of exhaust characteristics. The use of longer-chain alcohols is known to improve the cold flow properties and oxidation stability of the produced biodiesel (Huang et al., 2015). Fatty acid ethyl esters (FAEEs) have better cetane number, oxidation stability, and cold flow properties than fatty acid methyl esters (Reddy et al., 2014).

10.6 OPTIMIZATION STUDIES ON HETEROTROPHIC MICROALGAE

10.6.1 OPTIMIZATION FOR LIPID PRODUCTION

Microalgae, when grown heterotrophically yield higher biomass and result in higher lipid productivities. For the microalga *Chlorella protothecoides*, as compared to the autotrophic mode, heterotrophic cultivation is known to produce 15.5 g L^{-1} biomass in a fed-batch mode in bioreactor and at the industrial level (10,000 L), biomass production of 14.2 g L^{-1} was achieved.

TABLE 10.2 Fuel Properties of Biodiesel Fuel (FAME) Derived from Different Microalgae*

Source and Reference	Fuel Property						
	Density at 15°C (g/cm^3)	**Viscosity at 40°C (mm^2/s)**	**Cetane Number/ Cetane Index**	**Pour Point (°C)**	**Flash Point (°C)**	**Cloud Point (°C)**	**Calorific/ Heating Value (MJ/kg)**
Chaetoceros gracilis (Wahlen et al., 2013)	0.885	3.4	51	–	–	–	39.51
Crypthecodinium cohnii (Islam et al., 2015)	0.912	5.06	46.5	–	95	16	39.86
Chlorella sp. (Makarevičiene et al., 2014)	0.883	4.73	–	–	179	–	37.06
Nanochloropsis sp. (Haik et al., 2011)	0.869	4.19	–	–6	144	7	40.72
Chlorella protothecoides (Al-Lwayzy and Yusaf, 2017)	0.900	4.22	52	–	124	–	40.04
Spirulina platensis (Nautiyal et al., 2014)	0.860	5.66	–	–18	130	–	41.36
Soybean oilseeds (*Glycine max*) (Wahlen et al., 2013)	0.884	3.9	54	–	–	–	39.97

**Compared with commercially available soybean biodiesel.*

This represents a value of 7.15 and 6.36 g L^{-1} for lipid productivity, which in turn is greater than the usual values achieved through autotrophic cultivation of this microalgal culture by 10 and 20 folds. Hence, optimized growth and nutrient conditions are being increasingly investigated for heterotrophic microalgae to attain a higher lipid content.

Different approaches have been used to enhance lipid yield, one of them being response surface methodology (RSM). Instead of the conventional one variable at a time (OVAT) experiments, wherein, a single variable is varied and all other parameters are kept constant. RSM adopts a statistical and mathematical approach to design a minimal number of experiments that allows us to understand the interactions among multiple variables that result in output values also called responses (Wang et al., 2015).

In a study on *Chlorella pyrenoidosa* NCIM 2738, the increase in lipid productivity when single and multi-response optimization was employed was 2.3 and 2.9 fold compared to the control (Kanaga et al., 2015). Photo-autotrophic cultivation in *Nannochloropsis salina* yielded lower biomass (0.21 g L^{-1}) and lipid content (22.16 mg L^{-1}) as compared to the heterotrophic mode wherein, a higher biomass (0.91 g/L) and lipid content (37 mg L^{-1}) was obtained. Optimization by RSM resulted in biomass of 1.85 g/L and total lipid content of 48.6 mg L^{-1} (Marudhupandi et al., 2016). For the microalga *Micractinium* sp. M-13, RSM was used to determine the effect of sugarcane industry effluent and citric acid on biomass and lipid yield. Lipid productivity increased by 1.5 fold from 365.5 mg L^{-1} in unoptimized to 580.5 mg L^{-1} in the optimized medium was observed (Karpagam et al., 2015).

In a recent study, an axenic strain of *Ettlia* sp. YC001 was optimized for biomass and lipid production, and the results indicated that the optimum values of fructose and yeast extract were 72.2 g L^{-1} and 21.5 g L^{-1}, respectively, with lipid content of 13.8%. At the fermenter level (5 L), biomass, and lipid productivity reached 7.21 and 1.18 g L^{-1} d^{-1} in 6 days (Kim et al., 2019).

10.6.2 OPTIMIZATION OF THE TRANSESTERIFICATION REACTION/BIODIESEL PRODUCTION

Biodiesel is produced by transesterifying the extracted lipid to yield FAMEs. Conventionally, biodiesel production is a two-step process wherein cells are lysed to release the intracellular lipid. This lipid is then transesterified using methanol and catalyst to produce a mixture of FAMEs. Different factors like the alcohol used, biomass concentration, lipid content, solvent, and catalyst have been optimized to increase FAME yield.

Direct transesterification refers to all the reactants (biomass, solvents, and catalyst) being added in a single step. An *in situ* transesterification of *Aurantiochytrium* sp. KRS 101 employed K_2CO_3 as the alkaline catalyst. It was observed that ultrasonication immensely helped increase the recovery yield of FAEEs to 94.6% for 100 g L^{-1} biomass, 3% K_2CO_3, 70°C and 30 min (Sung and Han, 2018).

The biomass of *Chlorella vulgaris* was used for the production of FAEEs using immobilized enzymatic catalyst associated with pressurized fluid (propane). Optimization studies were carried out using different parameters like temperature (46.7–68.1°C), pressure (59.2–200.5 bar) to obtain an ethyl ester conversion of 74.39%. Further, the oil: ethanol and enzymatic concentration were also varied and an ethyl ester conversion of 98.9% was achieved (Marcon et al., 2019).

In a study on *in situ* supercritical methanol transesterification (SCMT) using *Chlorella* sp. FC2 IITG, different parameters like lipid content, biomass water content, methanol loading were varied. It was found that optimum values of 52% (w/w), 5.75 ml g^{-1} and 115 ml g^{-1} of the parameters as mentioned above resulted in a maximum FAME yield of 96.9%. The results indicate that the wet biomass of the microalga can be used with a minimum quantity of alcohol (Chauhan et al., 2019).

10.7 GENETIC ENGINEERING OF HETEROTROPHIC MICROALGAE FOR BIODIESEL PRODUCTION

As discussed in the previous sections, microalgae are a rich source of triglycerides that can be easily converted to biodiesel. Although microalgae have been known to produce an oil content of up to 70% of their dry biomass, attempts have been made to improve the oil yield using genetic and metabolic engineering (Majidian et al., 2018).

Single gene overexpression or deletion in microalgal genomes have resulted in significant increases in lipid production. In *Chlamydomonas reinhardtii*, the overexpression of a recombinant *E. coli* malate synthase resulted in increased cell biomass by enhancing heterotrophic metabolism (Paik et al., 2019). Targeted knockout of the phospholipase A2 using the CRISPR-Cas9 system in *C. reinhardtii* has also resulted in a 64.25%

increase in overall lipid productivities (Shin et al., 2019). The malic enzyme, a key rate-limiting enzyme in lipid accumulation in many microorganisms, was overexpressed in *Chlorella protothecoides*, which resulted in a 2.8 fold increase in total lipid accumulation compared to the wild-type (Yan et al., 2019).

Novel techniques are also being developed to identify genetic engineering targets to enhance TAG production as well as the transposon and CRISPR-based methods to edit microalgal genomes. Metabolomics, proteomics, real-time polymerase chain reaction, and lipidomics were used to study the differential protein, metabolite, and lipid expression profiles of *Scenedesmus* sp. IITRIND2 under fresh water and saline conditions (Arora et al., 2019). The study found that KCS (3-ketoacyl-CoA synthase), SAD (stearoyl-ACP desaturase), Alfin-like protein, and a putative salt-tolerant protein could be potential genetic engineering targets to generate oleaginous halotolerant species. CRISPR-Cas enzyme systems have been used for targeted genome editing (Naduthodi et al., 2019), and marker-free gene disruption (Poliner et al., 2019) in *Nannochloropsis oceanica* strains with remarkable success. Transposon-mediated random mutagenesis has been attempted in *Nannochloropsis oceanica* CCAP 849/10 using the Tn5 transposon which resulted in a 12% increase in total lipid content g^{-1} of biomass (Osorio et al., 2019).

Significant progress has been made in the development of metabolic engineering strategies to improve lipid yields in microalgae, but more intensive approaches have to be developed to realize microalgal biodiesel production.

10.8 SCALE-UP AND COST EVALUATION/TECHNO-ECONOMIC EVALUATION

Heterotrophic microalgae are easier to cultivate, unlike their autotrophic counterparts. This leads to reduced costs because the use of equipment to provide illumination and expensive photobioreactors for cultivation is eliminated. This mode is characterized by the relative ease in operation (Venkata Mohan et al., 2015). The heterotrophic mode also results in higher biomass and lipid productivities as mentioned earlier. But, the bottleneck is the cost of the carbon source which accounts for a major part of the production costs. This can be reduced by using various waste and renewable substrates like wastewater, non-glucose feedstocks wherein, the components of the waste can act as a carbon source, for example, in textile wastewater, the organic dyes act as the carbon source (Fazal et al., 2017).

Harvesting of microalgal biomass is another area of interest and different methods such as sedimentation, centrifugation, flocculation, flotation, and filtration are being used (Raut et al., 2019). Amongst these, flocculation is a very cost-effective method for harvesting and has tremendous potential for scale-up.

The use of supercritical CO_2 is a green technology used for lipid extraction from microalgae that can be scaled up. It was found that dynamic supercritical CO_2 extraction from *Chlorococcum* sp. yielded 0.058 g lipid g^{-1} dried biomass in 1 h 20 min. When hexane was used for Soxhlet extraction, it yielded 0.032 g lipid g^{-1} dried biomass in 5.5 h (Venkata Mohan et al., 2015). In another study, it was found that a surface of 7,500 m^2 would be required to manufacture 10,000 tons of biodiesel per year from the microalga *Chlorella protothecoides* (Tabernero et al., 2012).

Another strategy to reduce production costs is the use of wet microalgal biomass for direct transesterification. This approach not only avoids the otherwise costly drying process but also the two-step lipid extraction and transesterification is conducted in a single step (Chauhan et al., 2019). The process is economical, uses lesser solvents, requires lesser time, and is more feasible.

A study on large-scale production of the heterotrophic microalga *C. protothecoides* indicated that lipid contents of 46.1, 48.7, and 44.3% were achieved for 5,750 and 11,000 L bioreactors, respectively (Li et al., 2007). The 11,000 L commercial-scale cultivation was done in a stirred tank bioreactor which contained 8,000 L of medium. Harvesting was done by filtration and lipid extraction by the Soxhlet apparatus using n-Hexane as the solvent. For the same microalga, when waste molasses was used as a carbon source for its growth, lipid content of 57.1% was achieved in 178 h (Yan et al., 2011). Further, the cost analysis for this study revealed that the total cost of microalgal oil was 3.21, 1.65, and 2.21 $ per liter of oil produced for glucose medium, molasses hydrolysate nitrogen-limited medium (MHL), and molasses hydrolysate direct medium (MDH), respectively.

10.9 CONCLUSION AND FUTURE PERSPECTIVES

Biodiesel production from microalgae could ease the burden on fossil fuels and act as a partial solution to the energy crisis that we are facing today.

The need to supplement these heterotrophic microalgae with a carbon source can be used as an advantage by cultivating them in wastewaters and inexpensive substrates which cuts down the cost of production. Further work is needed to develop bioreactors with genetically engineered strains capable of rapid growth and increased lipid production that can be used for large-scale cultivation. The downstream facilities needed to process a large amount of biomass so produced and further transesterification at the industrial level are avenues for further research.

ACKNOWLEDGMENTS

Financial support from the DBT-RA Program in Biotechnology and Life Sciences is gratefully acknowledged by GR. MK would like to thank FONDECYT-CONICYT for the postdoctoral research fund (3180134). SK would like to thank DSKPDF, UGC, New Delhi for their financial support.

KEYWORDS

- **biodiesel generations**
- **fatty acid methyl esters**
- **fuel properties**
- **heterotrophic microalgae**
- ***in situ* transesterification**
- **optimization**
- **transesterification**

REFERENCES

Al-Lwayzy, S. H., & Yusaf, T., (2017). Diesel engine performance and exhaust gas emissions using microalgae *Chlorella protothecoides* biodiesel. *Renewable Energy, 101*, 690-701.

Ananthi, V., Prakash, G. S., Rasu, K. M., et al., (2018). Comparison of integrated sustainable biodiesel and antibacterial nanosilver production by microalgal and yeast isolates. *Journal of Photochemistry and Photobiology B: Biology, 186*, 232–242.

Andreotti, V., Solimeno, A., Chindris, A., Marazzi, F., & Garcia, J., (2019). Growth of *Tetraselmis suecica* and *Dunaliella tertiolecta* in aquaculture wastewater: Numerical simulation with the bio algae model. *Water, Air, Soil and Pollution, 230*(60), 1–14.

Arora, N., Kumari, P., Kumar, A., et al., (2019). Delineating the molecular responses of a halotolerant microalga using integrated omics approach to identify genetic engineering targets for enhanced TAG production. *Biotechnology for Biofuels, 12*(1), 1–17.

Azari, A., Noorpoor, A. R., & Bozorg-Haddad, O., (2018). Carbon footprint analyses of microalgae cultivation systems under autotrophic and heterotrophic conditions. *International Journal of Environmental Science and Technology, 16*, 6671–6684.

Baumgartner, T. A., Burak, J. A. M., Baumgartner, D., Zanin, G. M., & Arroyo, P. A., (2013). Biomass production and ester synthesis by in situ transesterification/esterification using the microalga *Spirulina platensis*. *International Journal of Chemical Engineering, 2013*, 1–7. https://doi.org/10.1155/2013/425604.

Cao, H., Zhang, Z., Wu, X., & Miao, X., (2013). Direct biodiesel production from wet microalgae biomass of *Chlorella pyrenoidosa* through in situ transesterification. *BioMed Research International, 2013*, 1–6. https://doi.org/10.1155/2013/930686.

Carpita, N. C., (1985). Tensile strength of cell walls of living cells. *Plant Physiology, 79*(2), 485–488.

Chauhan, D. S., Goswami, G., Dineshbabu, G., Palabhanvi, B., & Das, D., (2019). Evaluation and optimization of feedstock quality for direct conversion of microalga *Chlorella* sp. FC2 IITG into biodiesel via supercritical methanol transesterification. *Biomass Conversion and Biorefinery, 2019*, 1–11.

Cheah, W. Y., Show, P. L., Juan, J. C., Chang, J. S., & Ling, T. C., (2018). Enhancing biomass and lipid productions of microalgae in palm oil mill effluent using carbon and nutrient supplementation. *Energy Conversion and Management, 164*, 188–197.

Chen, J. H., Lu, L., Lim, P. E., & Wei, D., (2019). Effects of sugarcane bagasse hydrolysate (SCBH) on cell growth and fatty acid accumulation of heterotrophic *Chlorella protothecoides*. *Bioprocess and Biosystems Engineering, 42*(7), 1129–1142.

Chen, J., Li, J., Dong, W., et al., (2018). The potential of microalgae in biodiesel production. *Renewable and Sustainable Energy Reviews, 90*, 336–346.

Deshmukh, S., Kumar, R., & Bala, K., (2019). Microalgae biodiesel: A review on oil extraction, fatty acid composition, properties and effect on engine performance and emissions. *Fuel Processing Technology, 191*, 232–247.

Devi, P. M., Venkata, S. G., & Venkata, M. S., (2012). Heterotrophic cultivation of mixed microalgae for lipid accumulation and wastewater treatment during sequential growth and starvation phases: Effect of nutrient supplementation. *Renewable Energy, 43*, 276–283.

Du, Y., Schuur, B., Kersten, S. R. A., & Brilman, D. W. F., (2015). Opportunities for switchable solvents for lipid extraction from wet algal biomass: An energy evaluation. *Algal Research, 11*, 271–283.

Du, Y., Schuur, B., Kersten, S. R. A., & Brilman, D. W. F., (2016). Microalgae wet extraction using N-ethyl butylamine for fatty acid production. *Green Energy and Environment, 1*(1), 79–83.

Du, Y., Schuur, B., Kersten, S. R. A., & Brilman, D. W. F., (2018). Multistage wet lipid extraction from fresh water-stressed *Neochloris oleoabundans* slurry – experiments and modeling. *Algal Research, 31*, 21–30.

Ehimen, E. A., Sun, Z. F., & Carrington, G. C., (2010). Variables affecting the in-situ transesterification of microalgae lipids. *Fuel, 89*(3), 677–684.

Ehimen, E. A., Sun, Z. F., & Carrington, G. C., (2012). Use of ultrasound and co-solvents to improve the in-situ transesterification of microalgae biomass. *Procedia Environmental Sciences, 15*, 47–55.

Ende, S. S. W., & Noke, A., (2018). Heterotrophic microalgae production on food waste and by-products. *Journal of Applied Phycology, 31*(3), 1565–1571.

Engin, I. K., Cekmecelioglu, D., Yucel, A. M., & Oktem, H. A., (2018). Evaluation of heterotrophic and mixotrophic cultivation of novel *Micractinium* sp. ME05 on vinasse and its scale-up for biodiesel production. *Bioresource Technology, 251*, 128–134.

Fan, J., & Zheng, I., (2017). Acclimation to NaCl and light stress of heterotrophic *Chlamydomonas reinhardtii* for lipid accumulation. *Journal of Bioscience and Bioengineering, 124*(3), 302–308.

Fazal, T., Mushtaq, A., Rehman, F., Ullah, A., & Rashid, N., (2018). Bioremediation of textile wastewater and successive biodiesel production using microalgae. *Renewable and Sustainable Energy Reviews, 82*, 3107–3126.

Gifuni, I., Pollio, A., Safi, C., Marzocchella, A., & Olivieri, G., (2018). Current bottlenecks and challenges of the microalgal biorefinery. *Cell Press Reviews, 37*(3), 242–252.

Gujjala, L. K. S., Kumar, S. P. J., Talukdar, B., et al., (2017). Biodiesel from oleaginous microbes: Opportunities and challenges. *Biofuels, 10*, 45–59.

Günerken, E., D'Hondt, E., Eppink, M. H. M., Garcia-Gonzalez, L., Elst, K., & Wijffels, R. H., (2015). Cell disruption for microalgae biorefineries. *Biotechnology Advances, 33*(2), 243–260.

Haik, Y., Selim, M. Y. E., & Abdulrehman, T., (2011). Combustion of algae oil methyl ester in an indirect injection diesel engine. *Energy, 36*(3), 1827–1835.

Halim, R., Danquah, M. K., & Webley, P. A., (2012). Extraction of oil from microalgae for biodiesel production: A review. *Biotechnology Advances, 30*(3), 709–732.

Hidalgo, P., Toro, C., Ciudad, G., & Navia, R., (2013). Advances in direct transesterification of microalgal biomass for biodiesel production. *Reviews in Environmental Science and Biotechnology, 12*(2), 179–199.

Huang, R., Cheng, J., Qiu, Y., Li, T., Zhou, J., & Cen, K., (2015). Using renewable ethanol and isopropanol for lipid transesterification in wet microalgae cells to produce biodiesel with low crystallization temperature. *Energy Conversion and Management, 105*, 791–797.

Islam, M. A., Rahman, M. M., Heimann, K., et al., (2015). Combustion analysis of microalgae methyl ester in a common rail direct injection diesel engine. *Fuel, 143*, 351–360.

Jordaan, E., Roux-Van, D. M. M. P., Badenhorst, J., Knothe, G., & Botha, B. M., (2017). Evaluating the usability of 19 effluents for heterotrophic cultivation of microalgal consortia as biodiesel feedstock. *Journal of Applied Phycology, 30*(3), 1533–1547.

Kakkad, H., Khot, M., Zinjarde, S., & RaviKumar, A., (2015). Biodiesel production by direct in situ transesterification of an oleaginous tropical mangrove fungus grown on untreated agro-residues and evaluation of its fuel properties. *Bioenergy Research, 8*(4), 1788–1799.

Kanaga, K., Pandey, A., Kumar, S., & Geetanjali, (2015). Multi-objective optimization of media nutrients for enhanced production of algae biomass and fatty acid biosynthesis from *Chlorella pyrenoidosa* NCIM 2738. *Bioresource Technology, 200*, 940–950.

Karpagam, R., Raj, K. J., Ashokkumar, B., & Varalakshmi, P., (2015). Characterization and fatty acid profiling in two freshwater microalgae for biodiesel production: Lipid enhancement methods and media optimization using response surface methodology. *Bioresource Technology, 188*, 177–184.

Katre, G., Joshi, C., Khot, M., Zinjarde, S., & Ravikumar, A., (2012). Evaluation of single-cell oil (SCO) from a tropical marine yeast *Yarrowia lipolytica* NCIM 3589 as a potential feedstock for biodiesel. *AMB Express, 2*(1), 36.

Katre, G., Raskar, S., Zinjarde, S., Ravi, K. V., Kulkarni, B. D., & Ravikumar, A., (2018). Optimization of the in-situ transesterification step for biodiesel production using biomass of *Yarrowia lipolytica* NCIM 3589 grown on waste cooking oil. *Energy, 142*, 944–952.

Khot, M. B., (2016). Single Cell Oil of *Aspergillus terreus* IBBM1 as a Potential Feedstock for Biodiesel. PhD Thesis. doi: http://hdl.handle.net/10603/139691.

Khot, M., Kamat, S., Zinjarde, S., Pant, A., Chopade, B., & RaviKumar, A., (2012). Single-cell oil of oleaginous fungi from the tropical mangrove wetlands as a potential feedstock for biodiesel. *Microbial Cell Factories, 11*(1), 71.

Khot, M., Katre, G., Zinjarde, S., & Ravikumar, A., (2018). Single-cell oils (SCOs) of oleaginous filamentous fungi as a renewable feedstock: A biodiesel biorefinery approach. In: Kumar, S., Dheeran, P., Taherzadeh, M., & Khanal, S., (eds.), *Fungal Biorefineries* (pp. 145–183). Cham: Springer. https://doi.org/10.1007/978-3-319-90379-8_8.

Kim, M., Lee, B., Kim, H. S., Nam, K., Moon, M., & Oh, H. M., (2019). Increased biomass and lipid production of *Ettlia* sp. YC001 by optimized C and N sources in heterotrophic culture. *Scientific Reports, 9*(1), 6830.

Knothe, G., (2005). Dependence of biodiesel fuel properties on the structure of fatty acid alkyl esters. *Fuel Processing Technology, 86*(10), 1059–1070.

Knothe, G., (2008). "Designer" biodiesel: Optimizing fatty ester composition to improve fuel properties. *Energy and Fuels, 22*(2), 1358–1364.

Knothe, G., (2009). Improving biodiesel fuel properties by modifying fatty ester composition. *Energy and Environmental Science, 2*(7), 759–766.

Kumar, M., & Sharma, M. P., (2015). Assessment of potential of oils for biodiesel production. *Renewable and Sustainable Energy Reviews, 44*, 814–823.

Kumar, M., & Sharma, M. P., (2016). Selection of potential oils for biodiesel production. *Renewable and Sustainable Energy Reviews, 56*, 1129–1138.

Lardon, L., Helias, A., Sialve, B., Steyer, J. P., & Bernard, O., (2009). Life-cycle assessment of biodiesel production from microalgae. *Environmental Science and Technology, 43*(17), 6475–6481.

Lee, A. K., Lewis, D. M., & Ashman, P. J., (2012). Disruption of microalgal cells for the extraction of lipids for biofuels: Processes and specific energy requirements. *Biomass and Bioenergy, 46*, 89–101.

Lemões, J. S., Sobrinho, R. C. M. A., Farias, S. P., et al., (2016). Sustainable production of biodiesel from microalgae by direct transesterification. *Sustainable Chemistry and Pharmacy, 3*, 33–38.

Li, X., Xu, H., & Wu, Q., (2007). Large-scale biodiesel production from microalga *Chlorella protothecoides* through heterotrophic cultivation in bioreactors. *Biotechnology and Bioengineering, 98*, 764–771.

Liu, L., Chen, J., Lim, P. E., & Wei, D., (2018). Dual-species cultivation of microalgae and yeast for enhanced biomass and microbial lipid production. *Journal of Applied Phycology, 30*(6), 2997–3007.

Lü, J., Sheahan, C., & Fu, P., (2011). Metabolic engineering of algae for fourth-generation biofuels production. *Energy and Environmental Science, 4*, 2451–2466.

Maeda, Y., Yoshino, T., Matsunaga, T., Matsumoto, M., & Tanaka, T., (2018). Marine microalgae for production of biofuels and chemicals. *Current Opinion in Biotechnology, 50*, 111–120.

Majidian, P., Tabatabaei, M., Zeinolabedini, M., Naghshbandi, M. P., & Chisti, Y., (2018). Metabolic engineering of microorganisms for biofuel production. *Renewable and Sustainable Energy Reviews, 82*, 3863–3885.

Makarevičiene, V., Lebedevas, S., Rapalis, P., Gumbyte, M., Skorupskaite, V., & Žaglinskis, J., (2014). Performance and emission characteristics of diesel fuel containing microalgae oil methyl esters. *Fuel, 120*, 233–239.

Marcon, N. S., Colet, R., Bibilio, D., Graboski, A. M., Steffens, C., & Rosa, C. D., (2019). Production of ethyl esters by direct transesterification of microalga biomass using propane as pressurized fluid. *Applied Biochemistry and Biotechnology, 187*, 1285–1299.

Marudhupandi, T., Sathishkumar, R., & Kumar, T. T. A., (2016). Heterotrophic cultivation of *Nannochloropsis salina* for enhancing biomass and lipid production. *Biotechnology Reports, 10*, 8–16.

Naduthodi, M. I. S., Mohanraju, P., Südfeld, C., D'Adamo, S., Barbosa, M. J., & Van, D. O. J., (2019) CRISPR-Cas ribonucleoprotein mediated homology-directed repair for efficient targeted genome editing in microalgae *Nannochloropsis oceanica* IMET1. *Biotechnology for Biofuels, 12*, 66. https://doi.org/10.1186/s13068-019-1401-3.

Nagappan, S., Devendran, S., Tsai, P. C., Dinakaran, S., Dahm, H. U., & Ponnusamy, V. K., (2019). Passive cell disruption lipid extraction methods of microalgae for biofuel production – A review. *Fuel, 252*, 699–709.

Nagarajan, D., Lee, D. J., & Chang, J. S., (2018). Heterotrophic microalgal cultivation. In: Liao, Q., Chang, J., Herrmann, C., & Xia, A., (eds.), *Bioreactors for Microbial Biomass and Energy Conversion* (pp. 117–160). Singapore: Springer Nature. https://doi.org/10.1007/978-981-10-7677-0_4.

Nautiyal, P., Subramanian, K. A., & Dastidar, M. G., (2014). Production and characterization of biodiesel from algae. *Fuel Processing Technology, 120*, 79–88.

Osorio, H., Jara, C., Fuenzalida, K., Rey-Jurado, E., & Vásquez, M., (2019). High-efficiency nuclear transformation of the microalgae *Nannochloropsis oceanica* using Tn5 transposome for the generation of altered lipid accumulation phenotypes. *Biotechnology for Biofuels, 12*, 134. https://doi.org/10.1186/s13068-019-1475-y.

Paik, S. M., Kim, J., Jin, E. S., & Jeon, N. L., (2019). Overproduction of recombinant *E. coli* malate synthase enhances *Chlamydomonas reinhardtii* biomass by upregulating heterotrophic metabolism. *Bioresource Technology, 272*, 594–598.

Pavel, P., & Cepák, V., (2019). Screening for heterotrophy in microalgae of various taxonomic positions and potential of mixotrophy for production of high-value compounds. *Journal of Applied Phycology, 31*(3), 1555–1564.

Perez-Garcia, O., Escalante, F. M. E., De-Bashan, L. E., & Bashan, Y., (2011). Heterotrophic cultures of microalgae: Metabolism and potential products. *Water Research, 45*(1), 11–36.

Poliner, E., Takeuchi, T., Du, Z. Y., Benning, C., & Farre, E. M., (2019). Non-transgenic marker-free gene disruption by an episomal CRISPR system in the oleaginous microalga, *Nannochloropsis oceanica* CCMP1779. *Plant Journal, 99*(1), 112–127.

Qin, L., Wei, D., Wang, Z., & Alam, M. A., (2018). Advantage assessment of mixed culture of *Chlorella vulgaris* and *Yarrowia lipolytica* for treatment of liquid digestate of yeast industry and cogeneration of biofuel feedstock. *Applied Biochemistry and Biotechnology, 187*(3), 856–869.

Raut, G., Kamat, S., & Ravikumar, A., (2019). Trends in production and fuel properties of biodiesel from heterotrophic microbes. In: Meena, S. N., & Naik, M. M., (ed.), *Advances in Biological Science Research* (pp. 247–273). Academic Press. https://doi.org/10.1016/B978-0-12-817497-5.00016-1.

Ray, M., Kumar, N., Kumar, V., Negi, S., & Banerjee, C., (2018). Microalgae: A way forward approach towards wastewater treatment and biofuel production. In: Shukla, P., (ed.), *Applied Microbiology and Bioengineering* (pp. 229–243). Academic Press. https://doi.org/10.1016/B978-0-12-815407-6.00012-5.

Reddy, H. K., Muppaneni, T., Patil, P. D., et al., (2014). Direct conversion of wet algae to crude biodiesel under supercritical ethanol conditions. *Fuel, 115*, 720–726.

REN 21, (2018). *Renewables 2018 Global Status Report*. Paris: REN21 Secretariat. https://www.ren21.net/wp-content/uploads/2019/08/Full-Report-2018.pdf (accessed on 12 February 2022).

Salam, K. A., Velasquez-Orta, S. B., & Harvey, A. P., (2016). A sustainable integrated in situ transesterification of microalgae for biodiesel production and associated co-products – a review. *Renewable and Sustainable Energy Reviews, 65*, 1179–1198.

Sathish, A., Smith, B. R., & Sims, R. C., (2014). Effect of moisture on in situ transesterification of microalgae for biodiesel production. *Journal of Chemical Technology and Biotechnology, 89*(1), 137–142.

Shen, X. F., Hu, H., Ma, L. L., et al., (2018). FAMEs production from *Scenedesmus obliquus* in autotrophic, heterotrophic and mixotrophic cultures under different nitrogen conditions. *Environmental Science: Water Research and Technology, 4*(3), 461–468.

Shen, X. F., Qin, Q. W., Yan, S. K., Huang, J. L., Liu, K., & Zhou, S. B., (2019). Biodiesel production from *Chlorella vulgaris* under nitrogen starvation in autotrophic, heterotrophic, and mixotrophic cultures. *Journal of Applied Phycology, 31*(3), 1589–1596.

Shin, Y. S., Jeong, J., Nguyen, T. H. T., Kim, J. Y. H., Jin, E. S., & Sim, S. J., (2019). Targeted knockout of phospholipase A 2 to increase lipid productivity in *Chlamydomonas reinhardtii* for biodiesel production. *Bioresource Technology, 271*, 368–374.

Sivaramakrishnan, R., & Incharoensakdi, A., (2018). Microalgae as feedstock for biodiesel production under ultrasound treatment – A review. *Bioresource Technology, 250*, 877–887.

Sung, M., & Han, J. I., (2018). Ultrasound-assisted in-situ transesterification of wet *Aurantiochytrium* sp. KRS 101 using potassium carbonate. *Bioresource Technology, 261*, 117–121.

Suresh, M., Jawahar, C. P., & Richard, A., (2018). A review on biodiesel production, combustion, performance, and emission characteristics of non-edible oils in variable compression ratio diesel engine using biodiesel and its blends. *Renewable and Sustainable Energy Reviews, 92*, 38–49.

Tabernero, A., Del Valle, E. M. M., & Galán, M. A., (2012). Evaluating the industrial potential of biodiesel from a microalgae heterotrophic culture: Scale-up and economics. *Biochemical Engineering Journal, 63*, 104–115.

Tan, X. B., Lam, M. K., Uemura, Y., Lim, J. W., Wong, C. Y., & Lee, K. T., (2018). Cultivation of microalgae for biodiesel production: A review on upstream and downstream processing. *Chinese Journal of Chemical Engineering, 26*(1), 17–30.

Vasistha, S., Anwesha, K., & Rai, M. P., (2019). Progress and challenges in biodiesel production from microalgae feedstock. In: Alam, M., & Wang, Z., (eds.), *Microalgae Biotechnology for Development of Biofuel and Wastewater Treatment*, (pp. 323–345). Singapore: Springer. https://doi.org/10.1007/978-981-13-2264-8_14.

Venkata, M. S., Rohit, M. V., Chiranjeevi, P., Chandra, R., & Navaneeth, B., (2015). Heterotrophic microalgae cultivation to synergize biodiesel production with waste remediation: Progress and perspectives. *Bioresource Technology, 184*, 169–178.

Wahlen, B. D., Morgan, M. R., McCurdy, A. T., et al., (2013). Biodiesel from microalgae, yeast, and bacteria: Engine performance and exhaust emissions. *Energy and Fuels, 27*(1), 220–228.

Wang, H., Zhou, W., Shao, H., & Liu, T., (2017). A comparative analysis of biomass and lipid content in five *Tribonema* sp. strains at autotrophic, heterotrophic and mixotrophic cultivation. *Algal Research, 24*, 284–289.

Wang, Y., Yang, Y., Ma, F., et al., (2015). Optimization of *Chlorella vulgaris* and bioflocculant-producing bacteria co-culture: Enhancing microalgae harvesting and lipid content. *Letters in Applied Microbiology, 60*(5), 497–503.

Xue, Z., Wan, F., Yu, W., Liu, J., Zhang, Z., & Kou, X., (2018). Edible oil production from microalgae: A review. *European Journal of Lipid Science and Technology, 120*(6), 1–11.

Yan, D., Lu, Y., Chen, Y. F., & Wu, Q., (2011). Waste molasses alone displaces glucose-based medium for microalgal fermentation towards cost-saving biodiesel production. *Bioresource Technology, 102*(11), 6487–6493.

Yan, J., Kuang, Y., Gui, X., Han, X., & Yan, Y., (2019). Engineering a malic enzyme to enhance lipid accumulation in *Chlorella protothecoides* and direct production of biodiesel from the microalgal biomass. *Biomass and Bioenergy, 122*, 298–304.

Zhu, L., (2015). Microalgal culture strategies for biofuel production: A review. *Biofuels, Bioproducts and Biorefining, 9*, 801–814.

Index

B

C

D

E

I

O

Q

R

T

U

V